†J.M.J.†

Front Cover Images: Lucas Cranach the Elder. Adam and Eve. Located in Warsaw National Museum, Poland. 1510. Wikimedia Commons. Public Domain; Upper serpent from Cranach the Elder, Adam and Eve, located in Kunsthistorisches Museum, Vienna, Austria. 1510. Wikimedia Commons. Public Domain; UFO and AI neural networks: Adobe Stock File Numbers 538463882 and 269258088. Licensed; Earth from Space. Wikimedia commons. Rear Cover Image: The Creation of Adam, Sistine Chapel. Michelangelo (†1564) Wikimedia Commons, Public domain.

Hardcover, Paperback, Ebook, and Audiobook available from:
www.DSDOConnor.com
www.Amazon.com/author/DSDOConnor
www.TheFirstAndLastDeception.com

Published January 1, 2025.
Paperback ISBN: 9781957168104, eBook ISBN: 9781957168111, Hardcover ISBN: 9781957168128

AI Statement: This is an AI-free book. No part of this book was generated by AI. No part of this book was re-written, paraphrased, modified, or otherwise "optimized" by AI. AI-generated summaries were never consulted in preparing this book's content. Generative AI played no role in the research undertaken for this book, except for those cases in Part Four (and the Appendices) wherein doing so was necessary to illustrate generative AI's pitfalls.

***

This book uses content from the author's 2023 book, *Only Man Bears His Image.* Bible quotes will generally be taken from the Revised Standard Version Catholic Edition, but may also be taken from other translations. Uncommon abbreviations are avoided in this book, but the widely known ones are used: ET (*Extraterrestrial*), UFO (*Unidentified Flying Object*), AI (*Artificial Intelligence*), and, sometimes, NHI (*Non-Human Intelligence – an umbrella term referring to any idea of incarnate non-human intelligent creatures, including "aliens, rational AI, intelligent animals," etc.*), and AGI (*Artificial General Intelligence – a term referring to AI that has supposedly matched or surpassed humans*).

THE FIRST AND LAST DECEPTION

Aliens, UFOs, AI, and the Return of Eden's Demise

Daniel O'Connor

****

To The Seven Thousand

1 Kings 19, Romans 11

The Triumph of Christianity over Paganism. Gustave Dore. 1868.

Table of Contents

Part One: A Serenade for the Apocalypse

1. The First and Last Deception

As the sun began to set on the Eternal City one summer evening in the year 1972, the Bishop of Rome mounted the world's greatest pulpit to issue a warning not merely solemn, but apocalyptic.

"**The smoke of Satan has entered the Church of God**," he declared to the multitudes of now stunned lay faithful, priests, Bishops, and Cardinals there gathered before him.

It was the vigil of the ninth anniversary of his coronation as Pope Paul VI, and in the decades to follow, many Catholic pundits would trade barbs—each insisting (with varying degrees of credibility) that the enemies of his own tribe were the ones condemned by these pontifical remarks. But too few commentators concerned themselves with the remainder of Paul's preaching on that fateful day. For he explained just what this Satanic smoke consisted of, lamenting:

> We trust the first pagan prophet we see who speaks to us in some newspaper, and we run behind him and ask him if he has the formula for true life. I repeat, doubt has entered our conscience. And it entered through the windows that should have been open to the light: science.[1]

Who, then, were these *pagan prophets of science*, advertising their "formula for true life" and to whom, the Pope warned, men were so foolishly running? We needn't look far. "**The grand instruction book that tells creatures how to live forever**" —we were then being told—was about to descend from on high, though not from God. We'll get to the "era-inaugurating" scientist who issued that proclamation in a moment. First, we should consider the setting in which that claim emerged.

In the years leading up to the Pope's pronouncement, global headlines were aflame with reports of UFOs buzzing our skies. New religions were being formed and amassing devotees dedicated to occult communication with the "space brothers"—and these UFO religions quickly gave birth to the New Age movement itself. Some of the world's most renowned astronomers were promising the most unthinkable scientific breakthroughs they anticipated thanks to imminent contact with extraterrestrial civilizations. They pursued this contact with government backing and massive budgets. Hollywood's now explosive science fiction genre—suddenly the modern world's prevailing myth—was diligently ensuring that the masses were groomed to dismiss Christian eschatology and instead adopt unapologetically messianic ET aspirations as their own.

> "For the time is coming when people will not endure sound teaching, but having itching ears they will accumulate for themselves teachers to suit

their own likings, and will turn away from listening to the truth and wander into myths." (2 Timothy 4:3-4)

Most troubling of all, as one contemporary Ufologist* noted, was that in the decades leading up to Paul VI's reign, "*every* major" UFO incident had a Catholic connection.[2]

Today, new claims of the Catholic Church having long stood at the center of all-things-ET seem to go viral each day. These claims come from the lips of famous Congressional-testimony-providing "whistleblowers" like David Grusch and Luis Elizondo, top podcasters like Joe Rogan, famed "mystics" like Chris Bledsoe, presidential advisors, renowned authors, Apollo astronauts, (dissenting) theologians, and everyone in between. George Adamski, who in 1954 became the world's first famous "ET Contactee," even claimed he was honored for his work by Pope John XXIII—who in turn, others insisted, personally met an alien. (And in 2024, one theologian published a massive book which argued—with thousands of footnotes—that the current pope has been prepared, for many decades, to usher in for the world the Age of the Extraterrestrial.)

The nonsense of the claims themselves does not negate the veracity of one observation. Whatever—*whoever*—is ultimately orchestrating them is powerful, influential, and *fixated* upon fomenting expectation of the imminent "ontological shock of ET-Disclosure" through not only the US Government, but also the Vatican.

What is more, the years before Pope Paul's lamentation of Satan's smoke entering Catholicism saw these deceptions suddenly infiltrating the theology of the Church itself. One scholar who has chronicled the ET debate pointed out that in earlier decades, it was practically unheard of in the Vatican; however, "**In the early 1960s that situation changed.**"[3] The most renowned of unorthodox theologians such as Teilhard de Chardin,† Karl Rahner, and Hans Kung then began defending beliefs which had never before entered the heart of Catholicism—prompting whispers that echoed in the halls of St. Peter's about galaxies filled with other civilizations, each having their own incarnations of God, eclipsing Jesus of Nazareth. One Vatican clergyman and theologian later rose to celebrity status for declaring that humanity is pathetically unevolved, therefore in desperate need of being saved by the far superior aliens. Books began to be published insisting that all Marian apparitions—especially those at Fatima—were really just UFOs (this

---

* A "Ufologist" is an expert in the study of UFOs who endorses and promotes the view that nonhuman incarnate intelligent beings (extraterrestrials, "interdimensionals," AGI, or what have you) are piloting UFOs.

† Although de Chardin died in 1955, the popularity of his works exploded in the following decade. Their impact was so pronounced that the Holy Office decreed, in 1962, that de Chardin's writings contained "**serious errors which offend Catholic doctrine**." Unfortunately, this condemnation did little to blunt the influence of de Chardin's works. As we will see later, his thesis—which sought to turn Christianity into a "scientific religion"—remains among the most influential forces behind even contemporary Christian promotion of both the ET/UFO Deception and the AI Deception.

view continues to explode in popularity today). Even the design of church buildings evidenced the deception's infiltration, with many new parishes and cathedrals suddenly built to resemble something from *Star Trek*—imbued not with a sense of the sacred, but with a feeling of science fiction.

> But understand this, that in the last days there will come times of stress. For men will be lovers of self ... holding the form of religion but denying the power of it. (2 Timothy 3:1-2,5)

A dark consensus was emerging among those whose hearts had strayed from the Faith even while their bodies remained in the pews (and sanctuaries): a "higher power" would save us indeed, though it would not be Almighty God. Instead, the scientific and technological revelations from the aliens would grant us astounding new knowledge and even enable us to avert death itself. Where have we heard this before?

> Now the serpent was more subtle than any other wild creature that the Lord God had made. He said to the woman, 'Did God say, 'You shall not eat of any tree of the garden'? (Genesis 3:1)

Behold the first question ever asked; concise, though bursting at the seams with more seductive malice than any words spoken since. The Fall did not immediately transpire, but it was already a foregone conclusion. For Eve had dialogued with the Devil, though only while he assumed the disguise of an incarnate "Nonhuman Intelligence"—just as he is again doing today. And whoever dialogues with the Devil, regardless of the mask he wears, has already been defeated by him.

Eve's first response to the serpent prompted the latter's retort that what God had already revealed was actually *not* the full story.

> But the serpent said to the woman, 'You will not die. For God knows that when you eat of it your eyes will be opened, and you will be like God, knowing good and evil.' (Genesis 3:4,5)

The serpent promised Eve *immortality*; she ought to have remembered the hope she already had of this from God. She was promised her *eyes would be opened* with *new knowledge*; she should have considered that God Himself, not some secret category in the Great Chain of Being, would be the one from Whom any great revelation would descend. She was promised that she would be "like God;" she ought to have remembered that she already bore His Image and that this privilege was already so lofty as to surpass one's wildest imaginings.

Nevertheless, after hearing the Devil pronounce these words of "ontological shock," Eve was convinced. She giddily observed how "delightful" and "desirous" the fruit was (Genesis 3:6). She ate it, and so did Adam. The "whole world"—at that point consisting of just two humans—was deceived. Hell rejoiced.

Turning from the first book of Scripture to the last, one is confronted with the warning that the Devil—the "**deceiver of the whole world**" (Revelation 12:9)—will again embark upon the same expedition: "**The whole earth followed the beast with wonder**." (Revelation 12:3) Why? Because this Beast will be heralded by "**a strong delusion, to make [men] believe what is false**." (2 Thessalonians 2:11) Moreover, St. Thomas Aquinas wisely observed, "*As the end of a thing corresponds to its beginning, it is not possible to be ignorant of the end of things if we know their beginning.*"[4] The last deception will mirror the first. It, too, will entail a quest for immortality and new revolutionary knowledge; not from God, but from another "Nonhuman intelligence" which is only a *delusion*: another disguise worn by the Devil.

> "Do you think [Satan] is so unskillful in his craft, as to ask you openly and plainly to join him in his warfare against the Truth? No; he offers you baits to tempt you. ... He promises you illumination, he offers you knowledge ... He shows you how to become as gods."[5] —St. John Henry Newman. *The Times of Antichrist*

Alas, *how* could Eve have known not to engage in dialogue with the Devil? Satan did not come announcing his true nature to her. What, then, should have reminded Eve that she was confronted with something Satanic in this encounter? The answer is found in the same book of Scripture, though earlier still:

> Then God said, "Let us make man in our image, after our likeness…" So God created man in his own image, in the image of God he created him; male and female he created them. (Genesis 1:26, 27)

In the material Universe, man—male and female—and man *alone*, is made in the Image of God. Of all material creatures, only human beings have reason. Adam and Eve both knew this. They understood well what all their faithful offspring have, until recently, likewise understood; namely, that "**Of all visible creatures, only man is able to know and love his creator**" (*Catechism of the Catholic Church*, §356), and that **"Only man and woman, among all creatures, were made by God 'in his own image' (Gen 1:27)."** (*Compendium of the Social Doctrine of the Church*, §451) They knew intuitively what Pope St. Zachary and Pope Pius II later declared in their Magisterial teachings. These Popes denounced belief in nonhuman, rational, incarnate beings (fairies, space aliens, talking beasts, intelligent machines, interdimensional visitors, or what have you). They declared such propositions were a "**perverse and abominable teaching in opposition to God.**" They labelled such views the "**most pernicious errors,"** which consisted in no less than a **"sacrilegious attempt against the dogmas of the holy fathers,"** as we will see.

Reflecting upon this infused knowledge she enjoyed in the Garden, what should Eve have concluded? What should she have understood, upon seeing that a physical creature which was clearly not a human being began *speaking;* that is, displaying *rationality*? Answering that question in detail will be the topic of this book, and we will review the teachings above (and many others) in detail. Suffice it to say that a simple two-word response would have been appropriate: "*Begone, Satan!*"

Instead, Eve's curiosity defeated her wisdom. Her eagerness for exploring previously unconsidered nuance overpowered her trust in God and in the clear sense of His word. She engaged in dialogue with this apparently rational incarnate creature; this "Nonhuman Intelligence." The Fall of Man quickly followed when Adam, instead of correcting this folly, partook in it.

Returning to the 20th century, would even the boldest of scientists dare to propose something so audacious as did the serpent in the Garden? *Immortality*? Revelations of astounding new knowledge to render us God-like merely by reaching up our hand, taking it, and consuming it? Yes, they would; and they did.

"I fully expect an alien civilization to bequeath to us vast libraries of useful information, to do with as we wish."[6] This, the legendary Ivy League astrophysicist Dr. Frank Drake assured the world, would be the result of the "Search for Extraterrestrial Intelligence" (SETI) that he himself fathered and nurtured throughout the 1960s and beyond. What would these libraries include? Drake tells us: "*This 'Encyclopedia Galactica',*" as he calls it, will come to us from "*immortal civilizations*" of aliens, and includes "**the grand instruction book that tells creatures how to live forever.**"[7]

NASA declared that Dr. Drake "**inaugurated a new era**"[8] in the year 1960 at Project Ozma in West Virginia. With the dish of his 85-foot radio telescope thrust heavenward like a mystic with outstretched arms, Drake listened intently for the ultimate revelation that would enlighten us on how to finally fend off our most ancient foe. For all men rightly feel an existential need for deliverance from bondage to the fear of death. Those who do not seek it in Christ will seek it through Antichrist.

> "...[T]hat through death [Christ] might destroy him who has the power of death, that is, the devil, and deliver all those who through fear of death were subject to lifelong bondage…" (Hebrews 2:14-15)

With Drake, a new era had surely dawned, though with it only a black sun arose.

The movement Drake started swept the globe during the decade of which Ozma's creation marked the beginning. The next year, he devised the "Drake Equation," a purported means of mathematically proving the existence of aliens. Many referred to it as history's second most famous equation (beaten only by E=mc^2), and the increasingly secularized masses sought refuge in this equation like a Christian takes solace in *John 3:16*. Other SETI pioneers were similarly blunt in their aspirations. Dr. Jill Tarter is a prolific astronomer who worked for decades as a leader in SETI during its formative years. She proclaimed her eagerness for detecting an ET message by speaking of a scenario she excitedly foresees it heralding:

> Skeptics and true believers alike will be converted to the revealed, superior [ET] religion… [it] will be a missionary campaign without precedent in terrestrial history… The information will change our lives and our world view, and we will not be able to put the genie back in the bottle… it will be hard for humanity to resist the appeal of this universal religion and its God(s).[9]

The message everywhere was and is the same: ETs will soon provide everything man once sought in vain from almighty God.

The world's most popular scientist, the atheist but salvation-craving Carl Sagan, quickly became deeply involved with Drake's efforts to contact aliens. (Decades later, a major Hollywood film, *Contact*, proselytized Sagan's own pseudo-religious longings with SETI.) The Soviets also began their own search for extraterrestrial intelligence—and even claimed, in 1965, to have discovered an intelligent alien "supercivilization" in another galaxy. Newspapers announced SETI could bestow upon us a "**galactic heritage**" to end all wars.[10] This proved uniquely alluring for a world still entangled in the Cold War, populated by billions of souls terrified they would wake up one morning to find civilization itself obliterated. As never before in history, men were desperate for any *apparent solution* that would ensure such a fate could not arrive.

> Before Christ's second coming the Church must pass through a final trial that will shake the faith of many believers ... [it] will unveil the "mystery of iniquity" in the form of a religious deception offering men an apparent solution to their problems at the price of apostasy from the truth. (*Catechism of the Catholic Church*, §675)

Yes, people indeed began running to these Pagan prophets Pope Paul VI warned against—"prophets" of science who were promising the formula for true life.

Of no little consequence was the content of what had already become the prevailing myth of the day—and would only catapult to further prominence in that status during the decades ahead—namely, science fiction. The films of this genre were tantalizing their viewers with descriptions of advanced extraterrestrial civilizations visiting earth to save us from ourselves. Beyond the silver screen, too, UFO sightings and "ET-contact" experiences abounded, continuing an avalanche that began in 1947 and has not abated to this day. Just five days before Paul VI announced the infiltration of Satan's smoke into the Church, the Era of the UFO celebrated its silver jubilee, whereupon *The New York Times* observed in its June 25th, 1972, edition:

> Twenty-five years ago at 3 P.M. on June 24, 1947, Kenneth Arnold … ushered in the era of the flying saucer. … Since then, if these reports are to be believed, no part of the earth has been left unvisited by unworldly aircraft…For most of [that time], hardly a month has gone by without a deluge of publicity about new sightings, of the mysterious craft that some hoped and some feared were from space. …[Since 1947] there have been

> reports of at least 50,000 sightings, perhaps 100,000.[11]

June 24th—the feast of the birth of St. John the Baptist, the prophet sent by God to announce the coming of Christ. June 24th, 1947—the birth of the Era of the UFO, the false prophets sent by the Devil to announce the coming of Antichrist.

For although most of these "UFO sightings" are simply of misunderstood natural phenomena, human technology, hoaxes, or hallucinations, there has always remained a stubborn portion thereof—small in percentage, but large in magnitude—evading such explanations. As we will see in Part Three, these "non-debunkable" UFO sightings and "ET contact" experiences all bear hallmarks of the demonic and only include evidence of events that remain within the power of the Devil to bring about, even if their ubiquity would imply a "longer leash" than he has usually been given. Christians who know their Faith, however, would not rule out such a scenario. The Fathers certainly did not rule it out. St. Augustine warned against those who would succumb, "**being ignorant of the devil's power, and especially of such unexampled power as he shall then for the first time put forth**."[12] Yes, the Faithful know to affirmatively expect a time when an unprecedented Great Deception arises across the globe, prompted by Satanic illusions:

> The coming of the lawless one by the activity of Satan will be with all power and with pretended signs and wonders, and with all wicked deception for those who are to perish, because they refused to love the truth and so be saved. Therefore God sends upon them a strong delusion, to make them believe what is false. (2 Thessalonians 2:9-11)

Some were not fooled. They remembered Scripture, and they realized the Devil is not to be dialogued with.

The renowned Benedictine priest, theologian, and physicist, Fr. Stanley Jaki, has been described as one of the five Catholic scientists who "shaped our understanding of the world."[13] He traveled extensively among the pioneers and leaders of the Search for Extraterrestrial Intelligence, and he was perhaps the only insider to characterize it in the light of Faith. He wrote:

> Shortly after Frank Drake started his project in the 1960s various prominent people were interviewed as to what we can expect after we have detected radio signals from outer space. Some thought that this will mean the coming of Paradise on earth. What God failed to achieve, the science brought by extraterrestrials would make possible as they would bring with themselves the wisdom of their own evolution.[14]

In the same essay, Fr. Jaki shared details of his own trip to the Arecibo Telescope—where, sixteen years earlier, Dr. Drake had broadcast his famous 1974 message intended to alert extraterrestrials living in the constellation Hercules to our presence. Jaki wrote:

> As I visited [Arecibo] in 1990, the director of the observatory noticed my Roman collar and delivered to me a little sermon ... [whose meaning was clear:] Once we hear from extraterrestrials we can forget about Christ and Christian religion once and for all. Much the same was preached by the group of scientists forming the project SETI ...[15]

Nor has anything changed in the decades since. Dr. Seth Shostak is an American astronomer who currently heads SETI. He justified the importance of his efforts of finding aliens in the Milky Way by remarking that we must find aliens, for if they do not exist, then life on earth "is some sort of a miracle," which he regards as a scientifically abominable position.[16] In one 2003 tome, *Cosmic Company*, published by Cambridge University Press, Shostak wrote of the ontological shock he anticipates:

> ...[A] SETI detection will tell us something that will forever change the way we view ourselves... The barrier of isolation that has separated our planet from the Universe for billions of years, that has kept our eyes down and our thoughts parochial, will shatter. The world will change overnight, and we will wake to find that all our histories, and all our stories, are merely a small entry in an enormous book.[17]

The stakes have been set.

When one considers the whole gamut of the phenomenon—as we will do in this book—one conclusion becomes ever clearer. The quest to promote belief in—and then to discover and communicate with—extraterrestrial intelligence constitutes mankind's most herculean effort yet to once and for all be rid of religion, be rid of Christ; *be rid of God*. To be rid of all "medieval superstitions" and replace them with humanity's new object of awe, faith, and hope: a super-evolved extraterrestrial being who can both explain everything for us and solve everything for us. A "higher" being that comes to us without "dogmatic or moral baggage," and who can do for us men and women of the third millennium what once our unscientific ancestors considered the sole prerogative of Almighty God.

> Let no one deceive you in any way; for that day will not come, unless the rebellion comes first, and the man of lawlessness is revealed, the son of perdition, who opposes and exalts himself against every so-called god or object of worship, so that he takes his seat in the temple of God, proclaiming himself to be God. (2 Thessalonians 2:3-4)

As we will see in the chapters ahead, there is nothing hypothetical about this admittedly extreme warning. The whole UFO movement has already proven itself—repeatedly over the course of almost eight decades since 1947—to be a more obvious preparation for the Antichrist than even any author of fiction could have placed upon the page. Yet here, we are dealing with reality.

Fr. Stanley Jaki exposed this deception still more boldly than in his words quoted above:

> Behind these bravados about extraterrestrials you either see the strategy of the devil or you remain blind. How, you will ask yourself, can great scientific minds ignore the obvious? The answer comes only from the theology of original sin and of the devil. Leave them out of your equations and you will not understand anything ... And just as he was the father of lies in the first instance, in reference to Eve, the devil continues using the strategy of clever lies, which never look so clever as when wrapped in science.[18]

"**Just as he was the father of lies in the first instance**," this priest-scientist-prophet cautions. The parallels of this "first instance" with our own times are striking.

No reader of the third chapter of Genesis will struggle to recall who first delivered these promises that the originator of the Search for Extraterrestrial Intelligence was desperate to find the fulfillment of in extraterrestrials. The Father of Lies has changed the shapes of his appearances, but not the hallmarks of his tactics. Today, as Fr. Jaki observed, his "clever lies" are more beguiling than ever before, as they are "wrapped in science"—the myths of which have already become modernity's de-facto religion.

> "Seduced by the devil, [man] wanted to 'be like God', but 'without God, before God, and not in accordance with God'." (*Catechism of the Catholic Church*, §398)

Sixty-two years after his 1960 fathering of SETI, the ninety-two-year-old Drake discovered for himself that immortality cannot be found in the material Universe. But the living who now follow in his footsteps remain undaunted. Some are not as forthright as Dr. Drake in confessing their messianic motives (even though no other ones could account for their zeal in such a Quixotic undertaking). Most, however, are even more bold in stating their pseudo-religious aspirations.

We need only consider the man now carrying Drake's baton—not by heading SETI, but by founding and leading "The Galileo Project," a still broader and more zealous initiative (not to mention a billionaire-backed one) to discover aliens. His name is Dr. Avi Loeb. A renowned Harvard astrophysicist and bestselling author, he is now world-famous for announcing that the Age of the Extraterrestrial has arrived. Amazon lists his 2021 book, *Extraterrestrial: The First Sign of Intelligent Life Beyond Earth*, as their "Editor's Pick"—in the "Nonfiction" category. NBC News extolled Loeb as "*one of the country's best-known astronomers, [who] was the longest-serving chairman of the astronomy department at Harvard and has held prestigious appointments from the White House and top academic societies.*"[19]

Professor Loeb has not left us guessing about the depths of his mission's motivation or the heights of its aims. He declared in 2023 that **"Extraterrestrials could save humanity... the Messiah ... might be riding on a spacecraft from an exoplanet,"**[20] before contemplating the following year, **"...will there be anything else that God in religious belief systems can do**

and that science will not be able to do?"[21] In light of extraterrestrials, he is not reticent to give a resolute "no" to that question, and he boasts of the University lectures he plans to deliver a billion years in the future thanks to the secular immortality he anticipates.

In another missive, Loeb—an Israeli-born ethnic Jew—drew from knowledge of his religion to predict an impending "Cosmic Exodus," delivering not the chosen people from Egypt, but the whole human race from its hitherto earth-bound history. In this New Exodus, he writes:

> Extraterrestrial technologies could result in apparent miracles on Earth... Artificial General Intelligence (AGI), would impress us as much as the acts of 'God' impressed the Israelites...Would an Unidentified Anomalous Phenomenon (UAP) [UFO] serve as the burning bush moment for our new Moses?[22]

"**Who would serve as Moses for the new space community?**", Loeb next ponders, only before telling us, "**This baby may have been already born by now**." Loeb may be right about one detail, but that "new Moses" is not who Loeb thinks him to be.

> "...[The] time has come for the most astonishing wonders to take place on the earth and in the air ... because the true faith has been extinguished and false light illumines the world ... It will be at this time that the Antichrist will be born ...The demons of the air, together with the Antichrist, will work great wonders on the earth and in the air, and men will become ever more perverted."—Prophecy of Our Lady of La Salette

Loeb concludes his remarks with a simple exhortation: "**All we need to do is open our cameras and telescopes, as well as our mind, to the possibility that we might be saved from our terrestrial miseries by a superhuman entity**."[23] Scarcely more difficult than reaching up to pick a fruit from a tree.

Behold the Anti-Fiat.

Anything but "fringe" voices, men who share Loeb's objectives are of paramount importance and popularity in contemporary Ufology. Their missions define it. In the chapters ahead, we will see that they are joined by an army of colleagues and followers whose similarly messianic ET aspirations grow more popular by the day.

Moreover, we will chart out in detail the societal grooming process (through science fiction) that has brought us here. We will mine the depths of this deception's infiltration of Christendom, which—appropriately—goes all the way back to the birth of the Enlightenment. We will discern the fruits—as the Lord admonishes (Matthew 7:16) must be done in order to understand whether the tree is good or bad—and from this we will grasp the true nature of the ET and UFO *Strong Delusion*. We will consider demonology and relate its tenets to what is reported with UFO sightings and ET contact experiences. We will also consider multitudes of Scriptural, Magisterial, and scientific proofs which each independently refute the thesis that incarnate extraterrestrial intelligent life exists. And we will do all of this because the Word of God and the Signs of the Times have rendered the task urgent.

> But I am afraid that as the serpent deceived Eve by his cunning, your thoughts will be led astray from a sincere and pure devotion to Christ ... And no wonder, for even Satan disguises himself as an angel of light. (2 Corinthians 11:3,14)

Despite the gravity of the task before us, we are not—in fact—confronted with the "bad news" we may seem to be facing. It is news of impending chaos, to be sure—and a corresponding "**call for the endurance of the saints.**" (Revelation 14:12) Yet the words of the Lord are clear when such signs as these are observed: "**But when these things begin to come to pass, look up, and lift up your heads, because your redemption is at hand.**" (Luke 21:28) If indeed the apocalyptic persecutions, apostasies, and deceptions are at hand, then so too is the Triumph of the Faith.

We are promised that the time *will* come when, "**at Jesus' name, every knee must bend ... and every tongue proclaim to the glory of God the Father: JESUS CHRIST IS LORD!**" (Philippians 2:10-11) It will be a time when—according to Papal Magisterium—we shall see on earth, and not only in Heaven, the "**restoration of the human race in Jesus Christ**" (Pope St. Pius X, *E Supremi*, §14); in other words, "**that most auspicious day, whereon the whole world will gladly and willingly render obedience to the most sweet lordship of Christ the King.**" (Pope Pius XI, *Miserentissimus Redemptor*, §5) For the Immaculate Heart will Triumph and bring an Era of Peace to the world, as Our Lady of Fatima promised. This is all to say that the greatest prayer ever uttered will not go unanswered, for the Will of God shall reign on earth as in Heaven. (Matthew 6:10)

The French 19th-century seminary professor and missionary preacher, Fr. Charles Arminjon, wrote a book entitled *The End of the Present World and the Mysteries of the Future Life.* After reading it, St. Thérèse of Lisieux said that "**This reading was one of the greatest graces of my life**..." In the book, Fr. Arminjon wrote the following:

> Some have concluded from this passage [2 Thessalonians 2] that Christ is to come down in person to strike [the Antichrist], and that this will be the [end of time]. This interpretation is incorrect. St. Thomas and St. John Chrysostom explain … that Christ will strike the Antichrist by dazzling him with a brightness that will be like an omen and sign of His Second Coming … What is certain is that Satan will be hurled back into the darkness of the abyss, the reign of the man of evil will be utterly destroyed … The most authoritative view, and the one that appears to be most in harmony with Holy Scripture, is that, after the fall of the Antichrist, the Catholic Church will once again enter upon a period of prosperity and triumph ... These words are formal, and appear to leave no room for doubt.[24]

These are perilous times indeed, but—for those who love God—they are also the most exciting ones in history. We will discuss precisely why that is so in this book's concluding pages. For now it suffices to remember that while we must discern the times ahead in the Spirit, we must also approach them with no fear.

> "Choose today whom you will serve… As for me and my household, we will serve the Lord." (Joshua 24:15)

***

As the sun rises on the Eternal City one day near at hand, The Dawn from On High shall again break upon us. Announced by the resplendence of the Morning Star, God arises, and His enemies are scattered.

The Last Deception conquered like foam in the fire, as the First Deception was crushed by the Holy Cross. Eden's glory, rent asunder by the envy of the Devil, Now returns; the pride of his final onslaught is humiliated.

Fiat voluntas tua, Sicut in Caelo, et in terra.

Then will the lion lay down with the lamb; Every knee bent in adoration at the Name of the Most High. The Reign of God on earth veiled in the Sacred Mysteries. What He began with the Fiat, And He redeemed through the Fiat, So He concludes in the Fiat. Thus the Wife of the Lamb is adorned and made ready, To greet the Bridegroom when He comes.

~~~

Then, on That Day known to Him alone, As the sun rises, never again to set, He will descend as He ascended, lightning upon the clouds. His promises fulfilled, 'I am with you always; to the close of the Age.'

History concludes. The elements dissolve, the stars fall from their place, the dead rise, the Judgment is spoken. Those found in the Book of Life; Never to be separated from Him again. The Eternal Wedding Feast commences. Every tear wiped away.

God, all in all, In the World without end.

Amen. Come, Lord Jesus.
~~~

2. Leo XIII, Satan's Reign, and 1947 A.D.

"[Lucifer] will be unchained for a time fifty or sixty years before the year of Christ 2000."

—Prophecy of Blessed Anne Catherine Emmerich

Despite the shockwaves it generated, the 1972 sermon of Pope Paul VI announcing that the Smoke of Satan had entered the Church may not be the most consequential of all the papal prophecies. The mystical vision of his predecessor, who ushered in the 20th century, stands in a class of its own.

Before considering his vision, we first recall that most auspicious day exactly 33 years later: October 13, 1917. This day saw the most astonishing miracle (measured by the number who witnessed it and the magnitude of the marvel) since Moses split the Red Sea. The sun danced in the sky before a hundred thousand witnesses who had gathered precisely because a sign had been foretold. These pious souls had all been soaking wet from the downpour that the miracle interrupted, yet immediately found their clothing bone-dry upon the brief event's conclusion. Even secular newspapers widely reported on this display of Divine grandeur. It was the famous "Miracle of the Sun" during the apparitions of the Blessed Virgin Mary in Fatima, Portugal. (Later, the world's most renowned Ufologists would blasphemously insist this Marian miracle was actually a UFO sighting.)

Also on October 13—in the year 1884—Pope Leo XIII celebrated Mass in the presence of several Cardinals. This day, however—instead of proceeding to recite the customary three Hail Marys at the conclusion of Mass—the Pope turned white as a ghost and fell to the floor, appearing to be dead. Doctors were rushed into the room, but there was no work for them to do. Leo had revived as promptly as he had collapsed, though now with the task of sharing a "**horrible vision**" he had just received; "**a vision of Satan approaching the throne of God, boasting that he could destroy the Church**."[25] What this lion of a Pope saw was so immense that it compelled him to write a new prayer imploring the protection of St. Michael the Archangel, against the "wickedness and snares of the devil" who "prowls about the world," and order that it be recited after each Holy Mass across the globe. The words Pope Leo witnessed being exchanged between God and the Devil—reminiscent of the Book of Job's first chapter—are reported to be as follows:

> The voice of Satan in his pride, boasted to our Lord: "*I can destroy your Church.*" The gentle voice of our Lord: "*You can? Then go ahead and do so.*" Satan: "*To do so, I need more time and more power.*" Our Lord: "*How much time? How much power?*" Satan: "*75 to 100 years, and a greater power over those who will give themselves over to my service.*" Our Lord: "*You have the time, you will have the power. Do with them what you will.*"[26]

God allowed Lucifer this longer leash upon the earth, but only for a time.

Many Catholic commentators have since pondered when that period of Satan's greater power could have begun. Some contend those perilous years began immediately upon Leo's vision. They even pin the completion of the dark reign to Pope John XXIII's 1959 announcement of the Second Vatican Council seventy-five years later, as if the mere mention of a new congregation of Bishops within Rome was tantamount to a successful global exorcism of Satan's dominion. Other, less amusing, speculations appeared initially convincing. These proposed that the beginning of the period of enhanced Luciferian power coincided with the onset of the 20th century, or the start of World War I in 1914, or the birth of Communism in 1917. Yet not only the 75th anniversaries, but also the 100th of each, came and went. And they did so with no elimination or mitigation of diabolical action on earth, but only its increase.

Those speculations—now negated by history—overlooked the fact that Satan requires no unusually long leash to inspire political tyranny, war, and genocide. He has always proven competent in such tasks since the beginning of history, even though the 20th century's cultural, technological, and geopolitical developments rendered his efforts more destructive than ever before.

Satan does, however, need additional leeway for instituting an unprecedented seductive slew of false signs and wonders, appearing in new forms, deluding entirely new groups of people, communicating with men in new ways, tormenting them with new methods, deceiving them with new illusions, and so on. In a word, the Devil would need additional leeway to undertake precisely what we have continuously observed since—and only since—the moment the UFO Deception began in the year 1947.

> "Now the Spirit explicitly says that in the last times some will turn away from the faith by paying attention to deceitful spirits and demonic instructions." (1 Timothy 4:1)

Assuredly, the Devil needs "greater power" to begin his formal preparations for Scripture's prophesied and historically unique "Strong Delusion." Such onslaughts as these are not found within his usual arsenal.

The great Miracle of the Sun at Fatima was heralded 33 years earlier by Pope Leo XIII's prophetic vision. Similarly it was followed, 30 years later, by the birth of the UFO Deception, which promptly credited itself for all Marian Apparitions, Biblical miracles, and even the supernatural in general.

The year 1947 was, indisputably, the onset of the UFO Era. On June 24th of that year, Kenneth Arnold reported "flying saucers"—at Mount Rainier in Washington State—and within weeks, alien bodies and craft were reported at Roswell, New Mexico. With the arrival of those fateful months, the world was never the same. Before 1947, the UFO scene was practically nonexistent. One contemporary scholar of the phenomenon, Dr. Christopher Partridge, observed:

> Before 1947 there is not a single recorded episode involving mass sightings of saucer-like objects ... [1947] both began the modern wave of sightings and also ushered UFOs into the popular consciousness. ...By the end of that year 850 UFO sightings had been reported in America alone.[27]

What we see, here, is that a now explosive global sensation was almost completely absent on earth before 1947. Clearly, that year marked something extraordinary. For the ET promoters, that year marks the age of the space alien—an age they now say is about to reach its crescendo in the "Day of Disclosure." A discerning Christian will notice something altogether different emerging; though also "otherworldly" in its own right.

> "The summer of 1947... is when all hell broke loose...[and] in every major [UFO] incident in the '40s and 50s... there's some sort of Catholic connection..." —Richard Geldreich, Ufologist. [28]

Now, if Satan received this greater power in the year 1947—to "destroy God's Church"—it would be reasonable to expect that something which had proven an impediment to this power's increase was removed at that same time. This is exactly what we see in various cases; for now, we will examine just one. Although it will be strange for some to hear, there have always been "victim souls"—exceedingly holy men and women who have given themselves over entirely to Jesus Christ, hoping to imitate Him in all things, including His Passion. Eager to suffer whatever is necessary in order to obtain mercy for the world, these willing "victims" populate the pages of Church history in their spiritual "restraining" of various terrible events from happening.

One such victim soul was a certain 20th-century Church-declared Servant of God we will mention again later—and who was approved by the Vatican in August 2024. She was endorsed by multiple canonized saints who knew her—including Padre Pio. Despite the depth of her writings, Jesus told this victim soul that the thousands of pages of private revelations He gave to her are simply a commentary on those greatest of all words, "**Thy Will be done on earth as it is in Heaven.**" (Matthew 6:10) This mystic, who miraculously lived for decades on the Eucharist alone, was also repeatedly told by Jesus that her sufferings on earth were *holding something back*. Jesus made it clear to her: once those sufferings ceased upon her entry into Heaven (that is, upon her death), this "something" may not be able to be restrained any longer.

The UFO era began almost immediately after this victim soul's death in 1947. Only a few months later, the world was aflame with talk of little green men visiting us from outer space to usher in a new era. The rest, as they say, is history.

> "A 'preterhuman intelligence' ... announced [to Aleister Crowley] the overthrowing of the Old World Order, which he considered to be dominated by a repressive Christendom. A New Age in the evolution of humanity had dawned, which Crowley called the 'Aeon of Horus.'" —Dr.

Timothy Dailey[29]

If a victim soul holding back a diabolical deception died that year, what of the opposite: an especially perverse soul *in whom* this delusion dwelt? Indeed, just as a holy soul's sufferings can hold back chastisements—diabolical onslaughts which, upon that person's death, no longer find themselves fettered—so too we observe the reverse. The demons who inhabit a uniquely perverse man can find themselves freer to infect others when their prior host has died.

> "When the unclean spirit has gone out of a man, he passes through waterless places ... Then he goes and brings with him seven other spirits more evil than himself ... So shall it be also with this evil generation." (Matthew 12:43,45)

In 1947, the so-called "wickedest man on earth," the "Great Beast 666," died. A degenerate himself and an avid promoter of the most grotesque sexual perversion, Aleister Crowley was also the world's most renowned occultist. With all the zeal and influence of an anti-Apostle, he laid down the foundations of esotericism which would be built upon in the following decades.

His message was the veritable opposite of the one Jesus gave to the mystic described above. The principal commandment of the "religion" he founded (Thelema) is a direct inversion of the greatest prayer of Christianity. "**Do what thou wilt shall be the whole of the Law,**" this "Beast" exhorted his followers. While no one who recognizes that the Devil is real will struggle to identify him as the source of Crowley's teachings, too few are aware of precisely what form the demons acquired to instruct this chosen instrument of theirs.

The primary depiction of aliens that has deluged society for the last 78 years is that of "the Roswell Greys." But the image of their massive skulls rising above diminutive bodies came first from a 1917 illustration of the demon, "LAM," that Crowley interacted with, and which Crowley himself thereafter drew. Catholic expert in the occult, Ursula Bielski, explains:

> LAM appeared to Crowley while he was performing a ritual known as the Amalantrah Working, a ritual meant to open a portal between the seen and unseen worlds. ... Famously, Crowley remarked of the many entities he encountered in his supernatural travels, "*Today they call them angels and demons; tomorrow they'll call them something else.*" [30]

Bielski goes on to describe the equally demonic activities of Crowley's most famous student, Jack Parsons. Parsons was the founder of NASA's Jet Propulsion Laboratory, which, in turn, is heavily involved[31] in the Search for Extraterrestrial Intelligence (SETI); itself an effort to achieve immortality through contact with aliens. She writes:

> Parsons began attending gatherings of Aleister Crowley's Ordo Templi Orientis in Los Angeles... It was while operating [a] depraved hostel [in Pasadena] that Parsons met an aspiring writer and fellow occultist named

> L. Ron Hubbard, future founder of the Church of Scientology … Hubbard and Parsons became fast, close friends. After some minor occult misadventures, they decided to embark upon some major conjuring. The aim was to bring about the birth of an immaculately-conceived "moonchild" or Antichrist, the demon goddess Crowley called "Babalon." … Much like Crowley's earlier New York ritual that had, he claimed, manifested LAM, Parson's ritual was designed to manifest a collection of interdimensional entities … to bring about the end of Judeo-Christian civilization.[32]

A *New York Post* article even noted that "**Hubbard's son reveals that Hubbard claimed to be Crowley's successor: Hubbard told him that Scientology was born on the day that Crowley died.**"[33] Hubbard went on to declare himself the Antichrist as he built this massive, rich, and powerful UFO religion. Recall that not only was Scientology born[34] when Crowley died, but also Ufology itself. The Devil was on the move.

***

The year 1947 stands apart in history for additional reasons relevant to the present discernment. This was the year of the start of the Cold War; that decades-long global trial of existential dread which, for the first time ever, saw mankind seriously pondering whether his collective existence would continue another day. The incessantly looming specter of imminent nuclear war proved a possibility (and often an apparent probability) so terrifying that the masses began looking anywhere—and now, to the stars—for any purported means of averting the horrid prospect. The palpable dread also resulted in the creation of the now-infamous "Doomsday Clock" in the June 1947 edition of the *Bulletin of the Atomic Scientists*. The darkness that descended upon the earth along with the eponymous coldness of the war, however, did not depart when the Cold War "ended" in 1991. Indeed, the likelihood of nuclear war appears, today, higher than ever before. Many even argue—not unconvincingly—that the Third World War has already begun.

Today, the source of nuclear-apocalypse-angst derives especially from Israel and Russia (a primary cause of concern of the apparitions of Our Lady of Fatima, discussed above). As of this writing, the Holy Land is engaged in a war that risks spreading across the Middle East and even the whole world. The locus of this conflict, the modern state of Israel, also traces its history to the year 1947. On November 29th of that year, the United Nations adopted Resolution 181, and several months later, the Declaration of the Establishment of the State of Israel was proclaimed.

But a remarkable disparity presently exists between the two main Christian views about the question of Israel. Some contend this country remains the veritable Kingdom of God on earth (or at least some version thereof), as in the days of the Old Testament. Others insist the modern State of Israel has little or nothing to do with Biblical Israel and will only—upon the construction of The Third Temple—roll out the red carpet for the Antichrist. Regardless of the respective merits of each position, they remain in agreement on one basic premise. Each acknowledges that few things could be so prophetically significant as the rebirth of "Israel" in the year 1947 after millennia of its absence from earth. Each recognizes, therefore, that what the news reports about that globally and historically unique land stands as a harbinger of apocalyptic times.

Incidentally, today's largest explicit UFO Religion, Raëlianism (whose very name derives from "Israel"), holds as its supreme mission the establishment of an "intergalactic embassy" in Israel. The religion's founder—"Raël" (born 1946)—says that this embassy will serve multiple purposes. It will be the long-awaited "Third Temple" of the Jews, as well as the headquarters for the action of the "Elohim" on this planet (not to mention their literal landing pad). We are assured that these Elohim—advanced extraterrestrials who count among their ranks previous "prophets" whom earth has already seen (such as Jesus, Muhammad, and the Buddha)—will arrive by the year 2035 to usher in a global techno-utopia and a new world order with a single currency. No laughing matter, Raëlianism claims one hundred thousand adherents across the world, and has already amassed tens of millions of dollars for the construction of this "embassy." The primary impetus for their claims? The prospect of nuclear destruction that began with the Cold War, brought us into the "Age of Revelation." Now, only two options exist: witness the atomic extinction of our species, or accept the grand solution—brought by the ETs—to avert this catastrophe.

It is beyond clear. If the Great Deception will involve men desperately seeking an "apparent solution" to their problems at the price of apostasy from the truth—and we know, from the Catechism, it will consist in precisely that—then one fact emerges of paramount importance. The fact is this: no other "problem" society has ever faced in its history rises to a similarly apocalyptic stature as the one which began its formal phase in the year 1947.

> "You hypocrites! You know how to interpret the appearance of earth and sky; why do you not know how to interpret the present time?" (Luke 12:56)

Hollywood's grooming of society looms large here. The noble alien with his AI robot from *The Day the Earth Stood Still*, released only four years later in 1951, quickly emerged as a cultural mainstay that continues to speak to the aspirations of the masses. This ET came to earth on his craft in response to our nuclear weapons, and after much drama, his final message before ascending to the heavens on the same craft was blunt: "*Your choice is simple. Join us and live in peace, or pursue your present course and face obliteration. We shall be waiting for your answer.*" The alien's solution that we are seduced to accept? Robots. Artificially-intelligent machines that would immediately stop any potential source of aggression in its tracks.

We will see that the "Artificial Intelligence" (AI) Deception is deeply linked to the ET Deception. Therefore, we should not be surprised if both delusions were birthed like evil twins. Not coincidentally, the transistor was also invented in 1947. Although a great feat of engineering and obviously morally unproblematic in and of itself, this single event nevertheless heralded the birth of the Information Age; the "Digital Revolution." The rise of computing, more so than any previous Industrial Revolution's engine, prompted men to begin regarding technology—and now, "AI"—with a veritable *messianism*.

In both lies—the two main outlets of the "NHI" (Nonhuman Intelligence) Deception—society has only steadily descended to ever-new degrees of darkness in the intervening decades. Today, even the world's richest and most powerful elite serve as its prophets. The founder of Google, Larry Page, has excitedly announced the imminent rise of a "digital superintelligence" and a "digital god," which will replace humanity once AI reaches "The Singularity" to become "AGI" (Artificial *General* Intelligence). Similarly powerful elite have made identical claims, as we will see in Part Four. The most celebrated and popular pundits have also joined the bandwagon. Yet whether they giddily prophesy of an impending AGI-Utopia, or nervously fret of an impending AGI-Hellscape, both pronouncements derive from, promote, and hasten the arrival of the *Strong Delusion* (2 Thessalonians 2). For they both relay the supposition that "Nonhuman Intelligence" exists, and the exhortation to enter into dialogue with it. One does so by inviting it to replace us; the other by begging it not to. Just as there are Ufologists who caution about the arrival of malicious ETs and those who extol the arrival of benevolent ETs, so too there are AGI-heralds who caution about the "Singularity," and those who excitedly await it. What too few observers discern is that both camps—being built on the same lie—bring their devotees to the same tragic fate.

The priest-physicist quoted earlier, Fr. Stanley Jaki, was just as astute in describing the AI Deception as the ET Deception—noting that the former, too, threatens the most extreme destruction. He declared that by rendering "**homage not to God but to the Supreme Master Programmer**," thanks to man's own reckless and willful forgetfulness—choosing to become "**oblivious to his own uniqueness**" [35]—mankind capitulates to the doctrines of demons. This capitulation occurs just as terribly by believing dead and dumb computers are now sentient as by supposing aliens are arriving to save us. In both cases, mankind again listens to the voice of the serpent who ever whispers, "**Did God *really* say?**" (Genesis 3:2)

> "They forgot what he had done, and the miracles that he had shown them." (Psalm 78:11)

It is no coincidence that almost all those pundits now announcing the Age of the Extraterrestrial are likewise insisting that the Singularity of AGI is imminent. These two deceptions are as wed to each other as are the Beast and its image (cf. Revelation 13), as we will also see in Part Four.

***

Even the heavenly bodies themselves announced the change that the year 1947 brought with it. For what could better symbolize the emergence of a new spiritual darkness than the light itself—that is, the sun—being blotted out? Mankind has been observing sunspots for over two thousand years, but what remains by far the largest ever recorded is none other than "The Great Sunspot of 1947." It was "so large as to be seen easily with the unaided eye,"[36] according to reports from astronomers. Although dark in appearance, we know that these spots correspond to regions of great magnetic flux; so violent that they can wreak havoc upon earth by spawning coronal mass ejections that create geomagnetic storms. (Ironically, these are the very "storms" which could wipe out the entire infrastructure of the Computer Age that began also in 1947.)

As if to signal the imminent climax of the age 1947 initiated, sunspot activity rose massively in the year 2024. This surge even caused the Aurora Borealis to adorn the skies as far south as Mexico—astonishing those who had never before seen these Northern Lights.

If indeed Satan's 75 to 100 years of greater power began in the year 1947, then it could now reach its summit—and then conclude—at any point.

> "...The great questions, the formidable events of the present hour, the dangers that threaten the whole world ... The Year 1947! What will be the judgment of the coming ages?... Future generations will bless it or curse it, depending on whether it will represent for the great human family a starting point toward the awakening of fraternity...or instead a progressive decay ... of new and incalculable calamities" —Venerable Pope Pius XII[37]

The answer to the question the Pope here posed—"**The Year 1947! What will be the judgment of the coming ages?**"—is becoming clearer and clearer. In fact, it was rendered clear by the prophetic foresight of another bishop at the dawn of the same year. In January 1947, Venerable Archbishop Fulton Sheen delivered a radio address in which he described in great detail, just several months before the deluge began, the tenets of the ET Deception of the decades that would follow. Sheen prophesied:

> Progress, [they say,] is automatic thanks to science—education and evolution, which will one day make man a kind of a god as H.G. Wells said ... The anti-Christ will not be so called, otherwise he would have no followers ... This masquerade has helped the devil convince men that he does not exist, for he knows that he is never so strong as when men believe that he does not exist. When no man recognizes, the more power he exercises. ... How will [the Antichrist] come in this new age to win followers to his religion? He will come disguised as the Great Humanitarian; he will talk peace, prosperity and plenty ... he will invoke religion to destroy religion; he will even speak of Christ and say that he was the greatest man who ever lived ... He will set up a counter-Church which will be the ape of the Church because, he the devil, is the ape of God. It will have all the notes and characteristics of the Church, but in reverse

> and emptied of its divine content. It will be a mystical body of the anti-Christ that will in all externals resemble the mystical body of Christ. ... There are Times of Troubles and it is not so much a Third World War that is to be feared, as the rebirth of Leviathan, the coming of the Day of the Beast. [38]

Sheen's diagnosis of the times which then lay ahead has proven accurate. As we will see in Part Three, the ET Deception is precisely the religion of Antichrist—permeated, as this prophet described, by "**all the notes and characteristics of the Church, but in reverse and emptied of its divine content**."

Concluding his address, the Venerable Archbishop quoted the poet Heinrich Heine's warning, equally accurate: **"Once the taming talisman—the Cross—is broken, the savagery of the old battlers will flare up again ... Then the old stone gods will rise from forgotten rubble and rub the dust of a thousand years from their eyes; and Thor will leap up and with his giant hammer start smashing Gothic cathedrals**..."[39]

These words can be summarized simply: with Christianity no longer restraining the Devil as it once did, the Pagan "gods" now return. But the demons who once wore the disguise of gods now come clothed in far more apocalyptic and seductive attire.

***

Nevertheless, we must ask: how do we *know* we are dealing with demons here? A Christian must not be hasty in arriving at such a dramatic conclusion. Indeed, the Devil is not behind every rock. Therefore, before we take later chapters to trace out the origins of this deception before 1947—and follow its development since then, up to the present day—we must change pace. We must now apply the intellectual rigor needed to determine—from Scripture, Magisterium, theology, philosophy, and science—that aliens do not exist.

Thereafter, with "faculties trained by practice to distinguish good from evil," (Hebrews 5:14) we will be equipped to "discern the signs of the times" (Matthew 16:3) in the light of Faith.

Part Two: The Case Against Aliens

3. Origins of The Deception

"I thank you Father, Lord of heaven and earth, because you have hidden these things from the wise and the learned and have revealed them to the childlike." (Matthew 11:25)

Among the greatest achievements of modern science are its discoveries that what was once, in starry-eyed ignorance, seen as a surely impending achievement *of* science is, in fact, securely relegated to the realm of science fiction.

Before presenting the mature science on the question of extraterrestrials, we should review its inebriated adolescent phase. Teachers of chemistry, biology, and geology often begin their classes by exposing the fallacies that once gave rise to belief in phlogiston, spontaneous generation, and a flat earth—in order to ensure the premises which gave rise to these illusions do not infect their students' approach. Hence, in assessing the question of extraterrestrial intelligence from a scientific perspective, we must be aware of the spurious premises which led many early scientists to believe in other inhabited planets.

Hindus, Buddhists, Muslims, and other non-Christians have long believed in fables of aliens (their own religions teach as much). The brand of alien belief couched in scientific terms, however, first arose in Pagan ancient Greece several hundred years before Christ. Although the great philosophers—Plato and Aristotle—rejected this "plurality doctrine," Epicurus and Democritus, with their philosophies of hedonism and materialism, famously embraced belief in extraterrestrials. (Little wonder, then, that the postmodern West—which makes ancient hedonism look pure, and ancient atheism look pious—has turned back to the philosophies founded by these two.)

Despite the influence of Greek thought on early Christian history, ET belief was always unanimously condemned by orthodox Christians during the Faith's first millennium-and-a-half. This "sensus fidelium" (which, as we will see in later chapters, enjoys infallibility) *appeared* to change during the so-called Enlightenment. More truthfully called the "Great Darkening," the Enlightenment was conceived by the radical skepticism of René Descartes' 1637 proclamation—*cogito ergo sum*, or, "I think, therefore I am." The shockwaves swept the Christian globe in the decades that followed, and the 17th century saw academics begin rejecting the authority of Divine Revelation (even if not yet disavowing all its content—that would come later); reserving their ultimate deference for a purely secular account of all things.

The Enlightenment's chief premise for alien belief was called "The Principle of Plenitude." This pseudo-philosophical axiom held that any putatively positive thing which *could* theoretically exist, also *must* exist—at least somewhere in the Universe. Practical atheists of that era (including many nominal Christians)—discontent with placing supernatural hope in God and in Heaven—found in this principle an outlet for the eternally-oriented longings of the human heart. For Plenitude pointed, instead, to creation and to temporality. "**They exchanged the truth about God for a lie and worshiped and served the creature rather than the Creator...**" (Romans 1:25)

According to Arthur Lovejoy in his groundbreaking 1936 work, *The Great Chain of Being,* the rationalist Baruch Spinoza (born 1632) "expressed the Principle of Plenitude in its most uncompromising form."[40] Spinoza was a Dutch devotee of Descartes who had been a Jew before the Sephardi authorities excommunicated him for "abominable heresies." He espoused Pantheism, and his passion was to create a purely philosophical and scientific new world religion built upon his extensive treatises about "Plenitude," while remaining devoid of the supernatural. Unsuccessful during life in this quest, his legacy achieved its aims: postmodern secularism has diligently followed the instructions found in Spinoza's blueprints.

If Spinoza gave the boldest philosophical definition of Plenitude, Giordano Bruno—in the century prior—remained unsurpassed in zeal as the principle's chief apostle. Since Bruno's late 16th-century preaching—and emboldened still more by his execution at the dawn of the 17th century—the esoteric and Gnostic have remained fascinated by fantasies of aliens.

Contrary to the hasty assurances of certain 20th-century writers, this ex-Dominican priest and cosmologist was indeed burned at the stake precisely for obstinately preaching belief in aliens, as we will see in Chapter 9. Accordingly, contemporary ET promoters regard him as their ideology's "first martyr." Heretic he was; martyr he was not. Treating Bruno's legacy, Dr. Lovejoy explains that:

> [Bruno] must be regarded as the principal representative of the doctrine of the decentralized, infinite, and infinitely populous universe... [he] preached it throughout Western Europe with the fervor of an evangelist...[41]

Bruno did not merely preach a Universe filled with aliens; he denounced those who failed to agree with his thesis as being mentally deficient and levying a blasphemous insult against God. He insisted that there are worlds "*... no less inhabited and no less nobly [than earth]. For it is impossible that a rational being fairly vigilant, can imagine that these innumerable worlds, manifest as like to our own or yet more magnificent, should be destitute of similar and even superior inhabitants.*"[42]

But a theory divorced from concrete developments will generally fail to—for better or worse—change the world. A Marxism without a Lenin, or a Sexual Revolution without contraceptives, would have (happily) remained as much talk. Similarly, the theory of Plenitude—without new ways to peer into the heavenly bodies—would have only been another forgotten academic fad. In the years after Bruno's execution, however, other planets did begin to be observed—and described—in detail. Galileo's telescopes enabled this, and a new era was born. An unholy marriage was contracted between the new revelations from the science of astronomy and the Enlightenment's Principle of Plenitude.

Telescopes and Fallacies

Even without the benefit of hindsight, it is easy to surmise how this confluence of technology and academic thought would play out. No longer "shackled" by the need to believe that God did *what He said He did* (in Divine Revelation), Enlightenment thinkers were instead imprisoned by their own deductions. They were cornered by the juxtaposition of their presuppositions and their discoveries, and compelled to profess "absolute certainty" that *each* of the newly observed planets "must" contain alien civilizations. Even one planet without rational incarnate inhabitants meant the demise of their Principle of Plenitude, and risked undermining Enlightenment optimism itself.

By the later decades of the 1700s, academia had claimed for itself such certainty in the existence of aliens that even Immanuel Kant (†1804)—likely the most renowned of the Enlightenment thinkers—considered it an absolute. This so-called "Christian genius" appreciated his religion's ethics but had no time for its core. Although he harbored no real Faith in God (he denied we could even know God *exists*, much less that He became Incarnate), Kant did have great Faith in extraterrestrials. In his most famous work, *Critique of Pure Reason*, he wrote:

> I would indeed bet all that I own ... that there are inhabitants on at least one of the planets, which we see, [*in the year 1781; i.e., in this solar system*]. ... this view ... is not a mere opinion but a strong faith (on whose correctness I would surely risk many of life's advantages), that there are inhabitants in other worlds.[43]

Any scholar of Kant would be hard pressed to find a single other view for which this "Father of Modern Philosophy" expressed such conviction and enthusiasm—even declaring it a "strong faith" upon which he would readily bet everything.

> "... [Kant] claimed, among other things, that the inhabitants of Mercury must be very lively because of the great heat there, whereas the inhabitants of Saturn were superior philosophers, compared to whom Newton was a mere ape. The future celebrity of critical reasoning, or pure reason indeed, never disavowed these wallowings of his in pure fantasies." — Fr. Stanley Jaki[44]

A younger contemporary of Kant, the Reverend Thomas Dick (†1857), was another preeminent Enlightenment thinker. Whatever he lacked in stature compared to Kant, he made up for—in the context of this discussion—by also being a Protestant minister and famous astronomer. Rev. Dr. Dick taught that to fail to believe in aliens "**would be impious, blasphemous, and absurd**." He continued:

> We are, therefore, irresistibly led to the conclusion, that the planets are the abodes of intelligent beings ... This is a conclusion which is not merely probable, but absolutely certain, for the opposite opinion would rob the Deity of the most distinguishing attribute of his nature, by virtually denying him the perfection of infinite wisdom and intelligence.[45]

Alien belief promotion "became a core component"[46] of Reverend Dick's vocation, so much that he "*deluged both sides of the Atlantic with volumes in which ideas of extraterrestrial life appeared with great frequency*."[47] He was not only "absolutely certain" in aliens theologically, but also scientifically. In reference to the various planets in our own solar system, he argued we can easily "prove" they are inhabited. Among many other reasons, he insisted that their scenery "**forms a presumptive proof that both the planets and their moons are inhabited by intellectual beings**."[48] He even calculated the numbers of intelligent extraterrestrials that must dwell on our solar system's various bodies: on Venus, 53.5 billion; on Mars, 15.5 billion; on Jupiter, 7 trillion. On the list goes, providing rigorously evaluated alien census data for other moons and celestial bodies, totaling *about* 22 trillion aliens inhabiting our solar system. His count of the number of aliens on Saturn's *rings*, however, was especially precise: 8,141,963,826,080.

Still more amusing declarations made headlines in the days after Reverend Dick's passing. Some came from the man who next arose to be the world's most renowned astronomer, Camille Flammarion (1842-1925). The author of "**the most complete study of the history of observations of Mars ever written**,"[49] Flammarion had many exciting revelations in store for his global audience.

> "In two scientific treatises, *The Plurality of Worlds* and *Mars and Its Inhabitants*, Flammarion stated his belief that Mars not only housed life but also intelligent beings. Dwarfed in stature, positioned next to his fifteen foot-tall telescope, [he] described in detail the mountains, valleys, craters, lakes, and oceans of Mars in North American Review in 1896. 'It is obvious,' Flammarion concluded, 'the world of Mars is ... vigorously alive.'"
> —Dr. Michael Crowe[50]

Flammarion's works were so widely read that historians have noted they were "**unequalled among scientific books**," and he was certainly the "**most prolific of all writers on astronomy**..."[51] Like other contemporaries of similar persuasions, Flammarion *had been* a Catholic. He even studied for several years as a seminarian before forsaking his Faith to free his mind for the unbridled pursuit of Exotheology. One modern author noted:

> [Flammarion] dreamed of other worlds ... He had lost the Catholic faith of his youth... He substituted for it another faith, based on [extraterrestrial reincarnation]... His writings strongly endorsed the idea that ...[Mars] was almost certainly inhabited—so much so that when in 1891 a Madame Guzman, of Bordeaux, decided to bequeath 100,000 francs as a prize ... *"for the person of whatever nation who will find the means within the next 10 years of communicating with a star (planet or otherwise) and of receiving a response,"* Mars was specifically excepted.[52]

In the year 1891, contact with extraterrestrials on Mars was treated as such an obviously imminent accomplishment—such a foregone conclusion—that a prize in astronomy was not even deemed appropriate to bestow upon the one who first managed it!

Flammarion did not merely insist that Mars was inhabited. He assured his readers that, on Mars, "**these unknown brothers are ... active beings, thinking, reasoning as we do here**."[53] Only two years before Flammarion's death in 1925, *The New York Times* published an article entitled *"Flammarion Predicts Talking with Mars,"* and relayed that he anticipated *"'Psychic Waves' will be developed to the point of overcoming space."*[54] As with contemporary Ufologists, occultist mediumship (today classed "remote viewing") was a great interest of Flammarion.

Reminiscent of almost all ET promoters from Bruno onwards, who charge alien disbelievers with the crime of "limiting God," Flammarion employed the harshest rebukes against anyone who disagreed, saying that such people are simply blasphemers who are guilty of "**hurling a gross insult in the shining face 'of the infinite Power who fashions the worlds.'**"[55] Even the sun and the moon, Flammarion declared, *must* be inhabited. In all this, Flammarion was no lone ranger. He was only the chief exponent of the dominant scientific view of his day; one most sharply illustrated by the "empirical" claim we must consider next.

Mars and the Moon

In the year 1877, Giovanni Schiaparelli, an Italian astronomer, described *canali* (canals or channels) that he claimed to have observed on Mars. Contemporary scholar Dr. George Basalla notes that "**[Schiaparelli's] supporters called him a modern Columbus who had discovered a new world on Mars**..."[56] Basalla continues:

> Interest in Martian canali reached such a peak in those years that a historian has likened the Martian canal furor to mass hysteria. ... [Claims proliferated] that intelligent Martians had built enormous structures on the planet, sent light signals, and made plans to invade the Earth. Inventors Thomas Edison, Nicola Tesla, and Guglielmo Marconi gave credibility to claims of Martian signals when they offered their technical advice to facilitate radio communication between the Earth and Mars.[57]

One promoter of this thesis was Percival Lowell (born 1855), an American mathematician and astronomer who was also a powerful businessman. A 1905 newspaper article announced that "**Professor Percival Lowell is certain that the canals on Mars are artificial [i.e., alien-built]. And nobody can contradict him**."[58]

> "Lowell claimed that because Mars was smaller than the Earth, it evolved faster ... Mars was much older than the Earth, and hence, Martian life had advanced far beyond human intelligence ... Unlike their inferiors on Earth, Martians had risen above petty party politics and arbitrary national boundaries to govern on a planet-wide basis. Lowell concluded his tribute to the Martian intellect by speculating that Martian inventions surpassed our wildest technological dreams ..." —Dr. George Basalla[59]

These assertions—presented (and accepted) as scientifically verified truth—are not merely entertaining historical tidbits. Like the others we have considered, Lowell was not some science fiction writer, nor was he a social commentator or political columnist. He was a veritable astronomer with access to the world's best telescopes and who even founded, owned, and administered a major observatory (the Lowell Observatory in Flagstaff, Arizona) which is in operation to the present day. His arguments carried immense weight.

Later developments in telescope design (and, ultimately, NASA's 1965 Mariner mission to Mars) revealed the whole phenomenon to be nothing other than an optical illusion combined with a smattering of psychological deception.

Nevertheless, so ubiquitous were 18th and 19th century claims of certainty in the existence of aliens on the moon, and Mars, and every other planet, that the chaos caused by "The Great Moon Hoax of 1835" is deeply telling but not in the least surprising. In August of that year, a series of articles (intended to be satirical) in the *New York Sun* announced that a well-known astronomer observed humanoids and other aliens populating the moon's surface. Illustrating the impact of these articles, Dr. Michael Crowe collected some reactions published by other newspapers in response to the claims:

> '*It appears to carry intrinsic evidence of being an authentic document.*' The Daily Advertiser, it was reported, expressed its enthusiasm by stating: '*No article, we believe, has appeared for years, that will command so general a perusal and publication. ...*' The Albany Daily Advertiser [called] it a '*Stupendous Discovery,*' [and] told of having read the story '*with unspeakable emotions of pleasure and astonishment.*' *The New York Times*, according to the Sun report, pronounced the discoveries '*probable and plausible,*' while the New Yorker described them as creating '*a new era in astronomy and science generally.*'[60]

It should not have been necessary for any investigative journalism to take place to expose the obvious fraud. Nevertheless, the eagerness with which "respectable" mainstream media outlets like *The New York Times* devoured this manifestly fake news is both a testimony to the wholly illogical nature of alien belief in the 19th century and a preview of the still more credulous gusto with which ET claims are accepted in the 21st century.

Unlike academics and major publishers, ordinary folk were not so gullible. Dr. Crowe continues his description of what we can only fairly diagnose as a sudden explosion of mass-formation-psychosis:

> According to other contemporary reports, "*Some of the grave religious journals made the great discovery a subject of pointed homilies*" and an American clergyman warned his congregation that he might have to solicit them for funds for Bibles for the inhabitants of the Moon. … A person who had been in New Haven in 1835 described the situation there: "*Yale was alive with staunch supporters. The literati – students and professors, doctors in divinity and law – and all the rest of the reading community, looked daily for the arrival of the [moon alien reports in the] New York mail with unexampled avidity and implicit faith.*" … Edgar Allan Poe later reported: "*Not one person in ten discredited it, and (strangest point of all!) the doubters were chiefly those who doubted without being able to say why – the ignorant, those uninformed in astronomy, people who would not believe because the thing was so novel, so entirely out of the usual way.*"[61]

The "moon hoax" was, in fact, written with all the right verbiage to appear legitimately scientific, but the main reason for its success was found elsewhere:

> The main reason why thousands of Americans believed these fantastic fictions is that for a number of decades they had been prepared for them by the preachings and proclamations of such [alien belief promoting] authors as Paine, Chalmers, Emerson, and Dick.[62]

Rev. Thomas Dick no doubt felt humiliated, and he strongly rebuked the hoax's author, Richard Locke. The latter's response is also instructive for us today:

> So far from feeling that I deserve the coarse reproaches of Dr. Dick, I think it is quite laudable in any man to satirize, as I did, that school of crude speculation and cant of which he is so eminent a professor. My hasty speculation … proved how deplorably the public mind had been trained to gullibility on matters of science...[63]

Whoever today reads these same New York periodicals which promoted the Moon Hoax need not be reminded that no lessons were learned. Within their pages, a 21st-century reader will learn of far more farcical claims being taken at face value: "UFOs" regularly traversing our skies and – with quasi-supernatural prowess – breaking the laws of physics, only to crash shortly thereafter; alien explorers traveling millions of light-years only to probe the nether regions of a selection of unfortunate Americans who recall these "abductions" through hypnosis; and "proof" of massive 78 year cover ups of ET contact (and craft!) that never provide evidence sufficiently credible to be so much as admissible in court, much less convincing if presented therein.

Just a few decades after the "Moon Hoax," an equally absurd claim was made. This time, it was not intended as a hoax or a work of satire. It was the life mission of renowned scientists, though the content was no less fictional. Attention from ET promoters had shifted away from contacting aliens on the moon, and towards the same body astronomers claimed harbored engineered canals: Mars.

> "Whereas Flammarion only set out to describe the Martians, Tesla actually made plans to contact them."[64]

When pondering the great innovators of the 19th and 20th centuries, one name quickly comes to the fore: Nikola Tesla. This Croatian-born American inventor—whom we largely have to thank for the existence of the electric grid—was not merely an alien believer or alien belief promoter, but was rather a veritable extraterrestrial-obsessed man on a mission. Tesla expressed outright conviction that he was receiving electromagnetic signals from Martians, writing:

> Others may scoff at this suggestion ... [of communicating] with one of our heavenly neighbors, as Mars ... but I have been in deep earnest about it ... The character of the disturbances [I] recorded precluded the possibility of their being of terrestrial origin ... The signals consisted in a regular repetition of numbers, and subsequent study convinced me that they must have emanated from Mars...[65]

In fact, Tesla was only hearing the transmission of another inventor—Guglielmo Marconi—which is precisely what Tesla repeatedly ruled out as impossible.

Tesla was a troubled man. He did not believe in God, but was quite willing to embrace anything else. Having abandoned the Orthodox Christianity of his youth (his own father was a priest), he sought a new object for his faith and hope in sci-fi fantasies, which he pursued with all the vigor of a Don Quixote, though without noble intentions. It is unsurprising that the modern world has largely idolized this antihero. Tesla's biographer, the psychologist Dr. Marc Seifer, discussed the inventor's obsession with aliens:

> These outer space entities may have [for Tesla] symbolized beings existing in the afterworld. Certainly the need to believe in extraterrestrials is a powerful and popular one. It explains why so many people accepted Percival Lowell's "Canals of Mars" hypothesis and, in today's world, the extreme popularity of such movies as Star Wars, Star Trek, and ET...[66]

A true false prophet of dystopia, Tesla excitedly issued predictions of imminent eugenics, and **"robots and thinking machines [that] will replace humans."**[67]

> "As Tesla made headlines in the New York dailies and electrical journals for his bold prediction that he would "signal the stars" ... H. G. Wells

> gained notoriety with his serialized Person's magazine horror story *War of the Worlds* ... Although fictional, these stories were based on prognostications put forward by supposedly sober scientists." —Dr. Marc Seifer[68]

Again and again, the most internationally-renowned scientists of the 18th to 20th centuries only confirmed each other in their delusions that "the evidence" they discovered presented "certainty" in the existence of aliens.

One would then naturally ponder: after the spectacular failure of the assertions of these scientists, were lessons *finally* learned?

***

Nikola Tesla died a lonely and broken man in a New York City hotel room in 1943. Percival Lowell passed away a few decades earlier, in the year 1916, and was laid to rest on "Mars Hill," near the observatory he founded and to which he dedicated his life. Martian alien theory, however, was not yet similarly buried, though it was at that point moribund.

The admission that Tesla did not communicate with aliens, and the increasing acceptance that it was unlikely the Martian "canals" were canals at all, did not translate into an increasing acceptance that Mars did not have creatures upon its surface. Undeterred by prior failures, the ET promoters simply did the same thing they and their predecessors had always done. They modified their arguments slightly, doubled down on their fundamental fallacies, and rejected the opportunity to learn lessons.

Carl Sagan then entered the fray. Perhaps the most popular scientist of the late 20th century, Sagan was idolized across the globe for various teachings in astronomy ever intermingled with his own atheism, but the man himself was above all fixated upon contacting aliens, as we will see in Part Three. Informally, he posited aliens dwelling on Mars "as big as polar bears,"[69] but even in a scientific paper published in January of 1965, "*Martian landing sites for the voyager mission,*" he declared "**the present body of scientific evidence suggests**" that there is life on Mars. He went on to advocate for any spacecraft visiting Mars to be sterilized, lest our earthly microorganisms contaminate the pristine Martian landscape bursting with life.

Then came July of that same year. Only several months after Sagan published the claims above, Mariner 4 sent back images of the Martian surface as it flew by the red planet. As one publication reported:

> The 21 grainy images of the planet ... revealed no signs of water or life.[70] ... There were, alas, none of the canals seen by astronomers in the late 19th and early 20th centuries, nor evidence of senders of messages heard by Nikola Tesla or Guglielmo Marconi. Indeed, the hazy images of a barren, crater-strewn landscape ended speculation that Mars might plausibly be inhabited by higher life forms.[71]

Mars, like all bodies outside earth, is a "dead planet." Today's ET promoters are eager for their audience to forget just how existentially devastating these NASA missions are for their thesis. *Century after century* of repeated claims had come from the world's loftiest scientists; claims of "absolute certainty" in the inhabitation of *all* the planets (including Mars). Similar theologically "certain" claims came from the (non-Catholic) theologians. In one day, it was all definitively shattered. All of the premises that had been used for hundreds of years to justify ET belief evaporated overnight like a puff of smoke. Yet the belief—revealing itself as a false religion, not a scientific view—only persisted and expanded.

4. Science vs. Aliens

"The debate about extraterrestrial intelligence contains just one gleaming, hard fact..." —Dr. Stephen Webb

We have seen that all of "The Science" used, for centuries, to "prove" the existence of aliens was entirely bunk. What, then, does actually *good* science say today?

Contemporary promotion of alien belief has finally conceded the barrenness of our planetary neighbors, turning its attention beyond this solar system. Accordingly, tsunamis of popular hype have been generated in recent decades whenever news breaks that just one of a myriad of conditions needed for life *might* exist on some "exoplanet" orbiting another star, despite such "news" being entirely trivial. We first must address the fallacy upon which this hype is built.

The Impossibility of Habitability

Good science has now shown that it is impossible for any planet outside of earth to even be *habitable*, much less inhabited.

Two scientists recently presented an explanation of what is needed for complex life to flourish on a planet, and a review of why ours is the only possible contender for that honor. After hundreds of pages of rigorous analysis, the authors of *Rare Earth* (the joint effort of a renowned paleontologist and astronomer) declared, in the book's conclusion:

> Our planet coalesced out of the debris from previous cosmic events at a position within a galaxy highly appropriate for [life]... around a star also highly appropriate ... We became a planet where global temperatures have allowed liquid water to exist ... and for that, our planet had to have a nearly circular orbit at a distance from a star itself emitting a nearly constant energy output for a long period of time. Our planet received a volume of water sufficient to cover most—but not all—of the planetary surface. Asteroids and comets hit us but not excessively so, thanks to the presence of giant gas planets such as Jupiter beyond us. ... Earth received the right range of building materials—and had the correct amount of internal heat—to allow plate tectonics to work on the planet, shaping the continents required and keeping global temperatures within a narrow range ... [The] Earth's remarkable thermostatic regulating process [always] successfully kept the surface temperature within livable range. Alone among terrestrial planets we have a large moon, and this single fact, which sets us apart from Mercury, Venus, and Mars, may have been crucial to the rise and continued existence of animal life on Earth.

This assessment captures well just how much we take for granted about our planet. Consider that "space"—which is precisely what all planets, including earth, are floating within—is as hostile-to-life an environment as one can imagine. As any aerospace engineer knows, designing and building a craft capable of withstanding space while, within itself, providing an environment amenable to life, is an extraordinarily difficult feat. Success in this endeavor requires immense amounts of collective ingenuity, investment, and effort from gargantuan teams of experts. Yet, earth itself is precisely such a craft, and such crafts do not simply come about by chance, no matter how many innumerable trillions of opportunities chance is given.

Unfortunately, *Rare Earth* was largely ignored by a secular society eager to believe that extraterrestrials would soon come to save us from ourselves. Fr. Stanley Jaki noted:

> Nobody paid attention to *Rare Earth*, and certainly not *The New York Times*, which is most eager to publish the latest results about [the] search for other planets around other stars. And each time that daily puts, preposterously enough, the best spin on the findings. ... The script might have been written by Screwtape whose tactic is to bedazzle man's mind with specious non-sequiturs.[72]

(Incidentally, it was also similarly flawed reporting from *The New York Times* that, in December 2017, generated the present wave of UFO obsession.)

Some ET promoters today insist that *Rare Earth* has been "debunked," and they bemoan that it is "still quoted by those who disbelieve in aliens." In fact, the opposite is true. We have only since discovered far *more* requirements for a planet to harbor life, thus rendering earth exponentially rarer still (infinitely rare, that is) than the *Rare Earth* book asserted it was.

One astrophysicist, Dr. Hugh Ross, recently compiled a more thorough list of 153 prerequisites for any planet to be compatible with life. This list, too, is doubtless partial, but even from a mathematical consideration of the quantified probabilities affixed to their likelihood of arising, one can easily see it is mathematically impossible for any other habitable planet to exist, anywhere. Dr. Ross explains:

> The probability of a planet anywhere in the universe fitting within all 153 parameters [required for life to even be possible] is approximately 10^-194. The maximum possible number of planets in the universe is estimated to be 10^22. Thus, less than 1 chance in 10^172 (100 thousand trillion trillion trillion trillion trillion trillion trillion trillion trillion trillion trillion trillion) exists that even one such planet would occur anywhere in the universe.[73]

Less than one chance in 10^172 is scientifically equivalent to zero—there is not a single even theoretically possible real-world application wherein the two numbers would be distinguishable. Whoever doubts this can enter one divided by that number into his calculator, and see the result himself. There is *no* chance that any planet exists anywhere else in the Universe capable of harboring life. Contrast these authentically scientific expositions of truths about our planet to the bold and "certain" claim of the "world's greatest" astronomer, Camille Flammarion. He promised that "**Earth has no particular distinction among the planets ... Earth has not been at all favored over the other celestial bodies.**"

Thus far, astronomers have discovered over five thousand planets. No evidence has been found that even one of them enjoys more than a few of the hundreds of those conditions we already know are requisite for life. An alien belief promoter, of course, protests that the planets we have discovered are such a minuscule percentage of such a massive whole that some other planet must exist with the necessary conditions for life. But even Dr. Crowe, the preeminent chronicler of history's debates on aliens—who was by no means against belief in the same—describes this argument accurately: as the "**fallacy [of the] misuse of large numbers**," wherein proponents erroneously conclude that such numbers must mean "**some [planets] at least must be inhabited**."[74]

It is all too easy for a thoughtless idealogue to attempt to bolster his views by presenting arbitrary deferrals to enormous numbers, and supposing that these large quantities will give any situation, no matter how impossible, a chance—or even a certainty!—of happening.

But that is not how reality operates.

When massive quantities "compete," as it were, against statistical impossibilities, *the statistical impossibilities always win*. It does not matter how massive the quantities are, since all such quantities derive from the concrete constraints of the *actual Universe*. Statistical impossibilities, on the other hand, flow entirely from mathematical axioms and bear no such constraints, though they are just as accurately and scientifically quantifiable.

Today's astronomers often estimate that there are several hundred quintillion planets in the Universe. This number remains, in fact, so tiny as to be essentially non-existent in comparison to the statistical impossibilities evident—and quantified above—in proposing that any other planet exists which is even theoretically capable of supporting life. One might as well claim that the number of grains of sand on earth is so vast that it statistically guarantees that at least some have randomly coalesced on a beach somewhere in a pattern that writes out an engaging novel. In fact, the unfathomable number of grains of sand is a veritable nothing compared to the infinite improbability of that chance occurrence.

When, however, considering the notion of planets not just habitable, but *actually harboring* intelligent life, science still more clearly rules out aliens. But since we have already seen that the probability of habitability itself is zero, we needn't belabor the point here.[75] (In the Appendices, we consider another fundamental cornerstone of science—falsifiability—which is violated by ET promoters.)

(Note: Earth's existence does not refute the argument here presented. Indeed, it is not possible – materialistically speaking – for earth (or life therein) to exist. Yet, it does exist. Therefore, earth is a miracle. An ET believer, therefore, can only claim

aliens exist by asserting that they [and their planets] are also miracles. Yet this assertion transfers the discussion to the domain of theology – and in the following chapters, we will see that theology, too, rules out aliens.)

***

With the onset of the second half of the 20th century, the nature of the extraterrestrial debate radically changed. This seismic shift came with "UFO" sightings and the "Ufology" movement they spawned. Today, the entire conversation on aliens is invariably bound up with the insistence that such beings do not merely exist, but are already visiting earth in droves. Therefore, we must now scientifically scrutinize those assertions.

The Bedrock Premise of Ufology

The entire charade that is Ufology is built wholly upon one premise which is immediately and starkly self-contradictory, logically bankrupt, and even hypocritical.

This jarring contradiction is exposed by first noting that Ufologists unanimously insist upon the scientific nature of their mission.* They claim their belief in alien craft visiting earth is motivated simply by empirically verifiable criteria in accordance with established norms. In one breath – albeit a very long one – Ufologists recite a litany of scientific arguments in seeming support of their position. They speak of the scientific laws governing atmospheric phenomena, jet propulsion, electromagnetics, aerodynamics, optics, and so on. They presuppose the indisputable credibility of these scientific laws and they use it as the foundation upon which to build their later claims. For example, they ascribe trustworthiness to the readings of the various high-tech sensor readings they cite. They summon various scientific experts to give testimonies to further their own thesis. These presentations are made to support their case that the UFO sightings in question could not *possibly* be human in origin (or optical illusions, or otherwise "ordinary"), but instead are explicable solely by ETs – whether of the "traditional" space alien variety, or the so-called "interdimensional" pedigree.

In the very next lengthy breath, however, they castigate fully qualified scientists who, upon being given a description of some alleged phenomenon, point out that what is being claimed is mistaken because it is scientifically impossible due to violating *the most fundamental laws* of physics. (We will consider one such example later in this chapter.) Thus they carry themselves like some teenager who, clinging to childhood fantasies, declares Santa *must* have provided the Christmas presents because his parents had

---

* Recent global headlines even appear to lend additional credibility to the field as a scientific one. Official U.S. Government commissions have been established to investigate UFO phenomena (*now branded "UAP" phenomena*), Congressional hearings are being held, and an ever-increasing trend of "whistleblowers," who insist aliens are among us, have gained unprecedented notoriety. The latest whistleblower published a book, *Imminent*, which almost immediately topped global best-seller lists.

said so;[76] only to preserve that belief even when his own parents—the *sole basis* for the original claim—insist, much more strongly, that *they* provided the gifts.

Ufologists say, "*Yes, these craft are clearly operating in violation of the laws of physics ... As we know them. Don't you think super-advanced alien technology would do precisely that*!?" For example, in May 2022, U.S. Congressman Adam Schiff—during a House of Representatives intelligence briefing—implied ET belief by claiming that UFOs move in ways "**that are inconsistent with what we know of physics.**" He justified the claim by deferring to the measurements of "**multiple instruments**" (while also stating that examining these UFOs will "**shed light on one of the world's most enduring mysteries**").[77]

Yet, amid incessant claims like these—even from the highest levels of government—no one dares to point out that the emperor has no clothes. No one asks delusional men like Schiff the obvious question: if "our present understanding" of physics is so fundamentally flawed that these "alien piloted UFOs" can simply set it aside like a silly game's rules in undertaking their maneuvers, then why give "our" science any credibility whatsoever, *including its employment to supposedly prove that these sightings are of ET craft in the first place?* In other words, how can we defer to our scientific analysis of the readings of instruments—which, incidentally, are also themselves the products of that understanding of those same scientific laws—only to come to the conclusion that those scientific principles do not apply? That is as ludicrous as purporting to build an argument against the validity of the law of addition… *using algebra*. It is to saw off the limb on which one stands.

Scientific truths such as the Law of Conservation of Mass, Energy, and Momentum are so fundamental that if they are incorrect—if it is *even theoretically possible* for them to be suspended by some technological innovation—then modern science is as debunked as flat earth theory, and it should be discarded; not used as the very framework for solving "*the world's most enduring mysteries*." (Schiff) But Ufologists insist on having their cake and eating it too: deferring to the veracity of *our* science in their very insistence that *our* science is wrong.

My purpose is certainly not to argue that the laws of science *cannot* be broken. Most assuredly, they can be broken and they have been broken many times! When this happens, the typical course of nature is suspended, and there is a simple term for what has transpired: *a miracle*.

That situation, however, is entirely different. Ufologists never seek to ascribe truly supernatural capabilities to these UFOs, as this move would immediately contradict their cause's entire motivation (which is the creation of a secular pseudo-religion, as we will see in Part Three). This fact is supremely important in spiritually discerning the whole ET phenomenon; for in this and many of its other characteristics, it carries a uniquely evil seduction—namely, the suggestion that what was once considered the sovereign domain of God is now known to be within the grasp of alien technology. No clearer preparation for the lying "signs and wonders" of the Antichrist could be imagined. But the analogy they draw to defend this deception is always the same:

> "Imagine a medieval peasant observing the flight of a modern airplane. He would have no idea how that could happen, but it is still an entirely technological phenomenon. It is the same with these UFOs!"

Admittedly, such a sight would be quite astonishing for a man watching the skies a millennium ago. But even such an observed phenomenon as that would *not* be "contrary to the laws of science," *even as* they were then known.

The men of ages past did not describe the ordinary course of nature with the same jargon we use today, but they understood it quite well. They knew that there was such a thing as flight, and they knew it did not require a miracle or any other suspension of natural laws to achieve. They would have seen the wings of the airplane and realized their similarity to those of a bird. They would have noticed that, also like a gliding bird, this plane could not remain aloft with wings steady unless it was rushing forward at high speeds. It would not have been difficult for them to describe the phenomena that kept a massive plane in the air, and it certainly would have presented no paradox.

What if they were shown a helicopter? That, too, would be straightforward for them to understand. They could comprehend it in the same way they comprehended a bumblebee or a hummingbird's hovering. A rocket's propulsion would be even easier for them to grasp from the science they already knew. ("Newton's Third Law" was always understood long before it was attributed to a certain 17th-century Englishman.) Any serious thinker of any time in history would have found little difficulty attributing the forward motion of an object to the thrust of high-speed propellant in the opposite direction. One ancient Greek philosopher, Archytas, even built a device operating based on that principle in the year 400 B.C.! Kites and balloons date back millennia, and there were several ancient and medieval attempts to design aircraft. Although they were not technologically successful, the theory behind them was quite good.

The maneuvers claimed of today's UFOs, on the other hand, display absolutely no mechanism whatsoever by which they can be scientifically described, *even in (any) theory*. There is only one conclusion that can be drawn from those situations by a man who takes reason and science seriously.

The inescapable conclusion is this: what appears to be happening is not actually happening (for it is theologically absurd to claim that God would work a miracle like that). *It is a deception*. This deception, moreover—if it cannot be attributed to a hoax, an optical illusion, a psyop, or some other similar human phenomena—derives from the operations of demons. This takes us beyond the realm of empirical science and into spiritual discernment, so we will save it for later chapters. For now, it suffices to observe that

Ufologists take a fundamentally religious approach—though a perverse one—to their ideology. Although stating scientific motivations, the content of their claims and the messianism of their aspirations betrays something else. Instead of "miracles," they simply refer to "advanced science," even as they eject God. Instead of the "spiritual," they simply refer to the "interdimensional," even as they dismiss Heaven and Hell. Instead of "grace," they refer to "ontological shock" and "synchronicity," even as they mock Divine Providence.

Changing a few words, they have only proposed a new religion.

On Intergalactic and Interstellar Travel

Even if we counterfactually posit their existence, no ET could ever visit earth from another solar system or galaxy.

Beginning with the latter, many claims of extraterrestrials visiting earth involve the presupposition that, to do so, they have managed to travel faster than the speed of light. Indeed, only 0.000000000005% of the Universe's planets are said to reside in our own galaxy, and of the over five thousand planets in the Milky Way we have found thus far, none has been demonstrated to be *habitable*, much less *inhabited*. Accordingly, expectations of alien contact are often placed in aliens traveling from other galaxies. But even the closest galaxy to our own (if it deserves the title of "galaxy"), the "Canis Major Irregular Dwarf Galaxy," is 25,000 light-years away.[78]

For perspective, the manmade object that is farthest from earth is the Voyager 1 craft, launched in the year 1977. As of 2024, it is over 15 billion miles from earth, having left the solar system in the year 2012 traveling a dizzying 40,000 miles per hour. If it were perfectly aimed at the aforementioned galaxy, it would take almost a *billion years* to arrive. The next closest galaxy after Canis is 75,000 light-years away, and the distances continue expanding dramatically for other galaxies. Clearly, any hope of meeting aliens on earth must be placed in their ability to travel faster than the speed of light.

But this is absolutely impossible. If any object with a "rest mass"[79] were to travel at the speed of light, then its own mass would have increased infinitely. As noted in an explainer from the U.S. Department of Energy:

> ... as an object moves faster, its observed mass increases. This increase is negligible at everyday speeds. But as an object approaches the speed of light, its observed mass becomes infinitely large.[80]

If, however, the mass of any object anywhere in the Universe were to have *ever* increased infinitely, then so too would the force of its gravitational pull. An infinite gravitational pull would have caused the Universe to collapse into itself. As the entire Universe has not collapsed, nothing whatsoever (other than light)—anywhere in the Universe—has ever traveled at light speed, much less beyond it. Moreover, an infinite amount of energy—i.e., more than is contained in the entire Universe—would be required to propel so much as a single atom to the speed of light; therefore it goes without saying that no spacecraft has ever, anywhere, achieved this feat or anything remotely close to it. It is irrelevant whatever technology, from whatever alien civilization, one may posit. We are not dealing with technological limitations, but bedrock scientific ones.

Now, this absolute barrier does not prevent Ufologists from proposing all manner of alternative theories. Some conjecture quasi-teleportation via "wormholes," or describe the phenomena with other newly invented jargon. All of this is pure pseudoscience without the slightest evidential basis. Here as in all other realms, anyone can do a quick internet news search and immediately generate a long list of supposed experiments, from supposedly trustworthy periodicals, which in fact present only pseudoscience. "Scientists create wormhole!" is a headline that one can anticipate being repackaged more or less annually. It is always a lie, though promoted by media outlets in search of clicks.

A "wormhole" could, they say, allow extraterrestrials to essentially teleport past billions of light-years of spatial distance. "*We know this is possible thanks to relativity and quantum mechanics*," many ET promoters solemnly declare. Others make the same basic claim, but conjure up other silly notions related to "warp drive," "Krasnikov tubes," "Tachyons," or what have you. All, however, are refuted by the same scientific considerations that follow.

First, a common misconception must be addressed: wormholes are *not* a component of Einstein's Theory of Relativity. Although some writers dubiously claim that proposing wormholes is "consistent" with Einstein's famous theory, even this assertion is unhelpful. Flying cows and planets made of cheese are also consistent with that theory.

What, however, *is* a wormhole? A "wormhole" is an entertaining proposition wherein two black holes existing at far distant parts of the Universe are randomly "connected" in such a way that they form two end points of a single bridge. For no apparent reason—and reminiscent of the convenient irrationality of Pagan myths—this bridge "folds" a fictional sheet-like semi-substance they call "the fabric of spacetime itself." This folding causes the bridge endpoints to be very close together, enabling even relatively slow travel along this bridge to cause the spaceship to jump to another part of the Universe entirely. Finally—as the fable rambles on—these "wormholes" could also be constructed such that one black-hole-endpoint of the wormhole is located close enough to our own planet to enable aliens to visit us from distant parts of our galaxy or other galaxies.

Various wormhole-travel theories have been presented, all refuted by the insurmountable contradictions they invariably generate. Astrophysicist Dr. Jason Lisle noted that "**wormholes do not exist in the real universe**," but are instead simply a "**fun mathematical construct**." He explains:

> Such a wormhole would not be traversable, much to the disappointment of sci-fi fans. ... No journey from one end of the wormhole to the other

> would be possible. ... [It moreover would] require the black hole / white hole pair to have existed forever in the past and forever in the future. It is [merely] a fun mathematical construct. But since time has a beginning, this solution is disallowed in the real universe. Other types of wormholes have been proposed ... However, in all cases they are apparently unstable, collapsing before any particle could make use of the short cut. ... [W]ormholes do not exist in the real universe. Physicists and mathematicians may enjoy considering these interesting solutions to Einstein's field equations. But the real universe must obey not only the rules of general relativity, but other physical laws as well, such as thermodynamics...[81]

Wormholes do not exist; not even wormholes through which mere information or subatomic particles—much less physical craft or alien beings—could travel, and we can be certain that *there never will be* any such wormholes or other constructs to enable teleportation or any other brand of faster-than-light travel one might propose.[82]

In truth, the moment such explanations as "wormholes" are brought into the discussion, an ET promoter has demonstrated he has no desire to adopt a scientific approach to his views. He might as well simply say that Zeus transports aliens across the galaxies through black holes, albeit only by using "quantum strangeness and non-locality." Despite the flailing attempt, he has only again implicitly transferred the discussion into the domain of *spiritual* discernment, which we will undertake in Part Three.

***

Can, however, we at least hope for aliens visiting from other solar systems within our galaxy? The answer to this question is also a resolute *no*.

The closest solar system to our own, that of the star Proxima Centauri, is about 25 trillion miles away from earth. The Voyager 1 spacecraft we referred to above, hurtling through space at its phenomenal speed, would take over 70,000 years to reach that solar system (that is, if, by some miracle, its trajectory is perfect and it avoids destruction by mechanical failure, debris collisions, or radiation). As we can see, proposing the possibility of even the very easiest interstellar meeting requires positing travel at speeds coming within the relative vicinity of the speed of light.

Yet this too presents obstacles insurmountable not merely technologically but also scientifically. In his article, "**Interstellar Travel as Delusional Fantasy**" (in *Scientific American*), Dr. Ed Regis observed:

> Collisions in space are by no means rare: by the end of the Space Shuttle program, for example, more than 100 shuttle windows had been replaced after impacts with space debris, some objects being as small as [a] fleck of paint ... It might be thought that the interstellar medium is "empty space," or a vacuum. To the contrary, the space between the stars contains volumes of interstellar gas and dust, along with cosmic rays ... for a starship traveling at relativistic speeds, a collision with even a random small particle, according to Tom W. Gingell of Science Applications International Corporation, who did a study of the subject, would have the effect

upon the spacecraft of an H-bomb explosion.[83]

At speeds required to give even the faintest hint of palatability to interstellar travel, collisions with the very smallest pieces of interstellar dust would decimate any craft—regardless of its construction method or materials. For any spacecraft meaningfully larger than a molecule, countless such collisions would be statistically guaranteed to occur when traveling between stars.

Astrophysicist Dr. Paul Sutter put the matter in even more specific terms in a 2021 article for *Discovery* magazine. He accurately described the most "realistic" interstellar travel method as follows:

> ... [T]he Starshot Initiative, which aims to shoot a super-powerful laser on a lightsail (a giant nearly perfectly reflecting membrane), using the energy from the light to propel the spacecraft to a tenth the speed of light. That would enable it to reach Proxima [Centauri] in less than half a century. To make this work, the laser would have to use all the energy from every single nuclear reactor in the United States at once. And it would have to operate for 10 minutes, which is about a quadrillion times longer than we've ever operated our most powerful lasers ... and the spacecraft could weigh no more than a paperclip.[84]

Hypothetical alien civilizations could, of course, have more advanced inventions than those considered above. But it only takes a small dose of honesty applied to these observations to realize that it does not matter how much more advanced any civilization's technological prowess becomes: interstellar travel remains ruled out. Any sober assessment of the massive distances and intrinsic dangers involved, when combined with basic knowledge of the universal scientific laws governing the strength of materials, kinematic physics, thermodynamics, chemistry, and more (even leaving aside biology!) will quickly demonstrate that no technological object—much less one so sizeable as the UFOs reported today—will ever travel between stars. Astrophysicist Dr. Hugh Ross explained:

> At even half the velocity of light, the energy needed to propel an object is 170 million times greater than NASA's fastest spacecraft requires... any reasonably sized spacecraft transporting intelligent physical beings can travel at velocities no greater than about 1 percent of the velocity of light. At higher velocities the risks from radiation, space debris, leaks, and wear and tear simply become too great to prevent the extinction of the space travelers before they reach their destination. A spacecraft traveling at 1 percent of the velocity of light (nearly 7 million miles per hour) would need twenty-three thousand years to travel 230 light-years.[85]

No scientific grounds exist on which to posit the possibility of aliens traveling to us even from our closest neighbor, much less some other planet in the Milky Way beyond Proxima Centauri, still less from another galaxy. Whoever takes science seriously is compelled to securely file such propositions in the science *fiction* box. Yet we still have not reviewed the clearest scientific proof that aliens do not exist.

SETI's Great Silence

If aliens existed, one thing would be guaranteed: their presence would be detected through electromagnetic communications ("radio contact") *millennia* before there would be any hope of physical contact with these civilizations. Yet, no such detection has ever occurred. Therefore, yet again, even if we counterfactually posit the existence of aliens, it remains certain that none are on earth or anywhere close to earth.

Understanding this, "SETI" (The Search for Extraterrestrial Intelligence) has been an official government undertaking for several decades, devouring millions of dollars of investment and pinning all of its hopes in discovering aliens through radio waves.

It is irrelevant whether these hypothetical alien civilizations intended to make radio contact with us. They *would* have done so, despite themselves, if they had become technologically advanced. There is no use protesting that alien civilizations exist but simply wish to wait to alert us to their existence until we are prepared for contact or until they are desirous of it. They would not have been able to prevent their existence from becoming known; just as humanity, upon its first radio transmission (in the 1800s), lost the ability to conceal its own existence from hypothetical alien civilizations.

The first radio waves humans transmitted (by Guglielmo Marconi on May 13, 1897) are now 740 trillion miles away from earth. In their travels, they have swept 25,000 cubic light-years of the Milky Way galaxy; a volume of space likely including hundreds of thousands of planets—all of which have now been inundated with our transmissions. The "response" is only the sound of crickets ... on earth. Your conversation with your sister about favorite recipes, while holding your 1990s Nokia phone? It is now slamming into the planets of Gamma Pavonis—in the constellation of Pavo, visible in the night sky—having already passed innumerable other solar systems. The crickets remain. In the year 1934 (long before SETI), Cincinnati, Ohio's WLW radio station began belting out 500,000 watts of transmissions. These immense signals have been hurtling through space now for almost a century, likewise inundating thousands of solar systems with news from planet earth. And even this is dwarfed by the signals earth began emitting thereafter.* Assuredly, any "ET civilization" like those envisioned by today's Ufologists would have abilities to decipher—and respond to—such signals at

* A radio tower twice as powerful was built by the U.S. Navy two decades later, and its signals could scarcely be classed as undecipherable by alien civilizations. Several years after that, radio towers twice as powerful still (at two million watts) began their emissions in Maine, communicating with American submarines. For perspective, these astoundingly potent radio signals should be compared to what even an amateur ham operator can manage with a cheap radio emitting less than one watt of transmission: he can bounce signals off the moon itself, and communicate with another ham operator on the other side of the globe (ham operators call this "moonbounce"). The same inexpensive technology can receive signals in the order of a femtowatt, which is equal to 1x10^-15 watts (that is, a thousand times smaller than one-trillionth of a watt). Therefore, the signal

massive distances. Even the head of SETI himself, Dr. Seth Shostak, admits as much. In 2003, he noted that we have, for decades, already been "**irrevocably signaling our existence to other star systems**."[86]

What is also certain is that wherever technology of any sort exists, its own radio wave signatures provide indisputable evidence of its presence—evidence that will always be far beyond that technology's physical presence. No matter how fast any physical technology (a spaceship, probe, or the like) is traveling, it would not reach its destination for thousands of years after its radio emissions arrived at the same location. Electromagnetic waves (radio signals) are not merely one of a multitude of means of remote communication. (Some Ufologists make this false claim when they realize they should be troubled by "SETI's Great Silence," and then hasten to dismiss the absence of ET radio signals as inconsequential.) Instead, such waves constitute the most basic design elements of the Universe. It is impossible to prevent their emission once technology develops, no matter the route that development follows.

Where, then, are these electromagnetic announcements of alien civilizations?[87] SETI has scoured every crevice of the galaxy it can reach in search of them. Not one intelligently originated transmission has ever been detected. We need only consider what is by far the most heavily advertised SETI signal that, ET promoters claim, proceeded from aliens—the "Wow!" signal detected at Ohio State University's "Big Ear" telescope in 1977. Ufologists have written books and made entire documentary films about this signal, insisting it finally proved that aliens exist. One 2017 film about the event concluded by relaying that while some scientists even then claimed to have debunked the signal's putatively ET origin, "**no theory has ever proven true. The 'Wow! Signal' remains unexplained**." Inspiring music is then played to images of nebulas and stars before the Director's name is splashed on the screen.

The signal was debunked in 2024.

Scientists finally realized it was just an emission from a hydrogen cloud.[88] Claims of ET contact, UFO encounters, alien abductions, recoveries of alien corpses, and so on, are hereby demonstrated to be as much rubbish. They presuppose that aliens are already physically here—the contrary of which is already proven by SETI's Great Silence.

Drake and Fermi

The "Silence" of SETI also disproves what has become a prevalent pseudoscientific hypothesis presented in support of alien belief: the 1961 equation conjured up by the man we considered in Part One, Frank Drake. By multiplying various factors (many of which lack any empirical basis)—

---

the U.S. Navy's transmitter (noted above) was blasting into space was a thousand-billion-billion times stronger than what even we—with basic technology that is vastly inferior to the technology any "interstellar-space-traveling aliens" would have—can decipher.

including the average rate of star formation, the number of planets, the fraction of planets with life, the fraction with life that develop *intelligent* life, and so on—Drake proposed to thereby compute the number of advanced civilizations that exist in the Milky Way galaxy. He proceeded to declare that the number must stand between one thousand and one hundred million. As it happens, the lower bound of his estimate was only off by a mere 999.

Recall that Dr. Drake was an astrophysicist and astrobiologist who had dedicated his life to the search for aliens. He is regarded in today's official SETI literature as "*the Godfather of observational SETI,*" the man who "*turned science fiction into credible, respectable science.*" Seventy-three years of SETI's continual and zealous effort, however, has only solidified the status of its premises and aspirations as remaining thoroughly fictional indeed.

With its massive 85-foot-wide radio telescope, Drake's Project Ozma gave him the opportunity to listen for transmissions from those solar systems he thought might house advanced civilizations. He discovered no signs of intelligent life—aside from one false alarm, reminiscent of Tesla several decades earlier, when another human's technology on earth was thought to be extraterrestrial. (We should anticipate many other false alarms in the future—including deliberate psyops.) It was the year after the founding of Project Ozma—after who knows how many hours spent searching for alien communications and finding none—that Drake created the equation above and delivered its corresponding pseudo-revelation of thousands or millions of alien civilizations dwelling in our very own galaxy.

Ironically, the definitive rebuttal to this meticulously calculated "scientific" argument presented by Drake and his colleagues was implicitly given ten years earlier by a renowned Italian American physicist, Enrico Fermi. A Nobel Prize winner and the inventor of the first nuclear reactor, Fermi's genius status was unquestioned. He is often regarded as among the greatest scientists of the 20th century. But he dared to ask the obvious question about aliens that few, before him, had presented; namely: "**Where is everybody**?" In other words, if the planets really are so teeming with life as astronomers claim, then how is it even remotely conceivable that—despite decades of listening to radio communications—we have not heard from a *single* extraterrestrial, much less experienced any visiting earth?

> "Fermi's colleagues were in awe of him for his uncanny ability to see straight to the heart of a physical problem and describe it in simple terms. They called him the Pope because he seemed infallible."[89]

It is a misnomer to call Fermi's question a paradox, for there is nothing paradoxical about it. On the contrary, it is a classic rhetorical question which furtively illustrates its latent conclusion. This "paradox" has been among the most painful thorns in the side of ET promoters for the last century, and consequently it has solicited dozens upon dozens of supposed solutions. None are substantial, much less convincing.

One recent scholarly tome published by Springer Academic Press—dedicated to discussing SETI's work *favorably* and presented in honor of Frank Drake—was nevertheless honest in phrasing the present contradiction for alien belief, pointing out that "**The paradox has not been blunted... the deafening silence of the universe ... has sharpened**."

> At the very least we should see evidence of [ET civilizations] or their instrumentality when we look into space. But we don't. *Where is everybody?*... But it's not just that their craft aren't here; as mentioned above, astronomers have observed no signs of extraterrestrial intelligence anywhere. The paradox is not that their spacecraft aren't around when we might expect them to be here; it's that we neither see nor hear *any* traces of them [*emphasis in original*]. David Brin called this "the Great Silence". ... The paradox has not been blunted since Fermi asked his question or Brin pointed out the deafening silence of the universe. If anything, it has sharpened... Even if interstellar travel of any description is simply too difficult or too costly, an advanced extraterrestrial civilization could signal its presence ... And it is not as if we are deaf: in the years since Fermi asked his question our ability to search for electromagnetic signals has steadily improved by many orders of magnitude. The paradox thus remains.[90]

Several decades after Marconi's emission of unmistakably intelligently-originated EM waves into outer space, SETI began doing so *intentionally*. With extremely powerful radio signals carrying information so clear that no alien civilization could fail to recognize them as deriving from another civilization, SETI carefully directed their calls toward those areas of the galaxy thought most likely to harbor intelligent life. The crickets remain.

A conclusion emerges at this point from anyone interested in a scientific approach to extraterrestrial intelligence: *it does not exist*. One such scholar interested in the truth of the matter—the British physicist Dr. Stephen Webb—confessed that he very much "wanted" aliens to exist. He dedicated a large volume, published in 2002, to the task of rigorously examining each one of the dozens of "solutions" to the Fermi Paradox that had been proposed in support of belief in aliens. He found none of them to be convincing. At the end of his lengthy scientific analysis, he presents the following conclusion, "**The only resolution of the Fermi paradox that makes sense to me—is that we are alone**," and he asserts "**the uniqueness of mankind**" on purely scientific grounds. He continues:

> The Fermi paradox provides a shock that forces us to examine the widespread notion that the vast number of planets in existence is sufficient to guarantee the existence of extraterrestrial intelligent life. In fact, we need not be too surprised. The Drake equation is a product of several terms. If one of those terms is zero, then the product of the Drake equation will be zero; if several of the terms are small, then the product of the Drake equation will be very small. We will be alone. If one factor in the Drake equation is close to zero, then we can reasonably identify that factor as being

> the solution to the Fermi paradox ... I suspect there is a combination of factors—a product of various solutions we have discussed in this book—resulting in the uniqueness of mankind ... [of habitable planets,] only Earth remains. We are alone ... only one planet—Earth—has intelligent life-forms.[91]

To no one's surprise, Dr. Webb's sound analysis did not put SETI to rest. The fool's errand continued, and along with it arose dozens of additional supposed solutions to the Fermi Paradox. Therefore, over a decade later (in 2015), Webb published yet another book dedicated to the same theme; this time, not addressing fifty solutions, but rather the "seventy-five" solutions now given to the paradox. One might wonder: did Dr. Webb's additional thirteen years—time spent consulting many more scientists and other thinkers, his reading vastly more scholarly literature, and his learning about the deluge of new discoveries in astronomy—change his mind? Did he finally realize that aliens are likely after all?

No. The additional "solutions" to the Fermi Paradox only solidified his conclusion that human beings are the Universe's only rational inhabitants. He closed that more recent book by sharing that most of his friends assert aliens must exist, because—as he relayed their words—"*...space is big. Really big... surely we can't be the only intelligent species in such a large universe.*" He even confessed that most of his physical-scientist friends assert the same thing, albeit bedecked with more specific claims related to various large numbers. Most importantly, however, he observed that his biologist friends are those who realize that these large numbers (including even the most generous estimates for the number of planets in the Universe) are laughably tiny when compared to the chances of intelligent life emerging. Even if our Universe houses a trillion times as many planets as astronomers now estimate, amid a trillion universes each a trillion times as large as our own, the mathematical criteria that describe the statistical likelihood of intelligent life existing still render those incomprehensibly large numbers a mere nothing in comparison. Dr. Webb concludes the book by wisely reporting: "**My own view? Well, I side with my biologist friends. The debate about extraterrestrial intelligence contains just one gleaming, hard fact: we haven't been visited by [aliens] nor have we heard from them**."

Millions of "Encounters"

Ufologists are keenly aware that one of the weakest elements of their thesis is the total lack of evidence of any recovery of ET material. This deficiency is illustrated most sharply when contrasted with their claims of alien-piloted UFOs continuously buzzing our skies for 78 years while frequently crashing.

> "The simple fact is that no one has ever found a single artifact, or any other convincing evidence for such alien visits." —David Morrison. NASA Astrobiology Institute Senior Scientist.[92]

Of course, those same decades have witnessed a steady stream of claims, from various authors and "whistleblowers," that ET-originated material *is* in the possession of certain individuals (conveniently unnamed "for their safety"). Not one of these claims—including the most popular and supposedly reliable—has withstood even basic scrutiny.

A Christian must pause to consider that this is the one type of evidence which would grant some modicum of credibility to the belief that aliens are among us, and not merely demons doing what they always do. For demons have always been known to assume the disguises of a culture's prevailing mythology. And this evidence is not just scant or suspicious, but utterly nonexistent. Even Carl Sagan, whose faith in aliens was absolute, lamented:

> Where is the physical evidence? Some abductees allege that aliens stole fetuses from their wombs... But not a single medical record has been produced substantiating such claims. Some abductees say that tiny metallic implants were inserted into their bodies—high up their nostrils, for example. But no such implants have been confirmed by physicists or chemists as being of unearthly manufacture. No abductee has filched a page from the captain's logbook or a strange examining instrument, or taken an authentic photograph of the interior of the ship, or come back with detailed scientific information not hitherto known on Earth. These failures surely tell us something.[93]

Contrast this to the evidence in support of the miracles of Christianity, which is superabundant.* With Ufology, we have only an amalgamation of "evidence" that is immeasurably insignificant in comparison. The deficiency of any objective evidence of aliens is damning, as such evidence would be guaranteed to be ubiquitous if only a fraction of the claims of the Ufologists were true. For example, one of today's most prominent ET promoters, Dr. Steven Greer, claims that over three thousand separate UFO crash retrievals are currently in the possession of various organizations. But may I ask: where is just *one* of these thousands of pieces of evidence? The Ufologists have had over 78 years since Roswell to present it.

---

* There are scientifically impossible tilmas and shrouds, healing springs, and spinning suns. Whoever wishes to do so can, this moment, hop into a car or board an airplane and visit the bodies of incorrupt saints (whose un-embalmed corpses have not changed since death), see walls at shrines filled with crutches from those instantaneously and miraculously healed medically incurable impairments, and observe statues of Jesus and Mary continually and inexplicably exuding oil or even weeping blood. He can find rigorously documented examples of mystics surviving decades consuming nothing but the Eucharist. He can open both Scripture and the prophecies in private revelation since those times and find countless proven examples of an impossibly clear fulfillment of each—so clear that they exceed even what any demon could know about the future. He can read of not mere scattered reports of fighter pilots or men in their backyards observing something strange, but rather a full 70,000 witnesses describing obvious miracles at Fatima and elsewhere on par with Moses parting the Red Sea. Above all, he can read the direct testimony of eleven men who repeatedly witnessed—directly, with their own senses—the Risen Christ, and who each went to their death rather than deny this truth; a phenomenon that has no remotely close parallel anywhere else in the whole history of the world.

In accordance with the naked absurdity this situation presents for the belief that aliens are among us, nearly the entirety of the Ufologists' effort in this regard is found in their insistence that alien craft *do exist*, but are the object of history's greatest cover-up. They claim the surface of our planet has been littered with physical evidence of aliens for decades now, which thousands have seen. But they insist this ubiquitous evidence—wherever it is found across the entire globe—always is (and always has been) immediately and diligently detected and collected by shadow-government agencies possessing powers all but Divine. This, despite ever-changing administrations, ideologies, budgets, and ruling parties presiding over these programs as the decades dragged on. This, too, despite many different nations' governments (often very antagonistic towards each other) being allegedly involved. Such a feat would make the Manhattan Project—by far the largest secret government undertaking in history—appear negligible in comparison.

This deficiency of evidence must be compared to the magnitude of the claimed phenomenon. The existence of physical evidence of aliens visiting earth would not be just any intriguing top-secret matter. It would be the single biggest revelation in two thousand years. For a non-Christian, it would appear to be, by far, the biggest revelation in history. Who could resist feeling compelled to immediately shout that discovery from the rooftops, regardless of any risks? Let us suppose for a moment, however, that there are a few cowards who, if sufficiently threatened, would keep silent. Even so, one thing would remain certain. There are not *so many* extreme cowards as to allow for the Ufologists' scenario. These Ufologists are impugning thousands upon thousands of so-called "ET witnesses;" charging them with being accomplices to the gravest treason against humanity.

They implicitly claim that each one of the thousands of people who encountered alleged physical evidence of aliens also numbers among that remarkably degenerate group of superlative cowards. Even flat earth theory, outlandishly erroneous as it is, only requires a far less bold assertion. It alleges that each one of the over six hundred astronauts who have orbited earth are liars who endorse the "round earth narrative" even when they know better. Ufology, on the other hand, consists in a far graver calumny against a far greater number.*

---

* Contrast this dynamic to the actual greatest revelation in history: that of Jesus Christ—especially His Resurrection. The Jews and the Romans of those times did everything they could to silence this proclamation—including inciting genocide and administering a carefully crafted conspiracy (cf. Matthew 28:15). Countless Christians were murdered, from the start, for proclaiming their Faith in *what they saw*. (Where, we should ask, are the corresponding thousands of UFO martyrs?) The persecutors of the Faith could scarcely succeed in their nefarious machinations to silence testimony to the Risen Christ with any early Christians for so much as a day! Yes, despite the unfortunate fact that cowards exist, it remains true that most people will not shy from proclaiming *what they know they saw*, even in light of death threats, particularly if what they saw was extremely important for humanity. Yet, the Ufologists would have us believe this is precisely what thousands of individuals have been doing for three-quarters of a century.

Ufologists will shout, "*But that is all a straw man! We do have such testimonies! We have the whistleblowers! Their testimony proves that the evidence is there! We will see it soon enough!*" They have been shouting those words for 78 years, but we still have never seen any evidence of "ET encounters" that could not be accurately understood in either earthly or demonic terms.

The more astute of the Ufologists acknowledge the fraud. One of them is Dr. Jacques Vallée—who likely knows more about UFO sightings and "alien contact" experiences than any other man alive. After years of rigorous study of the movement which claims evidence of space aliens visiting earth exists, he castigated it as an "**incredible conglomeration of lies and stupidity**." In the Appendices, we review some of his trenchant critiques.

We can also, however, delve directly to the core of the claims themselves in order to ascertain their deceptive nature. For it is not only the lack of evidence for these Ufology claims and the nefarious machinations surrounding their issuance which is telling. Even more importantly, the content of what is popularly attributed to UFOs tells us squarely what these "UFOs" really are, as alluded to earlier.

UFOs vs Reality

> "An object in motion stays in motion with the same speed and direction unless acted upon by a force. The rate of change of momentum of an object is equal to the force applied to that object. For every action there is an opposite and equal reaction." —Sir Isaac Newton

This is one way of phrasing Isaac Newton's famous Laws of Motion. The reality, however, is that these truths (especially the first and third) were known long before Newton's day. (Modern claims that Newton overthrew Aristotelian thought on this topic are at best exaggerated.) Unfortunately, contemporary Ufologists—like dilettante self-proclaimed specialists in "quantum physics," (two overlapping if not coterminous categories)—regard these Laws as rather blasé "*in light of advancements in 20th-century science that overthrew Newton.*" Such puerile statements as those remind serious people that we are not dealing with a rational inquiry here, but a fiction-driven myth desperate to find some basis to pretend its fantasies are scientific.

Far from overthrowing Newton's Laws, even the most advanced quantum particle physics research in existence today often presupposes their validity. Newton's Laws have not been—and never will be—"overthrown." In some rare cases they may need to be calibrated with considerations derived from relativity or quantum physics. Yet these calibrations have no effect on the question we must ask: what feats could *possibly* be accomplished by the types of physical craft described in relation to "UFO sightings"?

One thing we can be certain of is that no craft could be capable of a near-perfect right-angle turn at an extremely high velocity, or a massive and near-instantaneous change of velocity. Let us consider one case quantitatively. Similar claims are pervasive among UFO enthusiasts, but one author recounted some of the more famous and official accounts as follows:

> Like American observations, the Soviet Air Force had many instances in which UFOs were observed hovering, then departing at great speed. Some of these reports estimated the craft to be ... more than 600 feet [in diameter] with speeds ranging from hovering to triple the capability of modern fighters, yet then stopping again instantaneously. Other reports indicated that the UFOs displayed "startling maneuverability," yet made no sound... Approaching the UFO, [one fighter pilot] reported the size of the target ..."the proportions of a flying aircraft carrier." ... Then suddenly the UFO took off at amazing speed. [He] estimated that it went from stationary to Mach 10 almost instantaneously.[94]

What is described here is absolutely impossible, and in fact has never happened. When such things are truly observed—and, admittedly, we have troves of similar testimonies—only a demonic illusion is being witnessed. We will save discerning the demonic for Part Three; for now, we must settle the relevant science.

When considering the "flying aircraft carrier" above, which "almost instantaneously" accelerated from stationary to Mach 10 (7,672 miles per hour), we must determine what period of time describes an event (in this case, an acceleration) that appears instantaneous. Here, ten milliseconds emerges as the most appropriate number.[95]

This assumption leaves us to calculate how much force must be exerted during these 10 milliseconds and how much energy must be expended in the process. Using an aircraft carrier as our point of departure—since that is how the pilot above described the UFO in question—we are dealing with a craft weighing about 100 million kilograms. Accelerating such an object to Mach 10 would require an amount of energy (calculated by the equation, 1/2 * Mass * Velocity squared) in the order of 6 x 10^14 Joules, or about 167,000 megawatt hours. This is about ten times the amount of energy released by the atomic bomb dropped on Hiroshima. If such an extraordinary amount of energy was generated by *any craft whatsoever*—regardless of what hypothetical advanced alien technology was used to generate it—such an event would be overwhelmingly obvious (and destructive) to anyone within a massive radius. Just imagine ten Hiroshima's all at once—merely to enable *a single act* of acceleration of the "UFO" in question.

Nevertheless, this consideration scarcely begins to capture the ludicrousness of the situation and the certainty that no such event ever has happened or ever will happen. For we are not merely dealing with an energy equivalent to ten atomic bombs. We are dealing with that amount of energy required simply to cause a *single maneuver* of the UFO in question (*each* acceleration—and deceleration, *and each turn*[96]—would require similar emissions of this unthinkable energy quantity). We are, moreover, dealing with that massive amount of energy exerted over the course of a mere 10 milliseconds. Let us now calculate the power and force requirements for such a feat.

Power is simply energy divided by time. Releasing the amount of energy at play (6x10^14 Joules) over the course of 10 milliseconds entails a power consumption of 6x10^16 watts (sixty thousand trillion watts). This is more than 21,000 times as much power as is consumed by *the entire world*—and it would provide enough power to supply a world of 90 trillion inhabitants at current usage rates. Moreover, the Ufologists insist that crafts constantly exerting this amount of power are incessantly buzzing our skies; not merely doing so occasionally (even though just one such event would obliterate an area's population).* If even a fraction of their claims were true, the mere heat generated by these accelerations and decelerations would long ago have boiled off all our oceans and cooked the entire surface of the planet.†

Innumerable other proofs demonstrate the scientific impossibility of what is attributed to UFOs. They are diabolical deceptions, plain and simple. There is no "mystery" to solve here. It makes no difference what theoretical alien technologies could be posited. It is the laws of science, which undergird all technologies and all possible technologies, that render the definitive verdict.

We cannot within these pages scrutinize the physics of every reported UFO sighting (not all are as outlandish as the one considered above—though many are more outlandish still). My intent here was only to encourage the proper approach. If science is to be taken seriously—and it must be—then it also must be acknowledged when it determines that some reported description of events is scientifically impossible. This diagnosis does not necessarily entail calling the witness a liar or a lunatic. He may well have seen what he described, notwithstanding the scientific impossibility of his

---

* To decelerate an object, one must convert all that object's kinetic energy into thermal energy dispersed into the surrounding environment. This is why brake pads become so hot. Therefore, while the accelerations of these UFOs would impart astounding amounts of heat to the environment (even the most advanced theoretically possible means of propulsion would lose at least 10% of their energy to heat—your own car likely loses about 70% of its energy to heat when accelerating), the decelerations would impart *all* of their energy, as heat, into the surrounding environment

† Still, we are not finished quantifying the absurdity of these UFO claims. For we also must consider the *force* required by these accelerations—namely, 6x10^16 Newtons thereof. This amount of force, even if applied perfectly evenly over the entire surface area of an aircraft-carrier-sized craft (about 24,000 square meters) would generate a pressure of 2.5 trillion pascals. No material exists which can withstand so much as a fraction of that amount of pressure—even if the UFO in question were nothing but a solid block of the universe's strongest substance. Such pressures would immediately disintegrate even the strongest possible material object, and would likewise immediately liquify any pilot inside the craft, regardless of its own makeup or its own supposed "advanced stage of evolution."

report, for in that case he merely misinterprets it as technologically originated when in fact it is spiritually originated.*

In fact, only one who respects the ability of science to rule over (and to rule *out*) any and all suggestions proffered within the domain of any possible technology can successfully undertake the logical deduction that is an indispensable component of spiritual discernment. We need only consider that if early would-be Christians were as foolhardy as modern Christians, then Christianity never would have gotten off the ground. Upon observing a man walking on water, raising the dead, restoring sight to the blind, teleporting into closed and locked rooms, reading minds, knowing the future, and ascending upon the clouds, they would have thought to themselves, *"This could be the product of advanced technology; there is no need for me to worship this self-professed 'Son of God.'"*

What I present here is no mere hypothetical fear. It is already what is happening in the minds and hearts of millions. (For example, they are insisting that Fatima's Miracle of the Sun was just a UFO.) A more grave and concrete concern could not be imagined. While promulgated by the false prophets Christ warned against—the wolves in sheep's clothing—as if it were an act of humility, the "openness" to believing in ET-piloted UFOs actually undermines the entire Faith with such malevolence as to stand unrivaled.

Interdimensionals and the Multiverse

We have now arrived at the final strategy ET promoters employ, ostensibly within the realm of "science," though in fact repudiating science itself. When they realize that even wormholes, teleportation, and other related wild conjectures will not succeed in rendering palatable the notion that aliens may be among us, they resort to one last strategy. They present a neo-Gnosticism which unashamedly traffics in the very mysteries (arbitrarily called by other names) that are, by definition, within the exclusive domain of Faith. Next, they contradict themselves by insisting their views are "scientific," and thus exempt from theological (and even philosophical) scrutiny. Jacques Vallée (the pioneer of this view) and his ilk pooh-pooh traditional religion, while proposing theses that are—by any account that

* With this insistence, I am certainly not espousing materialism—only the opposite. (Some Christian ET promoters try to save face by pretending their Ufology pursuits simply consist in opposing such materialism, as if the embrace of anything "paranormal" somehow automatically consists in evangelization.) In truth, this reverence for science is necessary to *refute* materialism. Just as a family doctor must respect the norms which rule out some symptom arising from bacterial infection—so that he can admit when he must send his patient to the oncologist—so too must we defer to science when it declares some phenomenon physically impossible. For only thus can we hope to discern the supernatural or the diabolical when it is present. Whoever insists that any phenomenon could potentially be explained technologically has just driven himself down a one-way, dead-end street. He will never succeed in acknowledging the finger of God even in the face of miracles, for his ET belief consigns him to permanent radical skepticism. He will always be wondering if, in witnessing the supernatural, he is actually just confronted with some alien technology.

surpasses mere semantics—identical to those proposed by every other founder of false religions throughout history.[97]

They posit the beings visiting us are actually "Ultraterrestrials;" that is, some form of beings from "another dimension" who need not travel through space to arrive here, but need merely "intersect" or "phase" or "vibrate" with us at will. As the fable goes, these beings are not physically traversing the expanse of space to reach us, or even passing it via wormholes, but are rather somehow entering our "dimension" from another one. We will deal with the patently demonic spiritual realities that are signaled here later. Presently, our task is only to briefly note how devoid of any scientific basis such claims stand.

Consulting qualified scientists' analyses, we will see the silliness of arguing for "multiple dimensions" or "multiple universes"—two conjectures which, if they differ at all, do so only nominally. One recent scholarly tome, entitled *Universe or Multiverse?*, and published by Cambridge University Press in 2009, contains a chapter written—surprisingly—by Professor Paul Davies. Surprising, because Dr. Davies is a physicist who works in astrobiology and has held a prominent position within SETI (the Search for Extraterrestrial Intelligence). This scientist rightly describes the notion of other dimensions in the "multiverse" as a "**fantasy verse,**" saying that those who promote it **"abandon the notion of a rationally ordered real world altogether**" while simply proposing their own "**religious conviction**." After presenting many scientific considerations, he writes:

> ... [T]o follow the multiverse theory to its logical extreme means effectively abandoning the notion of a rationally ordered real world altogether, in favour of an infinitely complex charade, where the very notion of 'explanation' is meaningless ... [a] 'fantasy-verse' of arbitrary virtual realities, whimsically generated by a pseudo-Deity designer.[98]

Davies is right: the moment one posits such notions as "the multiverse" or "other dimensions," he has, *ipso facto*, declared a religious conviction—a perverse one, at that, devoid of any redeeming value.

Especially noteworthy is Dr. Davies' observation elsewhere in the same chapter that, in multiverse theory, "**there is no end to the hierarchy of levels in which worlds and designers can be embedded ... *ad infinitum*...**"[99] The trend we have seen and doubtless will continue to see is as predictable as the dialogue of a cheap soap opera. In the early 20th century, it was considered edgy to propose that time was literally a fourth dimension. Not only is that notion now blasé, it is today even unfashionable to describe a tired old slew of five, or seven, or a few dozen, "dimensions." One must, instead, propose not only the "multiverse" of dimensions, but also multiple "timelines" and "strings" superseding them. These catchphrases, too, are becoming overly commonplace for those eager to remain on the cutting edge of ET-belief "ontological shock" discourse. Therefore, talk is also now increasing of a "megaverse" or "superverse" or "holocosm" within which all the relatively puny, old-fashioned multiverses dwell. And who knows what neo-Gnostic phrases they will invent tomorrow to pretend that they have expanded *The All* to yet another level of empty esoteric and quasi-occult abstraction. The only thing we can be certain of is that whatever it is, it will be pure fantasy. There is no difference between the proponents of these ideas and children arguing about who can describe a bigger number, with iterations of the latter's debate consisting in exchanges of shouts: "*infinity!*," "*well, infinity plus one!*," "*okay, infinity times infinity!*," and so on. An adult should tell them to brush their teeth and go to bed.

A wise man will simply step back not only from individual claims like those above, but also from the entire discussion itself, and realize that the whole matter is ridiculous. Unfortunately, however, we are not dealing here with mere comical absurdities presented by pundits desperate for attention. Propositions like the "Interdimensional Hypothesis" and the "Multiverse Theory" are nothing other than Trojan Horses to beguile the masses into accepting the ministrations of demons without conceding that it is the fundamentally nefarious, *spiritual* entities—described accurately by Christian Revelation—who are at work; beings whose sole objective is the eternal damnation of souls.

> "The contemporary 'science fiction' myths of time-travel and multidimensional space, derived from imaginative speculation on Einsteinian and post-Einsteinian physics, and often applied to the UFO phenomenon, are in some ways replacing the world-view of the revealed religions, since they seem to transcend materialism and provide the 'miraculous' possibilities always associated with religious faith and spiritual experience. For God all things are possible—but if all things, or many strange things, are possible to UFOs, and will be possible to human science in the future, then who needs God? If space, time, matter, and even some mental processes can be manipulated by various subtle material energies, then who needs grace? If time-travel is possible, who needs eternity?... Those who become obsessed with these myths are making themselves available to satanic forces whose goal is to hide from us the reality of eternity by means of a counterfeit..." —Charles Upton. *System of Antichrist.*

At this point, we can no longer remain within the boundaries of scientific analysis. The time has come to delve into considering the outright demonic plans afoot in ET promotion. We will first theologically prove aliens do not exist, and then spiritually discern the global phenomenon which nevertheless heralds their imminent disclosure.

5. Salvation History is Missing No Chapters

"There is another heresy that says that there are infinite and innumerable worlds, according to the empty opinion of certain philosophers—since Scripture has said that there is one world..."[100]
—St. Philastrius, Bishop of Brescia (†397 A.D.)

What even the most nominal of Christians have known since their earliest Sunday school classes reveals the truth about our topic: there is absolutely no implication, much less indication, of the existence of aliens anywhere within the thousands of pages of Sacred Scripture. And Salvation History is missing no chapters.

The early Christian Bishop quoted above, St. Philastrius, condemned the brand of belief in extraterrestrials that was in his day prevalent among Pagans outside the Church—and heretics within it such as Origen. He declared, in contradiction to this "**heresy**," that in fact "**Scripture has said that there is one world**," therefore that is precisely how many exist. To claim otherwise is to accuse God—our loving Father—of resembling some deceitful man looking to entrap a young woman in an illicit marriage; having told her of innumerable intimate details regarding all the comparatively minor aspects of his past, while concealing from her that he is a polygamist, and is already "married" to several other women and has many children. For God Himself is both the Bridegroom of the Church and its very soul; the one soul of the one and only Mystical Body of Christ. One bride, one groom. One Church, one soul. One family of children, one Father. "**There is one body and one Spirit... one Lord, one faith, one baptism, one God and Father of us all**..." (Ephesians 4:4-6)

Public Revelation—the climax of which is found in the Gospels' account of the Incarnation, Death, and Resurrection of the Son of God Himself—*says it all* (in the most important sense). There, and there alone, is the existential foundation of the correct worldview; the correct cosmos-view. *Forever.*

> "In giving us his Son, his only Word (for he possesses no other), [God] spoke everything to us at once in this sole Word—and he has no more to say." —St. John of the Cross. Quoted in the *Catechism of the Catholic Church*. §65

To propose the existence of rational incarnate non-human civilizations is to denigrate Sacred Scripture; treating it as a pathetically unfinished work that barely contains a complete prologue, though it poses as a comprehensive epic. Worse, this proposal treats Scripture as a deceptive work—one that even depicts "**things into which angels long to look**" (1 Peter 1:12), while actually leaving out almost the entire picture of the Creator and His creation, save one infinitesimally small portion thereof pertaining to events on the "third rock from the sun."

Most Christian ET promoters begin defending their view by dismissing this analysis as an "*argument from silence.*" They insist that just as God did not, in Scripture, reveal microbes or other such things we have recently discovered, so too the absence of any mention of extraterrestrials within the Bible need not concern us in the least. For example, one prominent Christian clergyman, Rev. Dr. Andrew Davison (an Oxford theologian and ET promoter among those recently commissioned by NASA to study the question of aliens) proposed within a scholarly tome published by Cambridge University in 2023:

> ...[T]here is no more reason to expect mention of [aliens] in the historic revelation to Jews and Christians than to find mention of the duck-billed platypus ... [therefore] arguments from silence will bear little weight.[101]

Before Davison, however, other prominent Christian-theologian-ET-promoters made the same argument. One of them, Dr. Ted Peters—within his chapter included in a 1995 University Press scholarly publication—denounced this rejection of aliens as "*blatantly fallacious,*" because "**Big Mac hamburgers exist indisputably, but they are not mentioned in the Bible.**"[102] Any discerning Christian will perceive how ridiculous Peters' and Davison's argument is. In making this point, these two (and their ilk) label ETs as a Faith-trivial matter, so inconsequential as to remain analogous to that of the duck-billed platypus or Big Mac hamburger.

Yet they only do this when convenient for their thesis. (In the next breath, they contradict themselves by expounding upon and agreeing with what we will presently discuss.) A moment's reflection stifles their scoffing. In truth, the most enormous theological ramifications immediately follow upon the question of whether earth, man, the Church, the incarnation of Jesus Christ, the Immaculate Conception of the Blessed Virgin, and Salvation History in general, are unique and supreme; or, rather, if they merely number unremarkably among countless similar or superior extraterrestrial realities scattered across the galaxies (and which will perhaps soon come crashing into our world with unprecedented "ontological shock").

In fact, "arguments from silence" *are* indispensable for understanding the Faith, despite the prevailing mood in academia to demean these as so-called "arguments from ignorance." "Aliens" themselves being alien to Scripture is only one such example (albeit a particularly important and instructive one). Others abound: the Old Testament portrays only one chosen people, therefore we must not seek out alternate Salvation Histories aside from what is found among the Children of Abraham. Jesus ordained only men, therefore only men can even now be priests. He is never mentioned as having married and fathered biological children, therefore we know He had none. The New Testament speaks of no more than Seven Sacraments, therefore we know there are only that many and only ever will be that many. Scripture defines only Three Persons in God, so we know that no one can ever assert so much as a fourth (much less millions, as some recent "ET revelations" claim). Genesis speaks of human beings as male and female, but makes no mention of a "non-binary" option—nor does it indicate one can change his sex—therefore we know such things are ruled out. These truths are clearly so central to understanding the Creator and His creation that the faithful have always been *certain* that "arguments from silence" pertaining to them are, in fact, *definitive* arguments.

The last item, however, requires extra attention.

Transgenderism and ET Belief

> "To be made in the image and likeness of the Son not only means that human beings possess intelligence—the ability to know the truth and to will and love what is good—but also that our bodies must ... bear the image of God. For it is the whole of us as human beings that bears God's image." —Fr. Thomas Weinandy[103]

Today, one often hears transgender ideology defended with the following line: "*The Bible never explicitly says you cannot change your sex*,"[104] or, "God did not say, in Genesis, that He *only* made humans male and female; He did not say that there would only *ever* be male and female!"

This advocate for transgenderism has adopted an absurd view of Divine Revelation—and an identical one to that of the Christian ET promoter. He supposes that something so fundamental to creation as our sex might have received such a glaringly inadequate treatment within the pages of the Bible that, only with the help of modern scientific studies and psychology, could we at long last come to the "correct" understanding of sex; namely, that it is nothing but a "social construct" which we are free to modify at will! For example, in a 2023 article promoting transgender ideology, the Jesuit priest and Harvard-educated professor, Fr. Richard Clifford, wrote:

> The expectation that this single verse in Genesis ["male and female he created them" (Genesis 1:27)] can adjudicate modern controversies about gender is thoroughly misguided. There is no hint that the ancient author knew anything about the modern issue of gender identity.[105]

The Church's most famous promoter of transgenderism, Fr. James Martin, lauded the article, singling out for praise this particular quotation from it.[106]

The sophistry on display is no different from that of ET promoters claiming the Faith is perfectly reconcilable with aliens. Just as Fr. Clifford dismisses a Biblical foundation for understanding sexuality because the Bible's authors were apparently too ignorant as regards "*anything about the modern issue of gender identity*," so Christian ET promoters claim that, since the Bible was written long before telescopes, the jurisdiction of its own claims is necessarily deprived of the ability to rule on the realms these inventions observe. Fr. Clifford dismisses those who realize that Genesis speaking of humans as male and female means *humans are only male and female*. Similarly, ET promoters dismiss those who realize that the same verse of Genesis speaking of only humans as made in God's Image *means only humans are made in God's Image*. Another prominent Catholic theologian, Fr. Daniel Horan, issued a similar defense of transgenderism in 2020, condemning those who rejected it because they sided with the Catechism and the Book of Genesis.* (Note: in 2024, Horan left the Franciscan order and the priestly ministry.)

Little wonder that today's most popular UFO "Religion"—Raëlianism (which we will focus more on later, and which openly regards Catholicism as its chief nemesis)—is also deeply committed to the transgender deception. The preeminent scholar of Raëlianism, Dr. Susan Palmer, explained:

> Seeing gender as an artificial construct, Rael [the "religion's" founder] emphasises its fluidity, resulting from different possible combinations of X and Y chromosomes. This implies that all human beings are essentially androgynous ... [In Realianism,] men dress up as women and women dress up as men, an exercise which encourages participants to experiment with different shades of gender and "choose" the melange of male-female which will best suit them in their everyday life. ... [These are the] sensually awake who will achieve salvation when the UFOs descend. ... [It is] part of the process of "deconditioning" from Catholicism to Raelianism...[107]

The overlap between the transgender deception and the ET Deception runs deeper still. Both are apocalyptic (Pope Benedict XVI said that "**Gender ideology ... will be the ultimate rebellion against God the Creator,**"[108]) as both are assaults on the human body's *revelatory* value. This value is best seen by way of Pope St. John Paul II's Magisterium. He focused heavily on the fact that the Image of God is found not only in the human soul but also in the human body—and, among all visible things, *only* in the human body. The Pope declared:

---

* Horan wrote that those who oppose gender ideology merely employ "*outdated anthropological foundations,*" consisting of "*13th century pseudo-science and ancient philosophical theories*" which cannot account for modern knowledge. He next addresses those who would take the "ancient" view of the human person (i.e., that it is either male or female), and denigrates their choice to regard that view as "*somehow complete in itself apart from whatever we might learn about humanity or the world subsequently.*" ("*The Truth about so-called 'gender ideology.'*" National Catholic Reporter. June 24, 2020.) Three years later, Horan condemned the American Bishops for warning against transgenderism; castigating their "**references to the Catechism of the Catholic Church... [and] recourse to proof-texted passages from Genesis.**" ("*US bishops' document against transgender health care is a disaster.*" National Catholic Reporter. March 32, 2023) Here, too, we do not see a *similar* argument to the one used to defend Christian ET promotion, but an identical one. A theologian who commented on Horan's writings put it well when he said, "**Horan's central problem, ultimately, is not that he disagrees with the bishops (which he does), but that he disagrees with divine revelation**." (Dr. Alex Giltner. "*Franciscan friar's attack on bishops and doctrine is a theological disaster.*" Catholic World Report. April 5, 2023) We can equally say that the problem with ET promotion is not that it disagrees with St. Philastrius, Bishop of Brescia (or other saints, Popes, and Magisterial teachings we will cover later), but that it disagrees with Almighty God.

> Man appears in the visible world as the highest expression of the divine gift ... The body, and it alone, is capable of making visible what is invisible: the spiritual and the divine. It was created to transfer into the visible reality of the world the mystery hidden since time immemorial in God, and thus be a sign of it.[109]

Of all material things God made, only the human body is capable of *itself* rendering visible what is invisible in Him. This observation could be considered the essence of the Pope's vision in his *Theology of the Body*, and it is simply another way of saying "*only man bears His Image*."

It is clear how transgenderism accordingly distorts one's very understanding of God Himself. If one supposes the body's fundamental nature as male or female is actually not fundamental at all, but only a so-called "social construct," then one will equally have his understanding of God disfigured in relativist terms.

Yet even that disfigurement pales in comparison to the disfigurement ET belief brings. The Christian ET believer finds himself wondering if the human body—assumed by God Himself 2,000 years ago—is merely one more-or-less arbitrary way of designing an incarnate intelligent being just as a Volkswagen Beetle is just one way of achieving transportation. From that sentiment, his whole grasp of theology, approach to prayer, and perhaps even adherence to morality, risks being corrupted.*

The universal implications of this reality are reinforced in another teaching of Pope St. John Paul II:

> There is no doubt here that precisely in the whole visible world (cosmos) that one body which is the human body bears in itself the potentiality for resurrection, that is, the aspiration and capacity to become definitively incorruptible, glorious, full of dynamism, spiritual.[110]

John Paul II could not be clearer that he is referring to the entire material Universe, and he emphasizes that there is "**no doubt here**" about what he says. What is he saying? He is outright condemning any views which would assert that other beings—*anywhere*—are capable of resurrection. (And we should add that this itself rules out the possibility of aliens, even aside from the implications of ET belief we are here addressing.)[111]

The crux of the present matter is ultimately existential; *pre*-theological. It derives from how an individual Christian has decided to fundamentally regard the very idea of Divine Revelation. Does he—in union with all the saints—regard it as *the* foundation of our grasp of the Creator's creation? Or does he—with Horan, Clifford, Martin, and the ET promoters—view it

* The soul bears the Image of God by possessing intellect and will. If one claimed that "other types" of *souls* also reflected the Image of God—perhaps "souls" lacking freedom—then this would distort his own understanding of God. Likewise, if one claims that other types of *bodies*—just think of the countless extraterrestrial forms the ET promoters believe exist—likewise bear the image of God, then this cannot but similarly distort his grasp of God. For these other types of bodies would reveal contradictory insights into God Himself.

as just one of many other equal (if not superior) insights into the same; one which ought to be dismissed if doing so becomes socially or scientifically fashionable?

Massive Ramifications

> "Man is the perfection of the Universe..." —St. Francis de Sales, Doctor of the Church.[112]

On the question of extraterrestrial intelligence, we are dealing with a similarly crucial issue as sex. Is man *the* unique being of all material creation made in the Divine Image—*the* setting wherein eternal destinies are determined—or only one of multiple (even innumerable!) races of rational creatures throughout the galaxies? This question does not merely touch upon the essence of the creature's relationship with its Creator. It defines it.

Those who engage this question most laboriously, regardless of their answer to it, agree on the question's own immense magnitude and the enormity of its theological ramifications. Agnostic apathy about ET belief is only found among those who have given the matter no serious thought—but comment on it occasionally, nonetheless.

In fact, Christian ET promoters themselves openly acknowledge that the discovery of alien civilizations would cause a veritable theological earthquake of unprecedented proportions; one that would require us to re-examine (and, most of them admit, *overturn*) what has always been held as sacred and inviolable in the core of Christianity. Even the most supposedly orthodox-minded of them have already written books or produced media seeking to detail, in advance, this earthquake they foresee, and instruct us on the rebuilding process. One Catholic seminary professor even wrote a lengthy 2021 book, *Exotheology*, in support of alien belief; and in it, this theologian gleefully announced that contact with aliens could "**call into question certain cherished religious doctrines**," since "**for two millennia theology has [had] ... an anthropocentric myopia**." Discovering aliens, he foretells, "**may wholly redefine many of our conceptions of God and creation**," because:

> ... evidence of a second Genesis could drastically call into question certain Christian foundational theological teachings regarding creation, the Incarnation, and the Redemption. ... [this] would inevitably result in a profound reformulation or recontextualizing of theology..."[113]

Such ET-derived apostasy as this—masquerading as authentic Christianity—is exploding in popularity today within the Church, largely unopposed. But this particular theologian is only reiterating what scholars of Ufology have conceded for decades. Let us consider a few others.

Charles Davis was a prominent Catholic theologian and priest who studied at the Pontifical Gregorian University. In the year 1960 he wrote in support of ET belief while serving as a seminary Professor—even as he noted this view would cause us to "**qualify our statements about [Christ's]**

primacy," and that the "**Second Coming of Christ... [would be] a subordinate factor in some higher scheme**."[114] (Six years after making these statements, Davis left not only the priesthood but also the Catholic Church—doing so to much fanfare as he lamented the supposed shackles that Catholic orthodoxy placed on his theological freedom.)

Fr. John Saliba, in an academic tome published by the State University of New York, drew from various other theologians of the late 20th century, observing that they insisted belief in aliens would require "**radical rethinking of the common Christian doctrines of creation and redemption**."[115] Jesuit priest Fr. George Coyne was the director of the Vatican Observatory for almost three decades. He insisted ETs entail a duty to "**rethink some fundamental realities**" of Christianity:

> The existence of extraterrestrial intelligence must be taken as a serious possibility with all its consequences. ... There are serious theological implications here for our understanding of God. ... [Theologians] must accept a serious responsibility to rethink some fundamental realities within the context of religious belief. What is the human being? ... We cannot rely, even theologically, solely on God's revelation to us in the Scriptures and in the churches, since that revelation was to us and was received, therefore, in a very anthropocentric sense.[116]

"Serious" here is the name of the game, and we can hardly fault Fr. Coyne for repeating the phrase. Despite his tragic exhortation of openness (although according to at least one Ufologist, Coyne is not merely open to the idea of ETs, but is actually a believer in them[117]), Fr. Coyne is at least logically consistent and honest in admitting the massive repercussions of that openness: it entails radically rethinking God and humanity, and it fundamentally demotes Divine Revelation itself.

What has been said above can be summarized in the title of this chapter: *There are no chapters missing from Salvation History*, for we now have "**Christ the Lord, in whom the entire Revelation of the most high God is summed up...**" (*Catechism of the Catholic Church*, §75)

No one should pretend that God intends for the Bible to provide a direct answer to every possible question. Neither, however, should any Christian succumb to the opposite extreme. He mustn't treat Scripture like a text authored by humans which may well be guilty of leaving out large and necessary chapters. The Church has already been torn apart by false prophets who have proposed to present the hitherto "missing Chapters" of Salvation History: Gnosticism, Islam, Mormonism, various New Age "revelations," and too many others to list. Not in vain does Scripture warn: "**Even if we, or an angel from heaven, should preach to you a gospel contrary to that which we preached to you, let him be accursed**." (Galatians 1:8)

Addressing each such danger is well and good. But a thoughtful Christian, discerning a trend, will step back and consider what all these apparently disparate deceptions have in common. The similarity is obvious: each one treats what has been handed down—not only through Public Revelation but also through Sacred Tradition—as desperately deficient; as needing not merely development (Pharisees wrongly reject even that), but *fundamental* correction and revision, across the board, in order to give the *real* picture of God's plan with His creation and with His true children.

Plenty of important questions are not addressed in the Bible—because they do not have massive ramifications for the Faith.* But the fact that only man bears the Divine Image is as consequential to the Faith as any matter that can be imagined.

***

It is, however, not true that Scripture is silent about this question. The methods used above, while certainly disproving aliens, provide only some of the theological reasons ETs do not exist. Public Revelation is replete with teachings indicating man alone bears God's Image. It is filled with passages from which we can glean that there are no incarnate intelligent creatures other than human beings. To those passages, we now turn.

---

* When considering questions about microbes, or the geometric arrangement of the Solar System's bodies, or the amount of "time" that had passed while the earth was only dead rock, or the "setting of the sun" caused by the earth's rotation (or the sun itself revolving around the earth), or the shape of the earth, or the number of continents on the planet's surface, or countless other such questions, we are certainly dealing with ideas that we may struggle to find clear and complete solutions to within Scripture. Yet solutions like these have no impact on the Faith, therefore their absence from those pages is immaterial, and if Christians' view on the proper approach to them changes over time, that is not intrinsically problematic.

6. The Old Testament

"In a particular way must be deplored a certain too free interpretation of the historical books of the Old Testament. ... [The] first eleven chapters of Genesis ... [do] pertain to history in a true sense, [its authors wrote] with the help of divine inspiration, through which they were rendered immune from any error..."
—Pope Pius XII, *Humani Generis*, §38

Divine Revelation often describes *creation in its entirety*, that is, "heaven and earth." In doing so, it always indicates that "earth," where rational incarnate creatures dwell, is *singular*.

Scripture refers to "earth" over one thousand times. Not once does it imply that earth may be anything but unique in harboring intelligent incarnate beings. Yet, "heaven and earth" constitute *all that is*. The Catechism teaches, "**The Scriptural expression "heaven and earth" means all that exists, creation in its entirety**." It continues:

> "The earth" is the world of men, while "heaven" or "the heavens" can designate both the firmament and God's own "place"—"our Father in heaven" and consequently the "heaven" too which is eschatological glory. Finally, "heaven" refers to the saints and the "place" of the spiritual creatures, the angels, who surround God. (§326)

No Scripturally cogent options exist, therefore, for extraterrestrial habitations. Individual passages also make this clear.

The first words of Scripture do not pick up at some undefined point *within* the history of time, as if God chose to leave undisclosed to us the nature of the Universe's origin. Revelation begins, rather, at *the* beginning itself: "**In the beginning God created the heavens and the earth.**" (Genesis 1:1) We can see that the very first verse of the Bible rules out any conjectures about infinite time, an eternal world, or other universes, timelines, strings, dimensions, or what have you. On the contrary, Christians know that there was a singular beginning, and that beginning—of everything—was none other than the point at which God called *the* heavens and *the* earth ("**all that exists**") from non-being into being.

Moreover, the billions of trillions of stars and planets are all gathered, in the Word of God, into the simple umbrella term of "the heavens," whereas only one planet merits any attention among the quintillions estimated to populate the great expanse: "*the earth*." Faithful Christian exegesis will pause on the definite article used here by God: "the." Scripture does not refer to this planet as "an" earth or as "our" earth but, rather, as "*the*" earth. St. Thomas Aquinas points out that this definite article means exactly what it indicates: *there is only one*.

Despite being a mere nothing, quantitatively, in relation to "the heavens," earth is given its own *category* in Scripture's opening line. Having considered only one sentence of Scripture, we can already see that it would be illicit for a believer in the Divine origin of this verse to assert that earth is only one of many planets whereupon theaters of eternal destinies are played out.

Faith in just the first ten words of Scripture, therefore, collapses all the most important premises of alien belief like so many houses of cards. The first chapter of Genesis even teaches *why* those bodies which populate the night sky exist.

> And God said, "Let there be lights in the firmament of the heavens to separate the day from the night; and let them be for signs and for seasons and for days and years..." (Genesis 1:14)

The Bible teaches that the heavenly bodies exist *for us*—for man. Genesis does not imply these heavenly bodies may also be "for" anything or anyone else. If they were, then God deceives in the Book of Genesis by indicating that these very things exist *for* marking *our* "seasons... and years." If in fact these otherworldly objects serve as dwellings of other creatures called to eternal life, that status as alien-theaters-of-salvation would, indisputably, be their *primary* purpose, which in turn would mean Genesis 1:14 misleads.

Since God already told us the reason for the existence of the "heavens"—that is, all that we see, or can possibly see, when observing the night sky—no Christian may validly protest, "*But why did God make all this if, despite its incomprehensible magnitude, it is uninhabited? Wouldn't that be a waste of space?*" He revealed that He made it for us—for what we can gain as we gaze upon it. No other disparate purpose of this immense creation should be postulated. Nevertheless, we can—as faithful Christian exegesis implores—delve more deeply still into pondering what *has* been revealed.

God *could* have made the earth become enveloped each sunset only in an endless black void. Instead, He populated the night sky with resplendent diamonds whose incomprehensible expanse we discover more and more about with each new telescope built. Whoever supposes that this "must" entail the existence of alien civilizations therein simply understands nothing of the magnanimity of God and His great goodness in imbuing creation with such endless beauty to be enjoyed by man. It is thus that the enormity of "the heavens" testifies to the eternity of *Heaven* itself. As astronomers rightly note, we can scarcely wrap our minds around the expanse of the galaxies. Similarly, we *cannot* grasp the expanse of eternal life. It is only right that, upon creating the wonders of the starry sky, God would enclose within those mysteries greater and greater expositions of His grandeur, in proportion to how much we discover hidden therein. What, pray tell, is lacking in this understanding of why God stretched out the heavens, like a beautiful tent for us to dwell under? Simply put, **"The heavens proclaim the glory of God."** (Psalm 19:1)

This verse, although used by some ET promoters to justify their thesis, only refutes it. We need not pretend the heavens "must" be inhabited, since in fact they *already* "proclaim God's glory"—even as the ancients understood them. St. Thomas Aquinas himself took pains to point out how absurd it would be to claim that this verse implies the heavens are actually rational (or, by extension, that they contain rational beings). The Angelic Doctor wrote:

> This argument is foolish despite the fact that Rabbi Moses [that is, Maimonides, a 12th-century Jewish philosopher] proposes it… the heavens are said to proclaim the glory of God … inasmuch as through them the glory of God is made manifest to men.[118]

Here we must note how absurd it would be to claim that acknowledging the truth of—and the corollaries of—teachings contained in the Word of God constitutes "limiting God." On the contrary, one is limiting God by implying that the Almighty is too shortsighted to mean exactly what He said in Divine Revelation. Jesus rebukes that shortsightedness: "**O foolish men, and slow of heart to believe all that the prophets have spoken!**" (Luke 24:25)

The Pharisees are guilty of limiting God by putting Him in a box of their own worldly making. But the Pagan, the Gnostic, the modernist, the apostate, and the heretic limit God by pretending that He might not really intend to stick with what He already revealed, whereas in fact, God "*cannot deny himself,*" (2 Timothy 2:13), and what God has said is "*…once for all delivered to the saints.*" (Jude 1:3)

The Mother, the Fall, and Aristotle

Scripture's very first chapter refutes all three of the foundational premises presented by Christian ET promoters.

We saw the following contradicted by God Himself in Genesis 1: first, the notion that earth (and, by extension, man) cannot be regarded as unique; second, the argument that disbelieving in aliens constitutes "limiting God;" and third, the idea that the incomprehensible multitude of stars and planets in outer space "must," in the Divine Order, have been created for their own sake and not merely for us.

Moving to the second chapter of Genesis, this begins with the assurance that, after the creation of the "heavens" and the various living beings who count earth as their home, "**…the heavens and the earth were finished, and all the host of them.**" (Genesis 2:1) The uniqueness of earth is reaffirmed and we find revealed that, upon the completion of its establishment along with the heavens, *God's overall work of creation was done.* The Creator, of course, would have much more to do *within* creation over the length of the ages to follow, but the fundamental categories of beings (that is, "all the host") were settled. No additional ranks would ever be inserted into the hierarchy of creation—the "Great Chain of Being" settled in both Scripture and Sacred Tradition. (Aliens not only do not exist, but also *will not ever* exist.)

This is a hierarchy, as we will see, which the faithful have always understood to consist in the following categories containing no gaps: inanimate matter, plants, animals, men, angels, and God. No place exists for "more evolved" aliens in between men and angels.

> In the act of creation, God calls every being from nothingness into existence. "Crowned with glory and honor," man is, *after the angels,* capable of acknowledging "how majestic is the name of the Lord..." —*Catechism of the Catholic Church,* §2566

The third chapter of Genesis follows, teaching that Eve is "**mother of all the living.**" (Genesis 3:20) Now Eve, of course, was not the mother of the animals, therefore "the living" here is to be understood in light of the preceding chapter: "**And the Lord God formed man of the slime of the earth: and breathed into his face the breath of life, and man became *a living soul.***" (Genesis 2:7) Here, the context clarifies that only those who have the "breath of life" from God are not merely "alive," but are "*living souls,*" which is to say, they are *rational* incarnate beings.

Trusting that Sacred Scripture does not erroneously use the qualifier "all," we yet again see the grounds for alien belief ruled out. Genesis 3 reveals that *all* who can be called "living souls"—due to receiving the *breath of God,* which alone bestows rationality—are themselves children of this specific human being: Eve.

> "God created everything for man..." (*Catechism of the Catholic Church,* §358)

Adam and Eve did not long remain in their state of Original Justice. Genesis 3 also recounts the Fall, whereupon the sin of Adam brought all creation down with it (thus ruling out the possibility of "unfallen aliens"). "**All visible creation, all the universe, bears the effects of man's sin,**"[119] as Pope St. John Paul II taught. This is established in Genesis:

> Because you have listened to the voice of your wife, and have eaten of the tree of which I commanded you, 'You shall not eat of it,' cursed is the ground because of you; in toil you shall eat of it all the days of your life; thorns and thistles it shall bring forth to you... (Genesis 3:17, 18)

Whoever might struggle to infer, from this passage, the universal impact of Adam's sin, should be reminded that the validity of such an inference is settled by Scripture itself: "**By the envy of the devil death came into the world,**" (Wisdom 2:24), and "**in Adam all die,**" (1 Corinthians 15:22), whereas even "**the whole creation groans**" (Romans 8:22) from the Fall, because "**through one man sin entered into the world, and death through sin.**"[120] (Romans 5:12)

It is just for Adam's offspring to die on account of the former's sin; he is our father. On the other hand, it would be radically unjust and entirely irreconcilable with God's goodness for *other* rational beings—not only lacking natural generation from Adam but even living on entirely different planets—to be subject to death, suffering, and other evils due to Adam's choice to eat the forbidden fruit.[121] Yet, precisely that unconscionable scenario would apply if aliens existed. As Fr. Thomas Weinandy explained:

> ...[B]ecause of sin, not only humankind was marred, but also the whole of creation. ... Since aliens [would be] members of the cosmos, they too would have to be affected by humankind's sin. The very absurdity of such a conclusion argues against the existence of alien intelligent life.[122]

Pope St. John Paul II made this clear in his teaching quoted above. It is simply a fact that the entire material Universe—he specified not only earth, but "**all visible creation**"—fell due to man's sin. The Pope, however, was merely reiterating what Scripture had already settled, in the verses noted above and elsewhere, along with reaffirming a theologically obvious truth. Spiritual effects are not limited by distance any more than mathematical truths are. If the Fall was real—and it was—then it would be ridiculous to strive to ascertain where, spatially, its effects somehow magically stopped. They did not stop. They brought down the Universe.*

Not only Biblical theology, but also purely philosophical truths speak to the illogicality of belief in aliens on a related point. For what transpires in "the heavens" is not a separate story from man's own, but rather an integral part of it. The stars have never been strangers to us. They belong to man.

> "Man is woven into the fabric of the universe. Time is first of all a cosmic phenomenon. Man lives with the stars."[123] —Pope Benedict XVI

Accordingly, what pertains to the stars can only be subordinated to what pertains to the *only* creature "**that God has willed for its own sake**" (*Catechism of the Catholic Church*, §356)—*human beings*. Put in another way, if the entire material Universe is a story—and history surely is "His (God's) Story"—then mankind is that story's sole plotline; the rest is simply the setting. Therefore, that plotline provides the story's only final end. As Aristotle taught, "*the chief good is evidently something final. Therefore... there is only one final end,*"[124] and "*nature makes nothing incomplete, nothing in vain*" [125] therefore "*she has made all things for the sake of man.*" This is why Aristotle rightly insisted—in contradiction to the confused philosophers who preceded him and whom he corrected on this and other crucial matters—that "...**the world must be unique. There cannot be several worlds... It is plain, too ... that**

---

* No passage in Divine Revelation suggests that Scriptural truths crash against an impassable barrier located somewhere in earth's stratosphere when they follow their Divinely ordained course. The notion itself is silly; akin to supposing that an astronaut should revisit the Pythagorean Theorem when walking on the surface of the moon. We all know that no matter how far any astronaut travels from the earth in future missions, he need not fret that his journey may exclude him from the same economy of salvation in which he has lived since his conception on earth. One can escape neither Original Sin nor Judgment Day if only he builds a fast enough rocket ship. Scripture affirmatively rejects any claim that its own teachings "only apply to earth," and any attempt to limit its jurisdiction to earth is materially heretical.

the universe must be one."[126] Not only history's greatest philosopher insists upon this; so does history's greatest theologian: "**For there to be another earth than this one is not possible,**" St. Thomas Aquinas declared.[127]

Why is Aristotle—who wrote hundreds of years before Christ and had no knowledge of God's Revelation to Israel—capable of such truth and clarity on this point? Because even reason alone proves that God exists, and is all-good, and is all-powerful:

> The Church teaches that the one true God, our Creator and Lord, can be known with certainty from his works, by the natural light of human reason. (*Catechism of the Catholic Church*, §47)

To suggest that the Supreme Being would exercise His Providential direction of all things to author a story so scatterbrained as the one proposed by ET promoters is to impugn either His existence, goodness, or power—all violations of those truths even a noble Pagan would readily grasp about the Almighty.

7. The New Testament

"Thy will be done on earth as it is in heaven."
(Matthew 6:10)

Jesus Christ *is* the New Testament. He is the Eternal Word of God; the only begotten Son of the Almighty Father; the Alpha and the Omega; the King of Kings and Lord of Lords. His absolute supremacy, primacy, and uniqueness is both unquestionable and unsurpassable, nor is it even possible to overestimate them. He *is* God. Scripture could not relay this point any more clearly:

> [Christ Jesus] is the image of the invisible God, the first-born of all creation; for in him all things were created, in heaven and on earth, visible and invisible, whether thrones or dominions or principalities or authorities—all things were created through him and for him. He is before all things, and in him all things hold together. He is the head of the body, the church; he is the beginning, the first-born from the dead, that in everything he might be preeminent. For in him all the fullness of God was pleased to dwell... (Colossians 1:3-4,15-19)

Because of this absolute supremacy, and because the Incarnation of Jesus Christ was in no way partial or incomplete as it transpired 2,000 years ago, we are left with the certainty that we await no greater contact with the Divine than what we have already been given in the Public Revelation of Jesus Christ—handed on "once for all" (Jude 1:3) to the faithful, by the Apostles, and rendered complete upon the death of the last Apostle, St. John.

The Old Testament, however, was Public Revelation in process and not yet complete. One could read it and perhaps understandably ask, "is this all?" Its own pages testify to its fundamentally anticipatory nature in awaiting the coming of the Messiah. But now that this long-awaited Messiah has come, and He has revealed not merely a *covenant* as in the days of old, but rather the "**eternal Gospel**" (Revelation 14:6). Accordingly, we know that our task, until the very end of time, consists not in awaiting a new public revelation, but rather living out—though ever more deeply—what is contained in the one and only, complete, Public Revelation.

Yet it is precisely this essential approach to the Faith that is rejected by ET promotion. Consider, for example, the Franciscan priest, Fr. Richard Rohr, who is *"one of the most popular spirituality authors and speakers in the world,"* according to PBS.[128] His books can be found on *New York Times* best-seller lists, he preaches to the world on Oprah Winfrey's television network, and Rohr even claims that Pope Francis has read, and encourages, his work.[129] Fr. Rohr proudly asserts that "when" we discover aliens, the type of Christianity that the saints have always lived and died for will be "shot to hell." In a popular sermon, he declared:

> We can't presume that God just got interested in God's creation 2,000

> years ago and left the first 13.7 billion years empty of revelation ... In our overemphasizing Jesus without understanding Christ we created a storyline ... that all depends upon a supposed sin that was committed between the Tigris and Euphrates River, and that just isn't a big enough storyline... When I was in college, they were saying there's 6 stars for each one of us, now it's six *galaxies*! ... you have to say, 'who is God?!' Well, he certainly isn't upset because someone bit into an apple ... Our theory of salvation is so tiny, so planet-bound ... When we discover life on another planet, which might be tomorrow, our little story line is shot to hell...[130]

The supposed "littleness" of the Incarnation—"earthly" as it is—emerges as Rohr's motivation for replacing Jesus with his (now viral) "Universal Christ." And this alleged "littleness" is most sharply illustrated within the context of his belief in galaxies filled with planets containing other civilizations of rational creatures.

Outlandish and blasphemous as these proposals are, they cannot be faulted for inconsistency. For they *are* consistent with a cosmos-view that includes belief in aliens. But the moment a Christian has begun to entertain the prospect that the Jesus we know from Scripture might not enjoy absolute supremacy, and rather might be only one Divine Incarnation of many—or, equivalently, that we must in some way replace Jesus of Nazareth with descriptions of a "Universal Christ,"—is the same moment this Christian's Faith has begun to die in his heart.

We need not scour the crevices of obscure documents to grasp the absolute supremacy of Christ. We need only look to Scripture's most important teachings and the most central and ancient formulations of Christianity, for which the faithful over the millennia have willingly endured the most gruesome tortures and deaths rather than deny a single word.

This truth is simple: Jesus Christ is God, and He has—and always will have—two and only two natures: a human nature and a Divine Nature. No Christian may entertain the possibility that He had any more natures than this, or that God Himself had any other nature anywhere else before *The* Incarnation two thousand years ago. No believer may speculate that those two natures of Christ are "the only ones *we* need to know about," but that "*maybe He also has a Martian nature, a Neptunian nature, an Andromedian nature, and innumerable other natures*!" In contradiction to such wild speculations, we have the dogmatic truth as expressed at the 5th-century Council of Chalcedon: "**We confess, therefore, our Lord Jesus Christ, the Only Begotten Son of God ... became a union of two natures**."

From its earliest centuries, the Church has always declared what was already clear from Scripture: whoever would fail to assert that the Incarnation of God consists of exactly two natures (*human* and Divine) is guilty of a "**perverse, foolish, and extremely blasphemous**" assertion—according to the wording of the Council of Chalcedon's declarations. Virtually all Christian ET promoters assert precisely this, or at least testify to their proudly "humble" openness to it. Condemning this ET-belief premise as "**in profound conflict with the Christian Faith**," Pope Benedict XVI, as Cardinal Ratzinger, taught the following in an official document of the Congregation for the Doctrine of the Faith, wherein he reaffirmed the Magisterium of Pope St. John Paul II:

> In contemporary theological reflection there often emerges an approach to Jesus of Nazareth that considers him a particular, finite, historical figure, who reveals the divine not in an exclusive way ... These theses are in profound conflict with the Christian faith. The doctrine of faith must be firmly believed which proclaims that Jesus of Nazareth, son of Mary, and he alone, is the Son and the Word of the Father. ... In this regard, John Paul II has explicitly declared: "*To introduce any sort of separation between the Word and Jesus Christ is contrary to the Christian faith...*" [The value of Christ is] unique and singular, proper to him alone, exclusive, universal, and absolute.[131]

The "**Son of Mary, and he alone, is ... the Word of the Father**." This is dogmatic truth, and it repudiates any suggestion of multiple Incarnations of God or any substitution of the specific man born of the Virgin with a "cosmic Christ." An ET promoter may incoherently claim such "other Incarnations" are also the Word of the Father. Yet he obviously could not claim that those Incarnations are likewise the Son of *Mary*, and we know that the Son of Mary is absolutely alone in His status as the Word of the Father. Therefore, any theory that refuses to accord to the Man, Jesus Christ, a recognition of His unqualified singularity, supremacy, and exclusivity—or that introduces separation between the Second Person of the Trinity and the humanity of Jesus—is here dismissed as heretical.

To underline this reality, let us again defer to the wisdom of Fr. Thomas Weinandy, who is undoubtedly one of the greatest theologians alive today. A Franciscan priest who has taught at many Catholic universities, served on the International Theological Commission, and worked as the American Catholic Bishops' chief defender of orthodoxy, he has—especially since 2017—been distinguished as one of the most courageous of the internationally recognized ecclesial voices in critiquing the doctrinal chaos the Church has recently been experiencing. He aptly explained why belief in extraterrestrials cannot be reconciled with what the Faith dogmatically teaches about the Incarnation of Jesus Christ:

> ...[T]he existence of intelligent aliens would be contrary to God's revealed eternal plan for the whole of creation, for human beings, and above all for Jesus himself. … It's Jesus, as the incarnate Son, who "reconciles all things, whether on earth or in heaven, making peace by the blood of the cross." (Colossians 1:20) The entire cosmos was created for the incarnate Son of God, Jesus. It's as man that the Son of God reigns supreme. For him, as man, to reign over non-human aliens defies logic ... Intelligent aliens find no place within God's eternal scheme—his entire focus is on

> human beings and the humanity of his divine incarnate Son ... Jesus cosmically reigns supreme as man. He is not Lord of non-human aliens, for they do not exist.[132]

The existence of aliens, indeed, would violate the very deepest elements of God's plan as already definitively revealed to us. As Fr. Weinandy explains, the *Incarnation* of the Word—the man, Jesus of Nazareth—cannot be extricated from *the Word Himself* (the Second Person of the Trinity). If, as Scripture teaches, the Word holds all things together in the Universe, then it follows that *Jesus* holds all things together, and if Jesus holds all things together, then it is the God-*man* who does so. There simply is no place, in this great cosmic picture we are definitively given in Public Revelation, in which intelligent incarnate non-human creatures of any sort could possibly fit. If there was ever any question as to whether such a place could exist (there was not), that question was forever closed upon the Incarnation.

It is impossible for the human mind to even begin to comprehend how magnanimous it is that the infinite and uncreated God—the Eternal Word Himself—now and forever truly has a human nature like ours. The entire human race—in fact, the entire Universe and all it contains—is absolutely nothing compared to God. Nevertheless, the Infinite Himself incorporated our "nothingness" into His own self... *for all eternity*; even preserving five wounds He received on earth—those in His hands, feet, and side—in Heaven.

The impossibility of grasping so much as an iota of this glory, however, must not prevent a Christian from abiding in each conclusion that follows from acknowledging its truth. There is only one Son of God, and this one Son forever assumed *human* nature. If this assumption was just a show that God put on (a premise inextricable from "Christian" ET promotion), then He could have put on similar shows for innumerable alien civilizations. Whoever entertains this possibility, however, should admit he is no longer a Christian. For a Christian confesses unto death that the Incarnation is not and was not an illusion. He instead acknowledges that the Eternal One is now man—not any other created nature—and that He never will be any other created nature.

> I believe in one God, the Father almighty, maker of heaven and earth, of all things visible and invisible ... I believe in one Lord Jesus Christ, the Only Begotten Son of God ... begotten, not made, consubstantial with the Father; through him all things were made. For us men and for our salvation he came down from heaven... (The Nicene Creed)

This is the very core of the perennial and infallible Faith for which all the martyrs gladly died.

Christianity and "Other Christs"

> "For God so loved the world that he gave his only Son, that whoever believes in him should not perish but have eternal life." (John 3:16)

Behold what is perhaps Scripture's most oft-quoted verse—underlining the uniqueness of Jesus of Nazareth and subtly but no less powerfully testifying to the uniqueness of mankind. The same Gospel also relays Jesus' teaching that "**no one**" (John 14:6) comes to God the Father except through the Son of Mary, Jesus Christ. In Scripture's next book, we see this truth reiterated. The Acts of the Apostles (4:12) teaches "**there is salvation in no one else**"—Redemption cannot be found in anyone other than Jesus Christ "of Nazareth," thereby condemning any suggestion of extraterrestrial messiahs.

> "[Jesus] is the radiance of [God's] glory and the exact representation of His nature..." (Hebrews 1:3. NASB)

Confirming this condemnation, the Gospels of Matthew, Mark, and Luke all contain Jesus' stark warning that others would come, likewise claiming to be the Christ—and that we must *always* categorically reject any such proposal (cf. Matthew 24:5, Mark 13:6, Luke 21:8). St. Paul's letters contain an even more shocking warning:

> I am afraid that as the serpent deceived Eve by his cunning, your thoughts will be led astray from a sincere and pure devotion to Christ. For if some one comes and preaches another Jesus than the one we preached ... you submit to it readily enough. (2 Corinthians 11:3-4)

A resounding lesson for all Christians throughout time emerges: any presentation of "another Jesus" (aside from the one handed down in the Deposit of Faith) must be immediately and outright rejected. Scripture contains this admonition precisely because its author—God—knew a time would come necessitating it.

St. Paul goes so far as to link the evil of the Fall itself to the willingness of the Corinthians to submit to "another Jesus," and later in the same chapter he alerts us that "**even Satan disguises himself as an angel of light**." (v. 14) The faithful are hereby implored to let no apparent sign or wonder attract them away from the one and only Jesus Christ. That is, the Jesus preached by the Apostles and Who lived on earth 2,000 years ago. *Any* "other Jesus" would be none other than a machination of the father of lies, seeking to reenact in our souls the supreme evil of the Original Sin he inspired in the Garden.

Here we have another unmistakable instruction from Scripture that would be unconscionable if alien civilizations existed. For each would need their own incarnation—or at least "their own Christ." However, such a scenario would entail God condemning Himself by these teachings.

Recall that ET promoters not only assert the existence of alien civilizations, but are also unanimous in insisting that we can one day communicate with these aliens and perhaps even make physical contact. Most insist we have already done precisely that. If this scenario were true, it would resemble indigenous peoples on continents that were yet undiscovered by Christ's immediate audience in His time. (ET promoters frequently draw this very parallel while promoting their own position, supposing it supports their thesis, when in fact it only contradicts it.)

It is true that centuries had to pass before communication between those lands and Christendom would take place, but this meant nothing. Those lands did *not* have "their own Christs," much less "other Incarnations." The early European explorers and missionaries who first ventured to the Americas—men of strong Faith, unaffected by wild speculations disharmonious with Scripture—embarked upon their expeditions knowing they would not find "another Christ," *no matter where* they landed, including if they "sailed off the end of the earth."

If, however, other incarnations—"other Jesus Christs," if you will—existed on other planets, then we would be confronted with the same unconscionable scenario we would have faced if other incarnations existed in the once undiscovered continents. In any such scenario, Scripture's command that we categorically reject the very presentation of "another Christ" would be tantamount to God denying Himself—which the Bible expressly teaches God *cannot* do (cf. 2 Timothy 2:13).

A Christian who asserts the possibility of other Christs of *any* sort on other planets is thereby holding to a Faith which requires rejecting other Christs, even while believing that other Christs exist. This intrinsic contradiction prompts the one who holds it to discard either alien belief or Scripture. Hopefully it is not Scripture that he will discard.

Moreover, positing the possibility of other Christs in ET discourse succumbs to the same syncretism one hears propounded in New Age circles. If saviors other than Jesus of Nazareth exist anywhere (whether we are talking about different continents or different galaxies is irrelevant), then Christianity's repeated exclusive claims about Jesus of Nazareth are false. These claims contain no indications or implications that they cease to apply when one escapes earth's atmosphere, or that they expire in the third millennium when "ET disclosure finally happens." Although devotees of the New Age Movement readily dismiss exclusive claims in the Deposit of Faith—as they embrace their own falsified version of Christianity which regards Jesus as merely "one path of many up the same mountain"—faithful Christians must reject these notions wherever they are found.

> ...[W]e know that "an idol has no real existence," and that "there is no God but one." For although there may be so-called gods in heaven or on earth—as indeed there are many "gods" and many "lords"—yet for us there is one God, the Father, from whom are all things and for whom we exist, and one Lord, Jesus Christ, through whom are all things and through whom we exist. (1 Corinthians 6:4-6)

Abusing Aquinas

Catholic ET promoters are prone to twist the theology of Church history's greatest theologian, St. Thomas Aquinas, as they argue for the orthodoxy of belief in multiple (extraterrestrial) incarnations of God. Although Aquinas' texts obviously do not rise anywhere near the level of Scripture, their prominence in this debate requires that we address them here.

Catholic apologist Dr. Paul Thigpen's 2022 book promoting ET belief contains a chapter entitled "*A Million Alien Gospels,*" within which resides the section, "*Saint Thomas: Multiple Incarnations Are Possible.*" Therein, Thigpen relays a paragraph of the *Summa* in which Aquinas writes, "*the divine Person is able to assume [unite to himself], in addition to a human nature that has already been assumed, another numerically different human nature.*"[133] Next, Dr. Thigpen writes, "*If this is the case for divine incarnations on Earth, the principle would seem to hold for incarnations on other planets as well.*"[134]

Conveniently omitted is the *Summa's* question immediately following what Thigpen quotes, where Aquinas declares it is *not* possible for there to be incarnations on other planets. In fact, Aquinas twice insists there that "**only human nature can be assumed.**" (Summa, III, Q4, A1). Aquinas affirms that non-human incarnations are impossible, thus directly contradicting Thigpen's claim in the very title of this section of his book.

Throughout the thousands of pages of the *Summa Theologica*, St. Thomas Aquinas pondered countless questions from the perspective of their logical possibility—doing so apart from the consideration of their coherence with Scriptural and Magisterial teaching. To use Aquinas as Thigpen does (and as almost all Catholic ET promoters do) is no different than using Aquinas to argue for the Divinity of the Buddha.

Indeed, by the same token, one could argue for the orthodoxy of believing that Siddhartha Gautama also is an Incarnation of God, equal to Jesus Christ. This is so because Aquinas affirms that it is *possible* for God to assume *all* human natures! Despite the possibility of this, Aquinas says it **"... would have been derogatory to the dignity of the incarnate Son of God... as He is the First-born of all creatures**."[135] As we can see, conflating *logical possibility* with *theological legitimacy* is among the most destructive fallacies one can fall victim to in approaching St. Thomas' *Summa Theologica*. For it is precisely those things which God absolutely *did not do and will not do* that are also the most important to recognize He *can* do.

Only thus do we adequately appreciate the magnanimous mercy of His Will. God *could* have consigned all men to hell on account of Adam's sin alone. Instead, we know that God "*desires all men to be saved.*" (1 Timothy 2:4) God *could* have chosen to simply not bother with creation at all, for He is "*infinitely perfect and blessed in Himself.*" (*Catechism of the Catholic Church*, §1) Instead, the Catechism continues, "*in a plan of sheer goodness [He] freely created man.*" Only a deceiver would take this "possibility" as grounds to speculate that all reality is just an illusion, and that we don't exist at all (as many today do promote with respect to "Simulation Theory.") God *could* simply annihilate all things (as Aquinas teaches is possible for the Almighty). Instead, as Aquinas also teaches, *"we must conclude by denying absolutely that anything at all will be annihilated."*[136] We should recognize that these things, and countless other Biblically outlandish hypotheticals, are nevertheless logically within God's power—and, indeed, within His *rights*. More importantly still, we must acknowledge that He has chosen (and always will choose, for His promises are immutable) to refrain from exercising those powers and those rights. By understanding both the possibilities and the contrasting realities, we come to love God more and glorify His compassion.

Similarly, God *could* have created innumerable worlds filled with rational incarnate creatures like us. Dr. Thigpen also cites saints who noted this, such as St. John Chrysostom. (But no saint has ever taught that God might have actually done this.) The ET promoters miss the fact that it is precisely by recognizing that God *could have* done this—but *did not* do so—that we can begin to appreciate the incomprehensible value of this passing life on earth; this cosmically unique "Theater of Eternal Destinies."

As Catholic scholar Dr. Marie George noted, ET promotion citing St. John Chrysostom and others on this point merely arises from "**…a careless reading of certain patristic authors… [because these Church Fathers] are simply affirming that God *is able* to make many worlds…**"[137] Incidentally—here as with Aquinas—although Thigpen cites Chrysostom repeatedly in support of ET belief, the saint being quoted condemned that very notion. One historian, Dr. Alberto Martinez, recently noted:

> … St John Chrysostom had said that it is 'heretical and contrary to scriptures' to assert that there are 'many heavens and orbs'. … Jesuit Nicolas Caussin also cited Chrysostom to note that heretics believe in 'other starry heavens and worlds'.[138]

Relatedly, God the Father, or God the Holy Spirit, also "could have" become incarnate. Aquinas affirms this logical possibility as well (*Summa, III, Q3, A5*), and he does so for the same reasons noted above. Yet Catholic teaching is just as dogmatically* clear that this *did not occur*—and will not occur—as it is that there are absolutely no other Incarnations of the Son. The Vatican, as recently as 2024 (not a year during which it was especially known for zeal in going out of its way to find theological errors to condemn) formally denounced referring to the Father or the Holy Spirit as becoming incarnate. The Dicastery for the Doctrine of the Faith taught that this is "**not acceptable**," and "**contrary to the Catholic Faith… since only the Word became incarnate**."[139] In a word: one may not safely believe in the reality of some

---

* Scripture teaches that only the Son became incarnate, but this was also dogmatically established upon the earliest councils. Even history's most famous council, that of Nicaea (325 A.D.)—impelled by history's most famous defender of orthodoxy, St. Athanasius—taught that *only* the Son is incarnate. Aquinas, as aware of this dogma as anyone, would be deeply scandalized if he knew any Catholic dared to abuse his writings in such a way as to entertain the notion it may be licit to believe in any incarnations of God other than Jesus of Nazareth.

scenario merely because Aquinas discussed its *logical possibility*; doing so is a sure path to heresy and, ultimately, apostasy.

God *could* have become incarnate as many times as pleased Him. However, as we have seen—with dogmatic certainty—He became incarnate only once, and He never will become incarnate again. Only by understanding that can we even begin to honor Jesus of Nazareth, the Son of Mary, in the way He deserves. Only thus can we await His coming at the end of time with the unbounded joy and supernatural hope which our anticipation of the Resurrection of the Dead upon His Last Judgment demands.

Dogma on the End of Time

"Soon" (Revelation 22:7) the whole material Universe will come to an end. Scripture teaches:

> ... [The] heavens will pass away with a loud noise, and the elements will be dissolved with fire, and the earth and the works that are upon it will be burned up… (2 Peter 3:10)

The same passage reveals that this end will come in accordance with the "promise" of the Lord—to *us men* (cf. 2 Peter 3:9). St. Peter also writes, "**The end of all things is at hand. Therefore be alert and of sober mind so that you may pray**." (1 Peter 4:7) Not "some" things, but *all*. The entire Bible even concludes with this assurance:

> He who testifies to these things says, "Yes, I am coming soon." Amen. Come, Lord Jesus. (Revelation 22:20)

After this Second Coming of Christ, there will exist only "**a new heaven and a new earth.**" (Revelation 21:1) The arrival of this moment is "soon," and it is entirely centralized upon, flowing from, and concerned with, events transpiring with human beings here on earth.

It is unconscionable to propose that the galaxies are filled with rational creatures whose temporal lives will be abruptly extinguished at some point determined by the events experienced by earthlings billions of light-years away. This moment would be entirely arbitrary to these hypothetical aliens, for it would arrive by virtue of Jesus returning to a planet (or another so-called "dimension") that has nothing to do with their own civilizations.

Yet we must make no mistake about it, the absolute end of time—not only the conclusion of the civilized era of the "third rock from the sun"—is a thoroughly *human* affair. Jesus Himself teaches in the Gospels that tribulations on earth constitute the signs which will "immediately" precede the "stars ... falling from the sky." (Matthew 24:29)

> Then I saw a great white throne and him who sat upon it; from his presence earth and sky fled away, and no place was found for them. And I saw the dead, great and small, standing before the throne, and books were opened… (Revelation 20:11-12)

Here in Scripture's last book, we see the Second Coming of Christ in the Flesh, the Resurrection of the Dead, and Judgment Day. Upon the commencement of these events, the very earth and sky flee such that "no place was found for them." All those planets, galaxies, nebulas, stars, and other celestial bodies our best telescopes observe will vanish in an instant when Jesus of Nazareth comes back to planet earth.

A moment's meditation on this overwhelming reality compels whoever undertakes it to put to rest any speculation that these other planets or galaxies harbor other intelligent creatures like us. If they existed, they too would be God's children just as we are, and for God to allow their lives to be so fundamentally contingent on earth's happenings would be ludicrous.

The twentieth chapter of the Book of Revelation concludes by forever relegating those "not written in the book of life" to the "lake of fire," to be eternally forgotten. Such unfortunate souls are no longer part of the Story of God's Love. The following two chapters, however—the last of Sacred Scripture—describe the eternal joys of the elect; those whose names are "written in the book of life." The twenty-first chapter opens with a description of this "new heaven," and we are there taught that "**the dwelling of God is with men.**" (Revelation 21:3)

It does not say God's dwelling is with "*all the elect rational creatures of the universe,*" but rather only with the descendants of Adam and Eve. In their new and permanent blessed abode, "*there shall be no more mourning ... for the former things have passed away.*" (Revelation 21:4) Christ, at that time, will have finished making "***all* things new**" (v. 5)—ruling out the possibility that other "things" pertaining to non-human civilizations might still await their own renewal. Of this moment, the *Catechism of the Catholic Church* teaches, "**In this new universe, the heavenly Jerusalem, God will have his dwelling among men**." (§1044)

Following the end of time, the Universe we gaze upon in the night sky—whether using the naked eye, or binoculars, or the world's most powerful telescope—will no longer endure. The Catechism, faithfully conveying the message of Scripture and Sacred Tradition, assures us that the present Universe will be replaced by a "new universe"—not that this "planet, solar system, or galaxy" will be replaced by a new iteration thereof. On the contrary, all temporality—without qualification or exception—will pass away upon Christ's Second Coming to earth.

Some Christian ET Promoters take the line that each supposedly inhabited planet simply has its own Judgment Day. This thesis is contrary to the infallible truths reviewed above, but it also invites certain questions. For example, "*how far must we travel into outer space to escape Judgment Day?*" As of this writing, the Voyager 1 spacecraft is far beyond our entire solar system; over 15 billion miles away from earth.[140] If someone had hitched a ride on it, would he thereby be dispensed from the Last Judgment? The notion itself is ridiculous; not just for pragmatic reasons of bodily survival on such a craft, but for dogmatic theological reasons.

Nevertheless—taking again the example of "orthodox" Catholicism's currently most popular ET promoter—Dr. Paul Thigpen argues that each of the supposedly inhabited planets will "likely" have *its own* Judgment Day. He asserts it is merely the case that "Christ comes to *Earth* to draw *our* history to a close [emphasis in original]."[141] He concedes that the Faith teaches the whole material Universe will come to a close, and that it "seems" this end will "follow Earth's Judgment Day," only to then imply there could actually be untold billions of years left after "our" Judgment Day, before the end of time.

The same prominent NASA-commissioned Christian theologian we have already quoted, Rev. Dr. Andrew Davison, dedicated an entire chapter in his own Cambridge University ET-promoting book to this matter. There, he begins by cogently admitting that if aliens exist, **"... it becomes arbitrary to place human beings at the centre of the timing of the eschaton**."[142] Accordingly, he writes that belief in aliens "**might call for serious revision, rather than simply expansion [of Christian teaching].**" (Remember that this is the theologian who, earlier in the same book, claimed that ETs are as irrelevant to Scripture, and therefore as unproblematically absent from its pages, as a duck-billed platypus.)

Davison proceeds to attempt to work out a solution that involves a return of Christ "*involving the return of all the [other] ways in which God has been Incarnate*," a "*string of local ends, worked out on many worlds*," or (quoting another author) simply a dismissal of eschatology, holding instead that "*Nowadays ... [we] no longer look for a supernatural cataclysmic end to the world ... it is the existential significance of the myth that remains—as the element of permanent value in it.*" Yes, a rejection of Christian dogma on Eschatology looms large in the ET Deception, and this is one of countless qualities that render it such an apocalyptic ruse.

Leaving aside these destructive speculations, we now turn to sources we can trust. Authoritative teaching unanimously holds that Christ's coming to earth again, thus instituting the General Resurrection and Last Judgment, is itself *the* end of time, not some event that will transpire an unknown period *before* that ultimate end.

> "... [A]t the end of time, when Christ, our life, shall appear..." (*Lumen Gentium*. §2) "[The Mass] will continue to be offered without interruption till the end of time." (Pope Pius XI. *Ad Catholici Sacerdotii*. §14) "Within the dimension of time the world was created ... its goal [is] in the glorious return of the Son of God at the end of time." (Pope John Paul II. *Dies Domini*. §74) "[God,] in the times of Noah brought the deluge because of man's disobedience ... [and] will bring [His justice] on the day of judgment at the end of time." (St. Irenaeus, Father of the Church. *Against Heresies*. Book IV. Chapter 36. §4) "He has determined at the end of time to pass judgment on the living and the dead ... He defers it, however, until the end of time, when He will pour forth His wrath in power and heavenly might." (Lactantius, Father of the Church. *The Divine Institutes*. Book

2. Chapter 17) "… [May we] hasten gladly to meet him when he comes in glory at the end of time." (Roman Missal. Mass for the Consecration of Virgins. Eucharistic Prayer I) "[In] Advent ... [our] minds and hearts are led to look forward to Christ's Second Coming at the end of time." (Roman Missal. Universal Norms. V. §39) "When the resurrection occurs, it will not be time but the end of time." (St. Thomas Aquinas. *Summa Theologica.* Supplement. Q77. A3) "When [Christ] comes at the end of time to judge the living and the dead..." (*Catechism of the Catholic Church,* §682)

Entire books could be filled with similar teachings, but what has been shown above already makes one thing clear. It is an infallible truth of the Faith, based on Sacred Scripture, that "earth's Judgment Day" (which is the *only* Judgment Day), is simultaneous with the end of time. Christ's Coming in the Flesh is *at* the end of time, not "before" it. This event defines the conclusion of the entire material Universe. This moment is, moreover, not merely the end of "humanity's time," or "earth's time;" it is simply the end of *time itself,* whereafter only eternity remains.

The only way to rescue belief in aliens from sheer absurdity is to contradict this dogma.

Limiting God, Aloneness, and the Our Father

At this point, ET promoters are sure to again pull their go-to arrow from the quiver and lament how one who abides in these dogmas is thereby "limiting God." To address this, let us again return to The Beginning.

After forming Adam, God said, "**It is not good for the man to be alone**." (Genesis 2:18) He wasted no time resolving the deficiency, saying immediately thereafter, "**let us make him a help like unto himself**." God made Eve. Adam gazed upon this magnificent creature—our mother—and said, "**This at last is bone of my bones and flesh of my flesh**." (Genesis 2:23) From that moment, man was no longer alone—and never again would be.

Today, as man has abandoned God, he has become a stranger even to himself. Therefore, he again thinks that he is alone, and instead of looking either to God or to his fellow human beings, he seeks company in the fables of his own creation. But when an adult crafts an idol, it does not remain an innocent story. "*Those who worship worthless idols abandon their hope for mercy.*" (Jonah 2:9) In every respect, the ET Deception parodies the deepest truths of Creation, Redemption, and Sanctification. And the ET promoters' favorite mantras underline this aping of the Faith.

We have seen that "aloneness" is regarded by no less authority than God Himself as a serious problem from the Bible's opening pages. Had Adam been the only human being ever to be created, there would indeed be a deficiency in creation; an "aloneness." After the creation of the animals, "*none proved to be a helper suited to the man.*" (Genesis 2:19) Therefore, God created Eve, and, as we saw, Adam immediately recognized that the prior deficiency had been resolved.

If Adam and Eve were the only two human beings to have ever existed, God's glory would nevertheless have attained the outlet it needed. Creatures would have been made who were rational and spiritual (made in His image—unlike the animals), while also being material (incarnate, unlike the angels). Therefore these beings could, in themselves, recapitulate all things—material and spiritual—and therefore both receive and reflect all the characteristics of God's goodness.

But God did not stop there. He commanded these two to "*be fruitful and multiply; fill the earth and subdue it.*" (Genesis 1:28). In the times since this command, such astounding wonders have been wrought that no one could behold them and still protest, without succumbing to impiety, "*that's it!?*"

Let us ponder, then, what we already know about our place in creation. We have billions of brothers and sisters—the Communion of Saints—filled with those elect who have gone before us in Faith yet remain present (though invisibly) in "**so great a cloud of witnesses**." (Hebrews 12:1) We have the Church on earth which is the very mystical Body of Christ. We have Heaven and earth (and the Universe) filled with more angels than we can fathom and whom we can call upon any time we like—each so unique as to truly constitute its own species. We have the millions of other corporeal *species* on this earth (here to serve us) that we are still discovering. We have the thousands of years of history that brought us to our day, and the vastness of the earth which no man (no matter how well traveled) will ever see the tiniest fraction of. We have the galaxies filled with stars and planets and nebulas that glorify God's greatness and beauty (perfectly well on their own, without being tainted by sci-fi myths). We have the depths of the sea which we have scarcely begun to explore. And of course, we have what infinitely surpasses this: the promise of union with God *for all eternity* in Heaven if we love Him.

To meditate on all this unfathomable wonder, which mankind—even collectively—has scarcely begun to discover or comprehend the surface of, and to nevertheless still insist upon asking, "*but are we alone!?*", is to reveal a profound deficiency in one's appreciation of the Creator and His creation.

This shortcoming can only be described as deeply irreverent and worse. Faithful Christians should never waste time speculating about (much less believing in) aliens on account of their absence supposedly rendering us "alone." Those things we *already* know about the existence of are enough to fill innumerable lifetimes—whereas we only get one. At the minimum, then, all suggestions of alien belief should be dismissed with the words of the Psalms:

> Lord, my heart is not proud; nor are my eyes haughty. I do not busy myself with great matters, with things too sublime for me. (Psalm 131:1)

As to the accusation that this contentment with what God *has done* constitutes "limiting God," we should remember that there are many "things," in fact, that God cannot do.

Scripture frequently describes them: He cannot deny himself (2 Timothy 2:13), He cannot lie (Hebrews 6:18), He cannot change (Malachi 3:6), He cannot violate His own words (Psalm 89:34). He cannot "deceive or be deceived."[143] He cannot bless sin.[144] Christians must ignore any accusation that they are "limiting God" by holding fast to these truths. If, even after being reminded of the many contradictions ET belief entails with what God has revealed, someone castigates others who disbelieves in aliens, saying they "limit God," he is only himself the proper recipient of his own rebuke. For he is accusing Almighty God of impotence: of being incapable of inspiring a Sacred Text that would truly withstand the test of time. He is implying that the Bible—the veritable Word of God—is suddenly unreliable now that two millennia have passed and we now have powerful telescopes and UFO whistleblowers. Whatever that supposition is, it is not Christianity.

So let us then turn to the words that *are* Christianity—even its very essence. The supreme prayer of Christianity—which Church Father Tertullian referred to as "*the summary of the whole Gospel*," and which St. Thomas Aquinas called "*the most perfect of prayers*"—is none other than the words committed to more hearts and minds than any other words in history: the Our Father.

St. Augustine said this prayer by itself contained all that could be found elsewhere in the prayers of Scripture. Assuredly, the essence of the Faith is found in the Lord's Prayer. The *Catechism of the Catholic Church* refers to this prayer as "**the fundamental Christian prayer**," (§2773) and "**the quintessential prayer of the Church**." (§2776). This summary of the Faith, however, can be summarized further still. Its central and climactic petition beseeching the accomplishment of the Will of the Father on earth as in Heaven encapsulates everything within itself.

> "'Your will be done on earth as it is in heaven.' There cannot be a greater prayer than to desire that earthly things should deserve to equal heavenly ones." —St. John Cassian, Father of the Church

These eleven words define the Christian religion for a simple reason: earth is the only place where the accomplishment of God's Will is even possibly in question, therefore earth is the only place that exists where this supplication has any non-trivial meaning.

In Purgatory and Heaven, the Divine Will is always done perfectly, and no risk exists that it might be contravened. In hell, God's Will has its exposition by way of justice, and none of the damned will ever lovingly embrace it, no matter what prayer we may offer to that effect. Finally, regarding the physical Universe outside of earth, there is no need to pray for God's Will to be done in such places—because no creature exists there who can reject it. Any rational being (outside of Heaven or Purgatory), however, is definitionally capable of rejecting God's Will. Therefore, there are no rational beings in the material Universe outside of earth. If extraterrestrials existed, then the Our Father would not be the greatest prayer ever prayed, but rather an abject failure whose Author would be cluelessly incompetent.

If there were aliens, Jesus would not have "summarized the whole Gospel" with the words "Thy Will be done on *earth* as It is in Heaven," but rather, "Thy Will be done throughout the *cosmos/galaxies/dimensions/universe* as It is in Heaven." Certain patently false and even diabolical "private revelations" today, which announce imminent alien contact, do seek to "improve" the Our Father by changing "earth" to "the galaxies."

The Our Father is perfect as it is. It cannot be improved, and Exotheology's attempt to improve it derives from the evil one.

8. Catholic Teaching

"Man, then, is unique within the cosmos. He alone of all kinds of beings is the object of God's personal solicitude both for his own sake and for the sake of his species."
(*New Catholic Encyclopedia*. Entry on "Man.")

We have seen that the Holy Bible—both in its Old and its New Testaments—teaches us that human beings are the only rational incarnate creatures that exist; it repeatedly rules out alien belief. Catholics, however, will find it even easier than other Christians to rule out the existence of aliens. For they also can discover truths, with certainty, from Sacred Tradition, and from the Magisterium which authoritatively interprets both Scripture and Tradition. Looking at these sources, we see aliens ruled out even more clearly than they are in Scripture alone.[145]

When considering the immense slew of texts in this great corpus, it is usually best for a Catholic to begin this journey by consulting—and submitting to—the simplest authoritative summary of it all: the *Catechism of the Catholic Church*.

The Catechism

"Man occupies a unique place in creation: (I) he is 'in the image of God'; (II) in his own nature he unites the spiritual and material worlds ..." (*Catechism of the Catholic Church*, §355)

Man's uniqueness has always been taken for granted in Catholic teaching; a fact which the Catechism—faithfully summarizing 2,000 years of Sacred Tradition—settles in the opening paragraph of its teaching on *Man*, quoted above. This text, moreover, uses the word "**unique**" because that is precisely what it means—"**being the only one.**" (*Merriam Webster Dictionary*)

Catholic ET promoters are known to protest that, even if innumerable alien civilizations populate the galaxies, "*mankind can still be considered special in another sense!*" This protestation, however, is a straw man; no one is claiming that the human race would be incapable of being "special" in light of aliens. The Church does not proclaim that man is merely "rare," or "special," or "extraordinary." It declares that, in the entire Universe—not just planet earth—man is the "**unique**" being made in God's Image: the only being who is more than just a creature of the Creator, but also a child of God.

If, however, anyone somehow fails to see how paragraph 355 of the Catechism rules out aliens, he will not likewise miss the teaching in the paragraph which follows: "**Of all visible creatures, only man is able to know and love his creator... and he alone is called to share... in God's own life**." (§356)

The Catechism here teaches that, out of the entire material creation ("*all visible* creatures"), *only* human beings were made in such a way that they *even can* know and love God. This is the same as saying that only human beings are *rational*. Moreover, this is not "an interpretation" of the Catechism (much less a "fundamentalist" one!), as Catholic ET promoters claim. It is simply *what the Catechism says*.

One corollary of this truth is that all non-human animals on earth are irrational; another equally sure corollary is that there are no intelligent extraterrestrials. For the Catechism does not refer to all "earthly" creatures, but it says, "**Of all *visible* creatures**..." Now, "all" means exactly what it says. This term cannot be arbitrarily restricted to mean simply "all *on earth*" by those who purport to read the minds of the Catechism's authors.*

"***Visible* creatures**" is an unambiguous phrase which refers to everything that God has made in the material Universe. The phrase encompasses the whole *corporeal* realm. If we cannot, pragmatically, see some individual material thing, that is irrelevant. In such cases, it remains true that this thing in question, whatever it may be, is nevertheless by nature "visible" if it is not pure spirit. Planets too distant for our telescopes to observe, depths of the earth that no human will ever lay eyes upon, undiscovered sea creatures, and all comparable things, are truly *visible*.

The purpose of this qualification, "visible," in the Catechism's teaching is to ensure angels are not misrepresented. Angels are creatures (God made them), they are not men, yet they can certainly know and love God! These creatures, however, are not in any way visible; they are pure spirit. They have power in the material realm and therefore can manifest themselves visibly and assume corporeal forms (as we will discuss later), but they *themselves* are immaterial; entirely "invisible."

Some ET promoters have even used this phrase to accuse the Catechism here of error, and in need of being "reformulated" in accordance with their own imaginative speculations. They make this claim by fallaciously describing angels as "visible" creatures (and rightly observing that angels do *know and love* God). Their claim that angels are visible contradicts Catholic dogma. If even angels must be categorized as "visible" (merely because they do, at times, *assume a visible form*), then the "visible vs. invisible" distinction—contained in the Nicene Creed itself—is incoherent. It is the views of these theologians, however, not Catholic teaching, which are incoherent. As Pope St. John Paul II taught, reiterating dogma, "**the angels ... are invisible, for they are purely spiritual beings.**"[146] And as the *Compendium of the Catechism of the Catholic Church* teaches, "**The angels are purely spiritual creatures, incorporeal, invisible, immortal, and personal beings endowed with intelligence and will.**" (Ch. 1 Q60)

---

* Even if they had this mind-reading ability, their insights would be irrelevant. It matters not what the personal views of the Catechism's authors were. It only matters what is actually taught by Catechism. The only relevant thing is what this text, which was promulgated as Magisterium by Pope St. John Paul II, actually states.

The Catechism's next clause states, **"... only man is able to know and love his creator."** Thus the Church teaches that of all physical creatures, "*only* man"—that is, only human beings*—are *even capable* of knowing and loving God. For all other corporeal beings, knowing and loving their Creator is ruled out by their very nature. *Any* creature with reason, however, is—by definition—capable of knowing and loving God.[147] We have already seen that the Catechism teaches that God's existence can be known with certainty by simple, natural reason and nothing more. (cf. §47) It is imperative for all the Faithful to understand that God can be known from the capacity of the soul which *any* rational being enjoys. If aliens existed, their intelligence would *guarantee* their ability to know and love God. Yet, according to this infallible teaching† of the Catechism, no beings other than man have that capacity. This rules out aliens.

The Catechism then declares, "**and he [man] alone is called to share... in God's own life**." This final clause is a reiteration of what came before it. Only human beings are called to be partakers of the Divine Nature. Observe that this is an additional teaching (the word "and" is used), not a qualification on or limitation of the prior teaching. Therefore, it is not only abhorrent but also incoherent to claim that aliens can exist so long as they are not called to share in God's life.

Nonetheless, Catholic ET promoters use several strategies to attempt to evade the Church teachings presented above. Their first is a simple rejection of grammar; they latch on to the next independent teaching—which declares that "**[Man] is 'the only creature on earth that God has willed for its own sake,'**"—and pretend that the phrase, "on earth," means that the prior teaching we discussed above is limited to terrestrial considerations. In fact, there is no "**on this earth**" restriction applied to the Catechism's insistence upon the absolute uniqueness of man, and no one who has learned to read could miss that.‡

---

* All of whom, Scripture and the Church teach, are descendants of Adam and Eve.

† It is infallible not because any clause in the Catechism is automatically guaranteed infallibility by inclusion within it, but because this particular teaching is reiterating 2,000 years of unbroken Catholic teaching.

‡ To expound: here is the entirety of the first two sentences of paragraph 356: "*Of all visible creatures only man is "able to know and love his creator."[GS 12 §3] He is "the only creature on earth that God has willed for its own sake,"[GS 24§3] and he alone is called to share, by knowledge and love, in God's own life.*" The attempt to reconcile belief in aliens with the Catechism here entails proposing that adjacent teachings in the Catechism are fundamentally in conflict, and proceeding to insist that Catholics are free to choose whichever conflicting teaching will serve as the basis for their own personal beliefs. The first sentence of paragraph 356, which rules out all aliens, is not restricted in scope by what is described in the second sentence. Indeed, the sentence we are now considering indicates that, "on earth," only man is willed for his own sake (plants, animals, etc., exist for the sake of man). But this condition is absent from the first sentence, which rejects the possibility of intelligent life among "all visible creatures"—with no qualification on "all" included. *Both* teachings remain true; only man is rational among "all visible creatures," *and* only man is willed for his own sake "on earth."

There is another conspicuously ignored fact which repudiates the "on this earth" defense ET promoters use to pretend they are abiding by the Catechism while still believing in aliens. For they claim aliens *already have* been on earth for decades, if not centuries or millennia! It is, therefore, nonsensical for them to defend this position even as they concede that belief in intelligent non-humans on earth is ruled out. We are left to conclude that man is—now and forever—the only material being that is a *person*. While paragraph 356's initial sentence rules out *all* aliens, the sentence following it additionally rules out any possibility of meeting aliens on earth.

Several still more slippery and theologically underhanded methods are used by ET promoters to evade this doctrine. Those who proffer them point to Pope Pius XII's teaching in his 1950 encyclical *Humani Generis*, wherein he taught that the Beatific Vision is a gratuitous gift from God, or they attempt to argue that the simple and faithful recognition that the Catechism rules out aliens is somehow actually the vice of "fundamentalism" or "proof texting." They assert that the Catechism "*must not rule out aliens, since the Pope who promulgated it, John Paul II, believed in aliens*" (in fact, he did not). These outlandish assertions and slanderous accusations are addressed and refuted in *Only Many Bears His Image* (pages 98-111 and 686-690).

> "[The Catechism] transmits what the Church teaches, whoever rejects it as a whole separates himself beyond question from the faith..." —Cardinal Ratzinger.*

***

Another weighty recent Magisterial document—not so lofty as the Catechism, but still deserving submission from all Catholics—is the *Compendium of the Social Doctrine of the Church*. With respect to its own content, it declares that "*...it is authentic Magisterium, which obligates the faithful to adhere to it.*" (§80) Within this Compendium, we are taught that:

> Only man and woman, among all creatures, were made by God 'in his own image' (Gen 1:27). The Lord entrusted all of creation to their responsibility... (*Compendium of the Social Doctrine of the Church*, §451)

This, of course, is yet another emphatic Magisterial rejection of ET belief.

Finally, yet another official Vatican document recently published addresses the same topic directly: "...**the uniqueness of the human person [is] incomparable to all other entities in the universe**." (Declaration of the Dicastery for the Doctrine of the Faith, "*Dignitas Infinita*," April 2024. §14)

---

* (*Introduction to the Catechism of the Catholic Church*, Ch 2). Note that the Catechism was promulgated by an especially weighty act of the Magisterium, which declared: "The Catechism of the Catholic Church, which I approved 25 June last and the publication of which I today order by virtue of my Apostolic Authority, is a statement of the Church's faith and of Catholic doctrine, attested to or illumined by Sacred Scripture, Apostolic Tradition and the Church's Magisterium. I declare it to be a valid and legitimate instrument for ecclesial communion and a sure norm for teaching the faith." —Pope St. John Paul II. *Fidei Depositum*. §IV

Ecclesiology and Mariology

In *Only Man Bears His Image* (2023), I considered that alien belief is ruled out by the simple recognition that God the Father is not hiding siblings from us, and God the Son is not a polygamist or an adulterer. For God is good, yet belief in extraterrestrials impugns that goodness with extraordinary audacity. We can similarly rule out ETs by acknowledging that God the Holy Spirit is not a monster. For the Holy Spirit is the soul of the Church—which in turn is a specific, concrete institution existing on (and only on) planet earth. "**The Church is one because of her "soul": "It is the Holy Spirit**," the Catechism teaches, before quoting St. Clement of Alexandria:

> There is one Father of the universe, one Logos of the universe, and also one Holy Spirit, everywhere one and the same; there is also one virgin become mother, and I should like to call her 'Church.' (*Catechism of the Catholic Church*, §813)

Expressly referring to the entire Universe, St. Clement (and the Catechism) insist there is one and only one Church. Pope Leo XIII taught, in his encyclical *Divinum illud munus*, that "**...as Christ is the Head of the Church, so is the Holy Ghost her soul.**" In the same document, the Pope declared:

> This being so, no further and fuller "manifestation and revelation of the Divine Spirit" may be imagined or expected; for that which now takes place in the Church is the most perfect possible, and will last until that day when the Church herself, having passed through her militant career, shall be taken up into the joy of the saints triumphing in heaven. (§6)

In seeing the sinfulness of so many of the Church's members and leaders, some Catholics may be tempted to doubt the astounding degree of union with God enjoyed by this seemingly temporal and imperfect "institution" into which they were baptized. Yet here as everywhere in the Faith, doubt only inclines one to error. In truth, the union the Catholic Church has with the Holy Spirit is, as Pope Leo XIII here taught, "**the most perfect possible**," so much so that it is not acceptable to "imagine or expect" a "fuller manifestation and revelation of the Divine Spirit." ET belief, however, is inextricable from that condemned expectation. Frankly speaking, everyone knows that extraterrestrial belief—wherever it is found—is an ideology built upon the messianic expectation of aliens revealing to us a whole gamut of realities (both technological and spiritual) far beyond what we have ever witnessed on earth. Even "imagining or expecting" such a thing, however, is illicit.

Moreover, it is not conceivable that *multiple* Churches could each have, as their very soul, the *One* Holy Spirit. This would render the Holy Spirit a monster akin to what is depicted in some horror movie, whereas in the Divine Order, one soul is to enliven one and only one body. Metaphors aside, it is certainly a scenario ruled out by the dogmatic teachings above.

Not only Ecclesiology but also Marian dogma rules out aliens. As contemporary Catholics are often quite poorly catechized on the glories of Mary, it is unsurprising that many neglect to see how irreconcilable belief in extraterrestrials is with what we know about the Mother of God. While we have already addressed why the notion of multiple incarnations—"other Christs" on other planets—is flatly contrary to Christianity, the impossibility of "other Marys" is just as clear an indication that the human race alone is made in God's image.

Wherever we see Christian ET promotion, we see it inextricably bound up with the denigration of the Blessed Virgin. For example, the prominent theologian we have already quoted, Rev. Dr. Andrew Davison, argued the following about the Blessed Virgin Mary's status in light of aliens. Addressing the perennial truth of the Faith that Mary is the Queen of not only earth and the Church, but also Heaven itself, Dr. Davison writes:

> ...[T]he best way to navigate these roles ... is to treat them as local or human matters: as part of our story, as a role in relation to our species ... [W]e may do well to set the language of queenship aside.[148]

Little would please the Devil so much as "setting aside" the Queenship of the Blessed Virgin Mary, whereas this theological mutiny against the Celestial Empress flows invariably from belief in extraterrestrials. Now we can, and should, lament the impiety inherent in such suggestions. But, here as elsewhere, we cannot find error in the internal consistency of one who realizes that if the Blessed Virgin is just one of "many Mothers of God" or "many Immaculate Conceptions" scattered about the galaxies, then we have no grounds to regard her as the Queen of Heaven. Refusing to accord this dignity to Mary, however, is contrary to Catholic dogma.

It is not merely the "fringe" ET promoters advocating for such views, but also the most mainstream Catholic ones. Davison is far from alone. For example, Dr. Paul Thigpen wrote:

> If the divine Son of God should become incarnate in extraterrestrial races, then His mother on other planets could also be rightly called the "Mother of God." He would, in fact, have several mothers ... [there may be] multiple incarnations with multiple Mothers of God and of the Church, who are also multiple Queens of the Universe through the divine Majesty of their Son...[149]

Dr. Thigpen insists that this proposal—which simple Catholic piety immediately perceives as blasphemous—is entirely unproblematic. He argues for the orthodoxy of this scenario, which sounds like the plotline of some degenerate modern sitcom, with the usual rhetorical device. He defers to "humility," as if this virtue demands that we embrace violations of the core truths of the Faith. But that is only disguised pride.

Again we see that the ET Deception carries within itself a revolution against all that is sacred. It seeks to undermine the Christian and Catholic understanding of our origin, destiny, Ecclesiology, Mariology, Christology, Eschatology, morality, and prayer.

Of the plethora of existing Magisterial teachings on Mary, one document gives some of the most relevant declarations. It is the same Apostolic Constitution by which Pope Pius IX infallibly defined the Dogma of the Immaculate Conception, *Ineffabilis Deus*. There, Pius taught that Mary was not only Immaculately Conceived, but also that "**...she approaches as near to God himself as is possible for a created being,**" and that "**she is above all men and angels in glory**."

The Blessed Virgin Mary is a "mere mortal" like us; an entirely *human* person (unlike her Son, she was not Divine); a woman who was conceived and born, on earth, to a human mother and a human father just as we are (though she had no stain of sin). It is nevertheless settled that she is *the* supremely exalted creature. One so exalted, in fact, that it would be materially heretical to even propose the possibility of the existence of a created being that matches or surpasses her, as Catholic ET promotion proposes.*

The Mariology of the Second Vatican Council's *Dogmatic Constitution on the Church* rules out aliens still more clearly: **"[Mary] far surpasses all creatures, both in heaven and on earth. At the same time, however, because she belongs to the offspring of Adam she is one with all those who are to be saved**." (*Lumen Gentium*, §53) Instead of asserting that Mary is "above" every other creature in the Universe, this teaching reminds us that the Virgin "*far surpasses*" them all, abolishing any expectation that alien civilizations could have their own creatures who are at least "close" to the Queen of Heaven born on the third rock from the sun.

More pointedly still, the Church here teaches that Mary is not only the Mother of the Church, but also that, "*because* she belongs to the offspring of Adam she is one with *all* those who are to be saved." This description of *why* she is one with *all* the elect is deeply revealing: it is "because" she shares our own biological heritage. The Council's decision to go beyond just *taking note of* Mary's common ancestry with us—by instead giving it singular explanatory power—is definitive on its own. The Virgin Mary is truly our sister in that way, which cannot apply to any creature, such as an ET, who could not number among Adam's offspring.

This exclusivity explicitly rules out the possibility of salvation for creatures who are not offspring of Adam. Because God is good, and would never create a class of intelligent creatures who could not be saved,[150] we find aliens ruled out yet again.

* Whoever would argue that Pius IX's teaching only applies to earth is contradicting the text itself. There are no limitations on the scope of his declaration. He bluntly states that the Virgin is as close to God "*as is possible for a created being*." The very next clause reads: "**[Mary] is above all men and angels in glory.**" This line also rules out the theory which would limit the document's application to earthly creatures, since angels are explicitly included, and they are not earthly. Instead, this declaration of Mary's superiority is expressly defined as universe wide.

We have seen ET belief refuted by Catechisms, councils, and by Magisterium on Mariology, Ecclesiology, and so on. What of direct Papal Magisterium explicitly addressing the matter? That also exists, though to understand it, we must first address the heresy of Polygenism.

Polygenism

Some ET promoters may insist that, even if we cannot posit the existence of *non-human* intelligent aliens, we can at least believe that aliens exist throughout the galaxies who are simply properly called "men," as we are. This claim still runs afoul of most of the dogmatic considerations we have already discussed, but there is yet another truth of the Faith that anathematizes this notion directly.

> Blessed are you, O God of our fathers… Let the heavens and all your creation praise you forever. You made Adam and you gave him his wife Eve to be his help and support; and from these two the human race descended. (Tobit 8:5-6)

"Polygenism"—the idea that men could exist who are not descendants of Adam and Eve—is condemned as a heresy by infallible Catholic Teaching. Every facet of Scripture, Tradition, and Magisterium unanimously declare that all human beings are descendants of Adam and Eve, who are themselves two literal humans who lived on this planet. The Papal Encyclical *Humani Generis* bluntly declares that "**the faithful cannot embrace**" Polygenism. Therein, Catholics are expressly forbidden from denying that Adam was the true "**first parent of all**." Obviously, it is not possible for Adam to be the parent of creatures on other planets, therefore we can immediately rule out the possibility of races of humans existing anywhere but on earth.[151]

Humani Generis is only one example of many Catholic teachings that condemn Polygenism. Eleven years before it, in a lesser-known encyclical that nevertheless bears equal Magisterial weight, the same Pope had already condemned it. Entitled *Summi Pontificatus* (On the Unity of Human Society), the document teaches against aliens by declaring that the creation of man in "**God's Own image and likeness**" was shown, in "**the first page of Scripture, with magnificent simplicity**," to be the "**culmination to His creative work.**" (§36) If aliens exist, however, then man is *not* the culmination of God's creative work; yet again we have an independent Magisterial rejection of alien belief. Pius XII also teaches:

> The Apostle of the Gentiles later on makes himself the herald of this truth which associates men as brothers in one great family… (§37) A marvelous vision, which makes us see the human race in the unity of one common origin ... (§38) These are supernatural truths which form a solid basis and the strongest possible bond of a union... (§41) ... [T]his unity of all mankind ... exists in law and in fact... (§42)

We are not, however, left with Encyclicals from one Pope alone in addressing the issue. Scripture clearly teaches that all human beings are descendants of Adam and Eve, and Sacred Tradition—ever faithful to its Biblical foundation—has always affirmed this truth, for **"[God] made from one man the whole human race."** (Acts 17:26)

Looking to another Encyclical promulgated several decades earlier, we see the same truth declared even more forcefully. In his encyclical entitled *Arcanum*, Pope Leo XIII teaches:

> Though revilers of the Christian faith refuse to acknowledge the never-interrupted doctrine of the Church on this subject... We record what is to all known, and cannot be doubted by any, that ... [God] decreed that this husband and wife [Adam and Eve] should be the natural beginning of the human race, from whom it might be propagated and preserved by an unfailing fruitfulness throughout all futurity of time. (§5)

The zeal exhibited by Leo in reiterating this "never-interrupted doctrine of the Church," along with his righteous indignation at those who deny it, is palpable and inspired. One cannot get human nature right if he gets its beginnings wrong. Accordingly Pope Leo uses the strongest language to denounce the idea that Adam and Eve were not the first parents of all human beings. He insists it "**cannot be doubted**" that Adam and Eve were the "beginning of the human race," and he prefaces this infallible teaching by denouncing those who would oppose it, saying they are not sincerely mistaken seekers of truth, but rather are "**revilers of the Christian faith**."

Popes Zachary & Pius vs. Aliens

The teaching of Pius XII and Leo XIII against Polygenism is just one example of Papal Magisterium ruling out belief in aliens. Another Pope's teaching from the 8th century all but explicitly denounces belief in extraterrestrial intelligence. It comes to us from Pope Zachary, who is venerated as a saint by both the Catholic and Eastern Orthodox Churches, and who was aflame with zeal to defeat Pagan practices. He famously built the church of "Santa Maria Sopra Minerva"—that is, "Saint Mary Above Minerva"—on top of the very ruins of the Pagan temple to the "goddess" Minerva, thus demonstrating the Faith's victory over idolatry. He was in regular formal correspondence with the great St. Boniface, who is famous for his skill with an axe as he chopped down "Thor's Oak," a large tree that became an object of idolatrous worship by locals. Having demolished their idol, Boniface proceeded to convert them to Christianity.

It is most fitting, therefore, that we have from Pope Zachary such a clear denunciation of the modern era's prevailing Paganism—alien belief—though, prophetically, issued 1,200 years before its popularity. In a letter to St. Boniface, Pope St. Zachary wrote the following about a certain cleric named Virgil:

> As for his perverse and abominable teaching, which he has proclaimed in opposition to God, and to his own soul's detriment—if the report of his having spoken thus be true—that is, that there are another world and other men beneath the earth, or even [*on?* *] the sun and moon ... take counsel and then expel him from the church, stripped of his priestly dignity.[152]

In this Papal Letter, the Vicar of Christ describes belief in aliens as not only errant, but rather, as "**perverse**" and "**abominable**;" so egregiously in "**opposition to God**," that for a Catholic priest to be found promoting such teachings requires that he be defrocked and excommunicated.

The move that the Catholic ET promoters make to attempt to evade this condemnation is predictable. They protest that they are not asserting that aliens exist on the "sun and moon" (if that is what Pope Zachary means here), nor on another world "beneath the earth," but rather that aliens exist on other *planets*. This position is mere gamesmanship. If indeed it is *abominable* and *perverse* to propose that "other men" live on "another world *beneath* the earth," then how exactly is this *abominable* teaching rendered licit just because one instead posits as their dwelling the planet Mars, or Alpha Centauri's planets, or those of distant galaxies or other so-called "dimensions"? There is no theological difference.

Their next move consists of pretending that "context" can invalidate this doctrine. The context of Pope Zachary's condemnation no doubt pertained to the discussion of the "Antipodes." These were the object of ancient and medieval speculations about "another side" of the earth, and discussions took place about whether it (though regarded as inaccessible to men of the civilization we have always known) harbored other intelligent beings whose origin was not Adam.

First, we must note that the Pope is *not* merely condemning the notion of rational incarnate beings living on the "Antipodes," but also the notion of such beings living on *extra*-terrestrial bodies: the sun and moon. Admittedly, however, debate among scholars exists as to what exactly the wording of the original text means in referring to those bodies, as the grammar is awkward. Yet writers of centuries past, who doubtless had access to better copies of the Pope's letter, *did* take Zachary's decree as ruling out the possibility of aliens on the sun or moon. One contemporary scholar noted:

> In 1631 Froidmont [a prominent theologian and scientist] cited Pope Zacharias [Zachary] at length to ... say that the notions that there is another Sun, another Moon or another inhabited Earth are 'heretical, or are nearly so'. Inchofer referred to the idea that there are worlds in the Moon

* As Dr. Marie George notes, "*Yet another difficulty in interpretation arises from the Pope's grammar: "Sun" and "moon" are in the nominative case, leaving one wondering what the Pope meant to say about them. Maybe he meant to condemn the notion of another sun and moon, and left out the word "another" ("alius," "alia") or maybe he meant to condemn the notion of human inhabitants on the sun and moon, and omitted the word "on" (L. "in")—in which case "sun and moon" should have been put in the ablative case.*" (*Christianity and Extraterrestrials?: A Catholic Perspective*)

> or the Sun as monstrous 'errors'. In his long manuscript he condemned it as 'heretical', following Philaster, Zacharias and others.[153]

Yes, Pope St. Zachary condemned alien belief—plain and simple.

Secondly—and most importantly—far from successfully defending their thesis, the context here only further clarifies how anti-Catholic ET promotion remains.

The debate of Zachary's day about the "Antipodes" is a perfect analogue to the present debate on extraterrestrials. As we can see from the letter, Pope Zachary is not at all concerned with the shape of the earth (i.e., whether—geographically—such a place as "the Antipodes" is possible). Recall that such questions as those have no effect on the Faith, therefore the Magisterium is not concerned with settling them. But the question of whether *rational incarnate creatures* other than descendants of Adam could exist (on these "Antipodes") has everything to do with the Faith, hence Pope Zachary's zeal in anathematizing it.

Like the condemned Virgil, Catholic ET promoters today insist that intelligent creatures exist in physical regions which are pragmatically inaccessible to us; beings who derive not from Adam, but from some other first parent. This position is, almost verbatim, the same thesis condemned by the Pope. The "**perverse anti-God abomination**" Zachary describes it as springs from the multitude of theological absurdities and blasphemies it contains, some of which we have already discussed. Not only is the substance of the position a perfect analogy, so too is its motivation. In its day, the Antipodean Thesis was presented on the basis of scientific conjecture and the prevailing Pagan mythologies of the time which posited all manner of fictional intelligent creatures living "under the earth." This is how belief in extraterrestrials is promoted today. It is supposedly derived from "astrobiology," but the real motive force behind the theories is, likewise, the day's prevailing mythology: science fiction.*

***

Several centuries later, one of Zachary's successors—Pope Pius II—promulgated a decree that is included in the most oft-consulted compilation

* Nevertheless, Catholic ET promoters unfortunately do not stop at these attempts to evade this Papal denunciation. Dr. Paul Thigpen, in his popular 2022 book on the topic, suggests Pope Zachary's condemnation is only a personal opinion expressed in a private letter. This is both fallacious and a category mistake. Fallacious, because this is not a mere letter: it is a clear instruction, written in response to St. Boniface's *Dubia,* on a formal canonical proceeding to be undertaken against a certain priest's "abominable" doctrine. This is not one friend writing to another to express personal opinions. Because the strict genres of Magisterial declarations we delineate today did not exist in the 8th century, Thigpen's contention is also a category mistake. Today, one could more validly regard a letter from a Pope as non-Magisterial. In the 8th century, however, Popes did not write such letters simply intending to chat. In the early Church, Popes exercised their Magisterial authority through letters precisely like the one here authored by Pope Zachary. One can even peruse *Denzinger,* where he will find several of Pope Zachary's letters *to St. Boniface*—just like the letter we are currently considering—and they are presented as true acts of the Magisterium. We are left with only one conclusion: this letter, too, is authentic Magisterium, as there is no indication that it substantially differs (in the intent of the Pope) from those already included in *Denzinger.*

of Catholic dogmas (Denzinger). Entitled "*Cum sicut accepimus*," it was published on November 14, 1459. The decree condemned what the Pope described as "**most pernicious errors ... a sacrilegious attempt against the dogmas of the holy Fathers**" taught by a certain Zaninus de Solcia. Among these "sacrilegious" errors was the following condemned proposition: "**God created another world than this one, and in its time many other men and women existed, and consequently Adam was not the first man.**"[154]

The popular Catholic scholar above uses the same strategy from addressing Zachary's letter in pretending to exempt ET belief from this Papal letter's condemnation. He dismisses the relevance of the condemnation, claiming that it only consists in Pius "*rejecting the more specific idea that God created a world inhabited by members of the human race before He created our world.*"[155] Here as there, one is struck by how oblivious such a dismissal is to the fact that no substantial difference exists between, on the one hand, modern ET promotion and, on the other hand, 15th-century belief in men on another world *before* Adam. One is likewise struck by how nonchalantly this dismissal ignores the nature of the condemnation, which the Pope was clear to promulgate not as addressing mere subtle theological mistakes, but rather as addressing "**pernicious**" and "**sacrilegious**" errors "**against the dogmas of the holy Fathers**." If indeed it is, as the Pope assures us, so exceedingly evil to believe in men on another world before Adam, then it is also evil to believe in aliens on another world before or after Adam.

Others play games by insisting one can still believe in aliens as being "other biological species of humans," while abiding by the Pope's decree. This move employs anachronistic fallacies with the universally accepted definitions of words.

> "MAN: Latin *homo*, a human being ... [The term] is part of the Church's official vocabulary and occurs in every major document of the Catholic Church ... In philosophical terms, "man" is a rational animal..." (Fr. John Hardon's Catholic Dictionary) "... man is a substance, corporeal, living, sentient, and rational... Being capable of reasoning, [man] verifies the philosophical definition of a person: 'the individual substance of a rational nature.'" (1914 *Catholic Encyclopedia*. "Man.")

The moment promoters of any thesis resort to playing such games with the meanings of words—meanings that have never once been in question—they have only demonstrated the bankruptcy of their view. "Man" is not a word about which the faithful have ever been confused. Philosophically, "man" is defined as "rational animal." Theologically, "man" is incarnate spirit; i.e., a "living soul" (Genesis 2:7), the (only) "Imago Dei." Both definitions reach the same destination; the former by rising from what is natural (as is philosophy's *modus operandi*), the latter by descending from what is supernatural and revealed (as is always the task of theology). Colloquially, "man" is simply a word that refers to a member of the human family, which is to say, a descendant of Adam and Eve.

> "How many a dispute could have been deflated into a single paragraph if the disputants had dared to define their terms." —Aristotle

Each of these commonsense definitions employed by Scripture and the entirety of Sacred Tradition are mutually confirming and refer to identical domains. Furthermore, each results in ET-belief being negated by the Scriptural and Magisterial teachings we have already considered (and still more). Confirming these truths, the great Father and Doctor of the Church, St. Augustine, should be given the last word here. He settled that **"no Christian can doubt"** that *anything* we can call human is in fact an offspring of Adam. This to say that all rational, material creatures which exist are children of Adam. The end. Attempts to complicate that simple truth are only ways of dialoguing with the Devil. The saint writes:

> But whoever is anywhere born a man, that is, a rational, mortal animal, no matter what unusual appearance he presents in color, movement, sound, nor how peculiar he is in some power, part, or quality of his nature, no Christian can doubt that he springs from that one [individual]... if they are human, they are descended from Adam.[156]

9. Sacred Tradition and the *Sensus Fidelium*

"On the contrary, It is said (John 1:10): "*The world was made by Him,*" where the world is named as one ... The very order of things created by God shows the unity of the world ... those only can assert that many worlds exist who do not acknowledge any ordaining wisdom, but rather believe in chance, as Democritus ... [It] is not possible for there to be another earth than this one..."

—St. Thomas Aquinas[157]

Although all Christians know (or should know) that Scripture teaches infallibly—even more so, inerrantly—and all Catholics know that the Magisterium of the Church also teaches infallibly, too few realize that a third source of infallibility exists. The *Catechism of the Catholic Church* teaches:

> The whole body of the faithful ... cannot err in matters of belief. This characteristic is shown in the supernatural appreciation of faith (sensus fidei) on the part of the whole people, when ... they manifest a universal consent in matters of faith and morals. (§92, Quoting *Lumen Gentium*)

To say that a given source "cannot err" is of course synonymous with declaring it *infallible*. It is also just as clearly the case that once a truth is ascertained to be infallible, that truth is permanent. Infallibly stated once, infallibly true forever. To say otherwise is incoherent.

Assessing the *sensus fidelium* on the question of extraterrestrials is uncomplicated. We need not carefully select one decade or century for our consideration; we need only look to any point whatsoever within the first millennium and a half of the Church's history, and there we will find the universal consent that manifests the *sensus fidelium*.

The substance of this "universal consent" is simple: *man alone bears the Divine Image*; a man is a rational animal, and all men are descendants of Adam and Eve (two literal people who lived on earth). Put another way, *angels and men are the only rational creatures*. Leaving aside Scripture and Magisterium for a moment, we are still confronted with the *sensus fidelium,* which has already infallibly settled the matter many times over.

Some ET promoters simply dismiss the entire matter as one outside the scope of the *sensus fidelium*. They argue that *"it is merely scientific and therefore not among those things that the Church is referring to when speaking of the faithful being incapable of erring 'in matters of belief.'"* However, as we discussed in earlier pages, alien belief has ramifications for the Faith that can only be described as seismic. All serious thinkers admit this, wherever they reside on the spectrum of views pertaining to extraterrestrial intelligence, and as such no one can say this is a purely empirical question. If anything is eligible for being determined by the sensus fidelium, it is the question of whether or not only man bears the Divine Image.* We have already quoted those Catholic theologians—Vatican priests and seminary professors, even—advocating for, in their own words, alien belief compelling us to "**wholly redefine many of our conceptions of God and creation**."[158] If this matter somehow stands outside of the domain of the *sensus fidelium*, then there is no such thing as the *sensus fidelium*.

Therefore, if we can determine that, at any point in the Church's history, the faithful were unanimous in their belief that aliens do not exist, then it is an infallible fact that aliens do not exist. And this is exactly what we will determine in the pages ahead.

Angels and Men in the Great Chain of Being

As Sacred Tradition developed, the Faithful remained unanimous in their understanding of the basic structure of the Universe. Traditional Christian art often depicts *all that is* from a supremely high-level view. For example, the 16th-century Franciscan missionary, Diego de Valadés', published the illustration entitled "The Great Chain of Being," contained within *Rhetorica Christiana*. From the Holy Trinity, to hell, and everything in between, within this "Great Chain" shows each of the rudimentary categories of beings. There is inanimate matter ("minerals"), living matter (i.e., plants and animals), spiritual-material creatures (human beings), purely spiritual creatures (angels and demons), and there is God. Notice that this and all such paintings lack one thing: empty spaces. There is no corresponding "altar to an unknown god" (cf. Acts 17:23) like ancient Athenians placed in their temples as they sensed the deficiency of their own theological worldview. Instead, Christian belief is on display, wherein no gaps exist between angels and men, and no gaps exist between angels and God.

Many other traditional masterpieces reveal the same belief. In all the paintings of Judgment Day, human beings are presented as the only incarnate creatures subject to judgment. (Whereas any rational creature would

---

* Comparing alien belief to the so-called "Copernican Revolution" illustrates well the present point. Let us say for the sake of argument that Catholics everywhere, in the year 1000 A.D., were in universal agreement that the sun revolves around the earth and not vice versa. An ET promoter would jump on this opportunity and argue, "*See?! Catholics everywhere were wrong about the Geocentric Model, so we can also grant that they were wrong about extraterrestrials.*" This comparison is a false one. The question of whether or not the earth revolves around the sun has no effect on any aspect of the Faith—theological or moral. It resides squarely outside of the domain of the *sensus fidelium*, even if universal consent among the faithful in answering it were to be found. This is proven by the fact that the adoption of the Heliocentric Model had no effect on a 16th-century Catholic's prayer, worship, morality, creed, sacramental life, or on his understanding of Scripture, mission, Redemption, the Incarnation, and so on. Ascertaining the applicability of the *sensus fidelium* by determining whether the subject constitutes Christian "matters of belief" (CCC, §92), we find that the "Copernican Revolution" was far less consequential than any one of the dozens of literal revolutions being fought with cannons and swords in the same time period. Belief in extraterrestrial intelligence, on the other hand, is a matter that resides at the extreme opposite end of the spectrum of Catholic-belief relevance. It is difficult to imagine a question more germane to matters of [Christian] belief than this one.

require a spot at Judgment Day.) Two thousand years of sacred architecture throughout the world delivers an identical testimony. These "sermons in stone," incarnating the *sensus fidelium,* have never given any place to belief in Nonhuman, non-angelic intelligences.

Even two thousand years of *authentic* private revelations—despite detailing hitherto undisclosed secrets of earth, Heaven, Hell, Purgatory, the cosmos, and everything in between—are likewise devoid of any teaching that aliens exist, and only confirm the completeness of the "Great Chain" in which only man bears the Divine Image. Some private revelations explicitly describe man alone as made in God's image, and no authentic ones contradict this description.

St. Bridget of Sweden was shown that, too often, "**while irrational creatures keep to their limits, rational human beings debase their original excellence ... by not making use of their reason**."[159] In the *Mystical City of God,* Venerable Mary of Agreda was shown clearly "**all parts of the creation**," which includes no space for extraterrestrials. She was shown that:

> All other creatures within the confines of the universe ... were to be subservient to Christ and most holy Mary, and through them to the rest of men. ... [God] overlooked all that could offend Him in the conduct of the men and angels that were to fall ... And before creating intellectual and rational creatures ... He created heaven for angels and men; and the earth as a place of pilgrimage for mortals. ... Thus [is] all parts of the creation...[160]

Another mystic is the one we referred to in Part One. Her writings contain many fulfilled prophecies, she experienced many miracles, and she was strongly endorsed by St. Padre Pio, Pope St. Pius X, and others. She was given private revelations by Jesus that do not merely touch on this matter, but render a definitive verdict on it. Jesus told this mystic, the Servant of God Luisa Piccarreta—in a volume approved, with a *nihil obstat,* by St. Hannibal di Francia, that: "**Each created thing holds out my love to man; and if it were not so, Creation would have no purpose; and I do nothing without purpose. Everything has been made for man...**" (January 9, 1920) Yet if everything is for man and creation is purposeless without man, there are no aliens. Lest any uncertainty remain, Jesus also told Luisa:

> There is great difference between the creation of the whole universe and the creation of man ... because man was created in Our [i.e., the Trinity's] Image and Likeness ... he is not only Our work, like all the rest of creation, but he possesses part of Our Life...(November 10, 1929) I did not love the heavens, but man in the heavens, and for him I created them... [the heavens] were to serve him as pure delight... (October 29, 1926) Only the human will, as We had created it with the great gift of the free willing ... [can] love Us and remain together with Us... It was the greatest honor and gift We gave. (May 12, 1934) We [The Trinity] made use of creation in order to form Our veils, Our vast residences; however, We held back from

giving reason to them, because they were created for man, not for themselves, and therefore We reserved giving to man the capacity and reason of the whole Creation..." (September 2, 1928)

Obviously, it is not possible to both believe in aliens—or any rational nonhuman, non-angelic creatures whatsoever—and believe in the authenticity of these messages.

A more recent mystic, Maria Simma, was also given private revelations on this matter—in her case, from the holy souls in Purgatory (who, being confirmed in grace, could never deceive). Many of the messages she received are compiled in the book, *Get Us Out Of Here*, which recounts an interviewer asking Simma, "*There is so much talk about and media attention given these days to UFOs. Do people confuse angels with UFOs or vice versa?*"[161] Simma responded, explaining what the Holy Souls had revealed to her:

> ...There is no intelligent life on other planets. But this does NOT mean that all those sightings and stories of abductions are not true. So many people would not wish to lie about this, but it does mean something else. If there is no other intelligent life elsewhere in our universe, and here I'd like to add that anyone with a deeper knowledge and faith in the Holy Bible would conclude the same, then all of those sightings are Satan's work. Satan wants us to be very curious and what better way is there than to lead us astray running after little green creatures out into the voids of space? Curiosity has killed far more than merely the cat. And all those films and television shows about civilizations out there can easily lead the young into danger.[162]

Simma is correct. It is not only the case that aliens do not exist: it is also true that the deception which claims they do is a diabolical ruse.

Two thousand years of Christian piety is also devoid of any recognition of extraterrestrial intelligence. No traditional piety can be found wherein Christians have prayed for aliens or incorporated them into their theological understanding of creation. This is evident even in the Old Testament. For example, in the Canticle of Daniel—recited continuously to this day in the Liturgy of the Hours—Christians bless God on behalf of *all things* in the Universe. **"All works of the Lord"** (Daniel 3:35) are included: from angels, to waters, to sun and moon, to stars, to frost and chill, to lightnings and clouds, to all beasts and cattle, to all "sons of men" (v.60), and everything in between. Despite such breathtaking comprehensiveness and detail, nothing in this canticle allows any room for aliens.

> "You formed man in your own image and entrusted the whole world to his care, so that in serving you alone, the Creator, he might have dominion *over all* creatures." (Eucharistic Prayer IV)

The supreme act of the Church's prayer, however, is found in her Liturgy—and one can simply glance at it to see it yields no space for belief in aliens. The Preface for the Feast of Corpus Christi even explicitly rules out the possibility of "human" aliens (i.e., aliens that are still "men," but a "different biological species; not *homo sapiens*"), by declaring that the whole human race is "**bounded by one world**." In the Preface, the Church prays:

> Nourishing your faithful by this sacred mystery, you make them holy, so that the human race, bounded by one world, may be enlightened by one faith and united by one bond of charity (*Preface II Of the Most Holy Eucharist*. §61)

The Church is never timid in her supplications. She does not limit their scope or water down their aspirations to pacify cynics or comply with the demands of Pharisees. Quite the contrary, she beseeches for God to give what "**prayer does not dare to ask**." (*Collect for the Twenty-Seventh Sunday in Ordinary Time*) Therefore, a Catholic ET promoter would only be grasping at straws by arguing that the Church is simply not yet bold enough to pray for the salvation and sanctification of extraterrestrials. While the Church frequently prays for the *entire* human race to explicitly acknowledge Christ, she never offers this petition directed to beings other than humans. The reason is clear: only rational creatures can acknowledge Christ. Angels already have done so (and will never fail to do so), and demons already have rebelled and never will repent.

Yet if aliens exist, then they are direly in need of these prayers, and the Church—whose Liturgy by nature encompasses the *whole* Universe and recapitulates *all* creation—has been miserably failing in its most basic mission for 2,000 years. Even implicitly supposing that this scenario might be plausible or possible (an accusation latent in all ET promotion) is derogatory to not only the Church, but also the Holy Spirit, Who is her soul.

> "[The Liturgy has a] Universal and, so to speak, cosmic character. Yes, cosmic! ... The Eucharist is always in some way celebrated on the altar of the world. It unites heaven and earth. It embraces and permeates all creation. The Son of God became man in order to restore all creation, in one supreme act of praise, to the One who made it from nothing." —Pope St. John Paul II. *Ecclesia de Eucharistia* §8

Finally, whenever *all* the Universe's rational creatures are referred to throughout Church history's great works, the object of reference is always identical: only "angels and men." That is, purely spiritual beings and descendants of Adam and Eve. Thousands of such references have accumulated throughout the centuries.

> "For God, wishing both angels and men, who were endowed with freewill..." (St. Justin Martyr. *Dialogue with Trypho*) "... [A] society composed of angels and men together; so that there are not four cities or societies—two, namely, of angels, and as many of men—but rather two in all, one composed of the good, the other of the wicked, angels or men indifferently." (St. Augustine. *City of God*. Book XII. Ch. 1) "After the resurrection, however, when the final, universal judgment has been completed, there shall be two kingdoms [Heaven and Hell]—both, however, consisting of angels and men." (St. Augustine. *The Enchiridion*. Chapter

111) "[God] hath set over all one and the same Head, i.e., Christ according to the flesh, alike over Angels and men. That is to say, He hath given to Angels and men one and the same government..." (St. John Chrysostom. Homily 1 on Ephesians) "All God's works are ordained to the salvation of men and angels; and the order of his providence is this ... out of the mass of that innumerable quantity of things which he could produce, he chose to create men and angels to accompany his Son, participate in his graces and glory, adore and praise him for ever." (St. Francis de Sales)[163] "... The number of the rational creatures, that is to say of Angels and men who were formed to see God, was lessened when man perished ... [but] the lost man was sought for on earth..." (Pope St. Gregory the Great) "...We affirm that God ... disposed before all things in his divine understanding all the principal parts of the universe which might render him honour; to wit, angelic and human nature..." (St. Francis de Sales "Of the Divine Providence in General") "Above all, I rejoice that God himself loves [Mary] more than all men and angels." (St. Alphonsus Liguori)[164]

Confirming this unanimous consensus of Sacred Tradition, St. Thomas Aquinas teaches there are exactly three types of rational creatures: men, angels, and demons.[165] The words of many Popes and their magisterial documents can be quoted as well:

> "...[T]he future [is] uncertain and full of cares, but do not lose heart. Rather so display yourselves, presenting 'such a spectacle ... to the whole creation, men and angels alike...'" (Pope Pius XII. *Orientales Omnes Ecclesias.* §60) "...[T]he Blessed Virgin is infinitely superior to all the hierarchies of men and angels, [she is] the one creature who is closest of all to Christ." (Pope Leo XIII. *Magnae Dei Matris.* §9) "Mary has by grace been exalted above all angels and men to a place second only to her Son, as the most holy Mother of God who was involved in the mysteries of Christ..." (Second Vatican Council. *Lumen Gentium.* §66) "'Christ,' [St. Cyril] says, 'has dominion over all creatures ... ' ... Christ is to be adored by angels and men... [and] to him as man angels and men are subject..." (Pope Pius XI. *Quas Primas.* §13)

Such references as these could be extended almost endlessly. The lesson they convey is clear. Of all creatures in existence, only men and angels are rational. Each time any saint, Father of the Church, Doctor of the Church, Pope, or Magisterial document, has spoken of "*all* rational creatures," they have attributed rationality exclusively to angels and men.

Nor will the ET promoters' *argumentum ad ignorantiam* fallacy defense work. For one cannot argue that the faithful of Church history had no ET belief merely because they knew nothing of the idea itself. First, the teachings reviewed here *expressly* rule out aliens—they do not just leave the matter untouched. They affirmatively state that all rational creatures *consist in* (that is, *include only*) men and angels.

Second, the idea of aliens *was* well known from the beginning of the Church, as we will see next. The faithful have always known exactly what

Pagan ideas they were rejecting by insisting rational creatures consist in men and angels.

The Ancients and the Fathers of the Church

Although Christian belief in extraterrestrials only became common during the Enlightenment, early Christians knew well of the thesis itself. They treated it as just another Pagan absurdity that was common in the cultures into which Christianity spread in its early days.

Centuries before Christ, among the most famous of the Pagan philosophers were the Atomist materialists and Epicurean hedonists. The men who founded these ideologies—Democritus (born 460 B.C.), Leucippus (born 510 B.C.), and Epicurus (born 341 B.C.)—left legacies of atheism and immorality. Additionally, however, they were famous promoters of belief in extraterrestrials. One contemporary scholar noted that, among the ancient Greek thinkers, "**They are the first to whom we can with absolute certainty attribute the odd concept of innumerable [inhabited] worlds.**"[166] The Hedonists of the lot—the Epicureans—were surely no strangers to early Christians; they are explicitly recorded in the New Testament, in the Book of Acts, Chapter 18. Their leader, Epicurus, declared:

> There are infinite worlds both like and unlike this world of ours. For the atoms being infinite in number, as was already proved, are borne on far out into space. ... [There] nowhere exists an obstacle to the infinite number of worlds.[167]

In another letter, this prototypical hedonist insisted that: "**We must believe that in all worlds there are living creatures and plants and other things we see in this world**."[168] Metrodorus, a follower of Epicurus, agreed. He taught that infinite worlds must exist because—as he wrongly reckoned—infinite atoms exist.[169] Early atomists were no different. "**In all these worlds [posited by the Atomists] there are living creatures. Under the influence of these Greek philosophers, this plurality of worlds idea eventually gained a following among certain Roman thinkers as well...**"[170]

Two and a half thousand years ago, therefore, we already see the position taken on this matter separating the wheat from the chaff. While those with legacies of moral and spiritual destruction (these atheistic Atomists and hedonistic Epicureans) believed in extraterrestrials, the nobler philosophers had no use for such theories. Plato and Aristotle condemned the notion of multiple inhabited worlds. Scholarly promoters of alien belief now concede we are re-living that same drama. Dr. Seth Shostak, the current head of SETI, wrote:

> Even today, the theological issues are not fully settled, and echoes of the ancient Greek debate between the atomists and the Aristotelians still reverberate in our time.[171]

We, today, are left with the choice of siding with Plato and Aristotle or with the hedonists and atheists.

The Fathers of the Church, however, were far more stringent in their opposition to belief in aliens than even the noble Greek philosophers. Indeed, early Christian writers knew well about these esoteric beliefs. The absence of the thesis from their own convictions was due not to ignorance of it, but condemnation of it. Summarizing her extensive and recent research, Dr. Marie George concluded:

> My position on the early Church Fathers (from the first through the eighth century) and the Doctors of the Church (from 170 to 1280, i.e., from Hippolytus of Rome to St. Albert) is that all of them who explicitly raise the question of whether there is one world or many, or who at least take an explicit stand on the question, reject pluralism [ET belief], with the exception of Origen... nothing in tradition supports the notion that a plurality of inhabitable planets exists ... [Church Fathers] who consider the question of a plurality of worlds plainly come down on the side of a negative answer... [E]very church Father and Doctor who paid some sort of attention to some version of many earths, always rejected the notion, with the single exception of Origen.[172]

Rightly does Dr. George observe that "**nothing in [Catholic] tradition supports the notion of [aliens]**," and that *all* the Church Fathers and Doctors who raise the question of aliens reject such belief.

The facts certainly bear out her analysis. St. Hippolytus also authored the *Elenchus*, or the "Refutation of All Heresies," a list of doctrinal errors—in it, he included alien belief.[173] St. Philastrius, Bishop of Brescia, expressly condemned the view as heretical. Particularly relevant is his denunciation of the use of certain apocryphal texts in support of alien belief—as the same strategy has again become popular among Christian ET promoters today, who defer to the Book of Enoch—and other wildly esoteric interpretations of the "Nephilim" of the sixth chapter of Genesis[174]—or other error-laden and non-Canonical ancient texts to ostensibly justify the orthodoxy of their position. St. Philastrius wrote:

> There is another heresy that says that there are infinite and innumerable worlds, according to the empty opinion of certain philosophers—since Scripture has said that there is one world and teaches us about one world.[175]

St. Jerome, a Father and Doctor of the Church—and originator of the official Catholic Bible (the Latin Vulgate)—was equally zealous in condemning the error of ET belief, writing that "innumerable worlds" were among the "**heresies in the books of Origen.**"[176]

Origen was an early Christian author whose writings on this topic provided a veritable sci-fi-like deception 1,700 years before its time. The Second Council of Constantinople was blunt in dealing with him. Including Origen in the same breath as the most famous heretic of the early Church,

Arius, it declared that those who so much as fail to condemn Origen are themselves thereby condemned:

> If anyone does not anathematize Arius, Eunomius, Macedonius, Apollinarius Nestorius, Eutyches and Origen, as well as their heretical books... let him be anathema. (Anathemas Against the Three Chapters. §11)

St. Augustine, in his greatest work, *City of God*, also denounced the multiple earths hypothesis. He rejected "**Epicurus' dream of innumerable worlds,**" (Book XI, Ch. 5) and he contradicted those who, "*though they do not suppose that this world is eternal, are of [the] opinion... that this is not the only world...*" (Book XII, Ch. 11) Later still, Augustine refuted the error of believing that men exist (on the "inaccessible underside of a round earth") who had not descended from Adam and Eve, in the context of the debate on the "Antipodes." (Book XVI. Ch. 9) Although he had no condemnations for belief in a globe-shaped earth, he denounces as lacking any credibility the notion that realms detached from the core of humanity are nevertheless populated by rational beings. For it would contradict Scripture, which, he reminds us, "gives no false information," to assert that such beings, though not offspring of Adam, nevertheless exist.

***

Following the condemnation of Origen's heretical views—both by the Church itself and by other Church Fathers—Christians, unsurprisingly, left this matter aside. ET belief was rightly regarded as yet another tenet of ancient Paganism from which Christ had rescued us; therefore, it was best off ignored.* Indeed, diabolical deceptions are only worth addressing when they become prevalent.

What strange thing, then, began to transpire as *The Katechon*—the "Restrainer" of 2 Thessalonians 2, which we will discuss in Part Three—began to be removed after a millennium-and-a-half of Church history?

---

* To confirm this, one may read any book in (or out of) print, peruse the archives of any conference, or even scour the primary sources himself: he will only discover that more than a millennium of Christian history passed without the extraterrestrial thesis receiving *any* support in Christianity. The books recently written—and there have been many—by the Christian ET promoters invariably pass on from Origen to the thinkers we will address next, who came over a thousand years later. They openly admit that they cannot find a single (non-condemned) Christian before the 15th century who so much as uttered a word in support of belief in extraterrestrials. Dr. Michael Crowe, a Professor Emeritus at the University of Notre Dame, may be the one scholar who has most extensively chronicled the history of the extraterrestrial intelligence debate. In a recent book whose scope includes ancient times to the year 1915, he proves that belief in aliens was alive and well in Pagan, pre-Christian centuries. After demonstrating that fact, however, he could only find enough material to fill up eight pages (of a 544-page book) with discussion of alien belief in an entire one-thousand-four-hundred-year period of history—most of the span covered by the book! Moreover, not one quotation provided within those eight pages demonstrates any alien belief whatsoever existed among Christians (only refutations of the same)—except, of course, from Origen. What this tells us is simple: Dr. Crowe did not find any alien belief for over a millennium of Christianity because *there was no* alien belief during that millennium of Christianity.

Oresme and Cusa: The Two Nicholases

The 14th-century Bishop of Lisieux, Nicole (or Nicholas) Oresme was one of the most influential thinkers of his era. A philosopher, physicist, mathematician, and economist, he is described by the old *Catholic Encyclopedia* as "**one of the principal founders of modern science.**" He was the actual originator of various concepts popularly attributed to Copernicus, such as relative motion and the earth's rotation. The secular *Stanford Encyclopedia of Philosophy* extols him in the loftiest terms, insisting that "**His work provided some basis for the development of modern mathematics and science.**"[177]

Suffice it to say that, if any Catholic thinker of those centuries was free of the so-called "shackles of tradition" or "enslavement to ancient unscientific thought," Bishop Oresme would be that man. Oresme even departed from some thinkers on the possibility of the plurality of worlds—insisting, contrary to them, that God *could* make multiple worlds. He was remarkably well positioned to accept belief in ETs if there were any grounds for regarding such a thesis as orthodox. Instead, Oresme wrote:

> God can and could in His omnipotence make another world besides this one or several like or unlike it. Nor will Aristotle or anyone else be able to prove completely the contrary. But, of course, there has never been nor will there be more than one [inhabited][178] corporeal world...[179]

This extraordinary Bishop was demonstrably unafraid of offending traditional thought. Yet he writes, "***Of course*, there has never been nor will there be**" aliens. Oresme's choice of words indicates that he knew full well he was merely reiterating what all Christians always believed: human beings are the only incarnate intelligences that exist. Commenting on Oresme, the 19th-century French physicist, philosopher, and historian of science, Pierre Duhem (also a Catholic), wrote:

> Someone like Nicole Oresme would have allowed that space can contain several [solar] systems ... [but] he would not have been tempted to place inhabitants on these worlds. As far as we know, no one defended such an assumption during the Middle Ages, when Nicholas of Cusa, whose imagination knew no bounds, proposed it.[180]

And thus we come to the other Nicholas, whereupon everything changes.

Nicholas of Cusa was a German polymath who was born just two decades after the passing of Bishop Nicole Oresme. He was named a Cardinal and Vicar General in the Papal States, and many considered him the second most powerful person in the Church. This Nicholas did indeed believe in aliens. He declared:

> Those on the earth are more material and rough, so that the intellectual beings found on the sun are more in actuality and less in potency, while the inhabitants of the earth are more in potency and less in actuality; the inhabitants of the moon are somewhere in between these two extremes ...

> The regions of the other stars are similar to this, for we believe that none of them is deprived of inhabitants.[181]

This was a radical teaching. For the first time in the history of the Church[182] since the condemned Origen (and Virgil), we see a thinker—a prominent clergyman, no less!—announcing his belief in aliens. Nevertheless, the 300 years that followed his own life saw no adoption of his thought among Catholics, save one man—the excommunicated and executed apostate ex-priest, Giordano Bruno, whom we will address shortly.

Considering Cusa's unmatched fame, we can be sure that—had there been any grounds for taking this novelty of his seriously—a deluge of ambitious writers proclaiming aliens would have followed like so many moths to the flame. Instead, we hear only crickets. A politically important cleric, Cusa was graced with his unorthodox views being politely ignored instead of openly denounced.

Nor will it work to claim—as Catholic ET promoters do—that Cusa's great power was itself a vindication of his beliefs. Neither unorthodoxy nor downright corruption have prevented ecclesial elevation in Church history. Christ protects the integrity of the Magisterium (cf. Matthew 16:18), not the prudence of ecclesial appointments. Moreover, Cusa's momentum in attaining ever more political clout in the Church was already established before his year 1440 A.D. announcement of his belief in aliens. The year before, the Pope personally sent him to the Council of Florence to heal the schism in the Church. (The attempt was unsuccessful.)

Not only ecclesial writers, but also secular ones, refused to welcome Cusa's beliefs. Dr. Michael Crowe, describing a 250-year period following Cusa's claims, wrote of the "resistance" shown to alien belief by the early heroes of science:

> In studying the period from 1500 to 1750, one is struck by the hesitancy, indeed resistance, shown by some of the founders of the scientific revolution to a full-blown pluralism ... [which is] defined as the doctrine that the earth is but one of the inhabited planets ... [Even Galileo], in at least six passages comment[s] on the question of extraterrestrial life. These range from rejection to cautious reserve concerning it ... [He wrote,] '*I agree with Apelles [Christopher Scheiner] in regarding as false and damnable the view of those who would put inhabitants on Jupiter, Venus, Saturn, and the moon, meaning by inhabitants animals like ours, and men in particular.*'[183]

Remarkably, even the man who is now regarded as the poster-boy for avant-garde propositions in Catholicism, Galileo Galilei, dismissed Cusa's belief in aliens as not only "false," but "**damnable**!"

Nevertheless, the era that followed changed everything. During the "Enlightenment," Cusa's views were indeed adopted. Not one who adopted those views is now venerated as a saint, blessed, venerable, or servant of God. None garnered any meaningful reception in the Church at large. But this trickle of Christians endorsing aliens, which eventually grew into a stream, was—tellingly—mostly evident outside of the Catholic Church (among those in newly formed denominations whose *raison d'être* was the overthrowing of 1,500 years of unanimous Christian teaching).

On the surface, Cusa should have been a prime contender for canonization. This prolific Cardinal, known and respected by Popes, is also referred to as "**arguably the most important German thinker of the fifteenth century,**" (*Stanford Encyclopedia of Philosophy*) and even "**the first modern thinker,**" according to the German philosopher, Erns Alfred Cassirer. Yet the Church has given him no exaltation whatsoever (though esoteric and Satanic groups have).* A serious impediment must exist which precludes bestowing any such formal honor—a problem so grave that it outweighs the mountain of biographical data seeming to support his canonization. To find that obstacle, we need look no further than Cusa's first-of-its-kind rejection of the fact that only man bears the Divine Image.

We must next examine Cusa's legacy; the "fruits" of his teachings—which, as Jesus admonishes, tell us the quality of the tree from which they derive (cf. Matthew 7:15). This examination brings us to the only significant thinker in the two centuries following Cusa who welcomed the latter's views on aliens: Giordano Bruno, whom we introduced in Chapter Three.

Giordano Bruno

The *Stanford Encyclopedia of Philosophy* notes that, "**In spite of his significance few later thinkers, apart from Giordano Bruno ... were influenced by [Cusa] until the late nineteenth century.**"[184] Cusa did not focus on his alien belief, and was not condemned for it, therefore Bruno is described by ET promoters as "the first martyr" of their pseudo-religion. Indeed, Bruno was excommunicated by multiple Christian denominations and was ultimately burned at the stake in the year 1600 for his heretical teachings.

Giordano's lurid tirades would have fit in well with 21st-century discourse. The old *Catholic Encyclopedia* entry on his life does not fail to testify

---

* One of the most famous Satanic "Churches," the Thelemites, is an occult group founded by the uniquely perverse individual discussed in this book's introduction, Aleister Crowley. Their central text, *The Book of the Law*, contains a "revelation" from the "angel" (demon) Aiwass. This false revelation is simply a slight paraphrase of Cusa's deeply erroneous metaphysics which referred to an infinite material universe. One recent tome from Oxford University Press observed that, "**Crowley saw the god in terms analogous to Nicholas of Cusa's description of the universe as a reflection of God: '*an infinite sphere whose centre is everywhere, circumference nowhere.*'**" (Henrik Bogdan and Martin Starr. *Aleister Crowley and Western Esotericism*. Oxford University Press. 2012. Page 183.) Additional problems exist. Cusa's work, *De Paci Fidei* (On the Peace of Faith) espouses religious syncretism. He refers to the multiplicity of (false) religions on earth as "one religion in a variety of rites." Bizarrely, he claims that Islam's basic text, the Koran, teaches the Divinity of Christ and the Trinity. Cusa is also often accused of espousing some pantheistic beliefs, along with the heresy of Conciliarism (which regards a council's authority as superior to the Pope's). Therefore, despite Cusa's fame, discerning Christians will have no trouble seeing that if a man like him is the origin of some proposal in Christendom, this is a mark *against* the proposal, not an argument in its favor.

bluntly to his antics, observing that he "**failed to feel any of the vital significance of Christianity as a religious system**." Bruno, in a word, was the example par excellence of a man who was *"holding the form of religion but denying the power of it,"* (2 Timothy 3:5) thanks to his infatuation with the notion of a cosmos filled with inhabited planets. Christianity appeared to him comparably insignificant.

Catholic ET promoters defend their view from the implications of Bruno's condemnation by claiming that he was not executed for his belief in a plurality of inhabited worlds, but rather for various heresies of other sorts. Recent scholarship has exposed the falsity of that claim. One professor of the history of science observed, in a 2018 article in *Scientific American*:

> To be sure, the Inquisitors didn't condemn [Bruno] for believing in Copernicus. ... [but] in the 1590s Bruno's claim [of multiple inhabited worlds] was considered heretical. Many authorities denounced it, including theologians, jurists, bishops, one emperor, three popes, five Church Fathers and nine saints ... [Bruno's belief in many worlds] was the most frequently recurring charge [in his trial]. ... Thirteen times, in 10 depositions, six witnesses accused Bruno of believing in many worlds. No other accusation was invoked even half as much. ... In 1597, Bruno was confronted by inquisitors, including the authoritative theologian [Saint] Robert Bellarmine. Bruno *"was admonished to thus abandon his delusions of diverse worlds."* ... Bruno was condemned for several heresies, but the one about multiple [inhabited] worlds was the strongest case against him.[185]

The many other heretical views Bruno held were likewise held by countless of his contemporaries who were not formally condemned, much less burned at the stake. Additionally, no one—not even in the most overly severe times of Church history—would be executed for a recanted view. Bruno, however, did recant of much: for example, he retracted various heresies against Christ, the Eucharist, the Blessed Virgin Mary, and so on. What he did *not* recant was his belief in multiple inhabited worlds. And this view is the principal one that set him apart from the many heretics who were not similarly punished. Belief in aliens, moreover, was not just one of Bruno's doctrines; scholars of his work have recently painstakingly shown that it was his overarching exhortation and his mission's primary motivation.*

Importantly, the Church has never apologized for the condemnation of Bruno, despite repeated calls for her to do so. As recently as the year 2000, Cardinal Sodano acknowledged that Bruno's *execution* constituted a "sad

---

* For example, one scholar recently noted: "**Bruno wrote: "...*[there] are innumerable globes like this on which we live and grow...*" [and] this was not just a passing theory, but part of a broader vision. "*The whole of Bruno's philosophy is based on his view of an infinite universe with an infinity of worlds,*" wrote one of his early biographers, Dorothea Singer. ... Bruno forcefully deployed the principle of plenitude... This was the cosmic perspective that colored all of Bruno's other works and guided his daily life. As another of his biographers, Ingrid Rowland, wrote, "*Bruno's enthusiasm for his real concern... was to show how the realization that we inhabit an infinite universe will transform every aspect of our lives...*"** (Steven J. Dick. *Space, Time, and Aliens.* Chapter 42.2)

episode" for the Church, but he stopped well short of apologizing for the condemnation. Instead, he praised the desire of the Inquisitors to defend the truth and reiterated the incompatibility of Bruno's teaching with Catholic doctrine.[186]

Giordano Bruno was executed at the onset of the same century that saw a movement which—despite its contradictory title—consisted in none other than a new darkness descending upon Christendom. In Bruno's own day, the fervor of the Copernican Revolution was in full swing and was electrifying scholars around the world. An unspoken competition soon commenced to see which academic could be the more "innovative," and the "least chained by tradition." The foundations of Modernism—"the Synthesis of All Heresies," according to Pope St. Pius X[187]—were being laid down by way of what came after Bruno: the "Enlightenment," more appropriately called the *Great Darkening*, as we saw earlier. This so-called Enlightenment, which began only a few decades after Bruno's death—born, as it was, in the *cogito* of René Descartes in 1637—sought to reject the supreme authority of Divine Revelation and replace it with a purely rationalistic account of all things.

How fitting, then, that the utmost illustration of the Enlightenment, in the French Revolution (1789-1799), saw "reason" symbolized by an idol and irrationally enthroned as a goddess in the desecrated Notre Dame Cathedral—amidst all manner of murders, blasphemies, and other outrageous evils. Even if it must assume new forms, Paganism always manages to fill any void created by men abandoning Jesus Christ. And it is, tellingly, in this time alone that we see the beginnings of the aforementioned trickle of Christian thinkers begin to espouse belief in extraterrestrials; a trickle which grew, throughout the swelling of the Enlightenment and its increasing darkness, into a steady stream in the deeply occult 19th century.

> "Who would not have been thought a madman or idiot, before that period [of the French Revolution], who had conjectured such a portentous approximation towards Paganism as actually then took place?" —St. John Henry Newman. *Discourses on the Antichrist.*

While it is true that some Catholic thinkers espoused belief in aliens during this century, today's ET promoters wildly mischaracterize that era. In fact, even as alien belief was considered a secular dogma in the 1800s, it was Catholic theologians who outshone the Protestants in remaining adamantly opposed to the idea of extraterrestrials. Most were of the view of Fr. Augustine Hewit (1820-1897), Superior General of the Paulists, who taught that **"belief in the past or present existence of other worlds inhabited by intelligent beings"** was **"unphilosophical, untheological, and unscriptural,"**[188] and Fr. Joseph Filachou, who condemned alien belief as "**absolutely rejectable.**"[189] The few Catholics who did seem to endorse the view were not actually advocating for belief in aliens (as today's Catholic ET promoters are), but were only insisting that alien belief cannot be used like Thomas Paine used it—to attempt to eradicate Christianity. (This matter of 18th and 19th-century ET promotion is discussed further in the Appendices.)

The Occult 1800s

> "The [19th] century has been described as both the democratization of Christianity and the feminization of it; the same could clearly be said of the occult and of the many places where those two systems of meaning overlapped and competed..." —Dr. Cathy Gutierrez[190]

Despite all that has been covered, some will doubtless still feel tempted to justify believing in aliens because of those 19th-century Catholics who did the same. Even a certain false traditionalism, reflexively glamorizing all times before the mid-20th century, may induce one to find utterly convincing the mere sight of any grainy black-and-white photograph of a 19th-century Catholic. Such an approach to Tradition contradicts its own premises.

The popes and Magisterial documents most often appealed to by traditionalists themselves were, of course, concerned with condemning the evils that had become pervasive in the Church in the time of their own promulgation. Consider the most famous of these "traditionalist" Magisterial documents—Pope Pius IX's *Syllabus of Errors*. It was not written to condemn the sexual revolution of the 1960s or the brand of Modernism peculiar to pre-war Europe. It was promulgated in the year 1864, directed against the errors that had *already* ravaged the Church in the decades leading up to its publication. There, the Pope rejected many notions commonly associated with 19th-century ET belief. For example, he condemned the view that "*Divine revelation is imperfect, and therefore subject to a continual and indefinite progress, corresponding with the advancement of human reason.*" (§5) [191]

The Pope concluded his condemnation of these errors with a grave warning that "the synagogue of Satan" was then gathering "its troops against the Church of Christ." Our own era being worse still must not preclude accurately assessing the evils of the 19th century. By that time, the Enlightenment had largely accomplished its goal of devastating mainstream scholarly thought, and almost all the popular thinkers had their minds clouded by a whole retinue of ungodly propositions. Then as now, it took a saintly scholar to protect himself against succumbing to this tendency.

The 1800s were also heavily dominated by the influence of freemasonry and the occult. Everywhere was found Theosophy, séances, Ouija, automatic writing, mediums, secret societies, table-rapping, spirit-photography, mesmerism, telepathy, divination, and on the list goes—all of which are simply portals for demons. Extraterrestrial fixation was just another item to add on to that list.

Occultism was so prominent in this period that multiple scholarly tomes exploring it have recently been published. One such work, published by Routledge, acknowledges in its opening preface that "**occultism was a**

point of entry for nineteenth-century thinkers," [192] explaining that it was among the "**central forces in nineteenth-century debates**."[193]

Catholics of that era were not immune. Nineteenth-century occultism infiltrated the Church just like the New Age Movement did in the following century. The same Routledge tome notes:

> [Occultism] was not a denomination, but rather an amplification or overlay on one's denominational beliefs and practices... it cut widely across churches and classes, from high to low and broad church Anglicanism and Roman Catholicism to virtually all of the Dissenting faiths, touching the heart of most major faith systems ... [Occultists] thought of themselves as expanding or improving their native religious practice with the séance, not replacing them.

Shockingly, many 19th-century Catholics supposed that séances "improved" their faith. Few now succumb to such folly. But it has only been replaced with something worse: the supposition that belief in extraterrestrials somehow "improves" Christianity.

The pattern these 19th-century occultist Christians advocated for also disturbingly resembles the pattern we see displayed in much of today's Christian promotion of extraterrestrial ideology. In the 19th century, the adoption of occultism (including ET promotion) meant a certain "bargain" for the Faith; one whose terms require that "**Christianity ... [be] stripped of most of its traditional doctrine.**"

> God, heaven, sin/redemption, and even the word 'Christianity' have been [muted]... Instead one finds euphemisms such as 'the Divine One' and 'the Great Religion', and the predominant focus is on the individual soul and a free-floating conception of the afterlife that is divorced from the traditional, dualistic geography of heaven and hell.[194]

We should not be surprised to see that today's Christian Ufologists deliver similar lessons as those found through the séances, mediums, and Ouija boards that first became popular in the 1800s. All derive from the same source.

The séance-organizers who dialogued with demons pretended they were doing good by "*demonstrating the reality of spiritual phenomenon,*" when in fact they were only ruining lives and endangering salvation. In the same way, today's Christian Ufologists pretend that their promotion of expressly diabolical teachings is somehow virtuous merely because the extraterrestrial phenomenon is concerned with "anomalous" events, whereas legitimate miracles are also extraordinary. Yet even the veneer of virtue that 19th-century occultism sometimes wore is absent from Ufology. For the former would at least often convince those it seduced to abandon atheism. Today's ET/UFO promotion usually has only the opposite effect.

***

In this chapter, we have seen total agreement throughout the entire testimony of Scripture and Magisterium; the unanimous consensus of all saints, Fathers, and Doctors; the unbroken development of Sacred Tradition; and the orthodox consideration of core components of Dogmatic Theology. Each rules out belief in extraterrestrials.

Are there aliens? The answer is *no*.

That answer is already clear, therefore it ought not be qualified with academic middle ways or so-called nuanced takes. If Eve rejected nuance when pondering, "*Could I perhaps justify eating the fruit?*", the Fall never would have happened. Let us not waste time lamenting the Fall, but let us also not re-enact it in our own lives.

God could have placed a large sign on the Tree of the Knowledge of Good and Evil, cautioning Eve that a serpent would approach her there and speak to her, and that his words would be nothing but a trap. Doing so, however, would have defeated the very purpose of the test. Eve already knew it was God's Will that she must not eat from that tree, no matter who told her to do so. However, she was allowed *just enough* wiggle room to pretend she could justify obtaining the promised hitherto secret "knowledge" (Genesis 3:5) by listening to the speaking serpent and eating the fruit.

Similarly, God could have inspired the authors of Sacred Scripture to proclaim—or moved a Pope to declare ex-cathedra—"*No incarnate intelligent creatures exist on any planet, in any galaxy, other than earth. Human beings are the only rational material beings God made. There are no aliens. Whoever says otherwise, let him be anathema.*" But we are in apocalyptic times, and as Eschatology (last things) mirrors Protology (first things), the Church is about to undergo a final test analogous to the original one our first parents failed in the Garden. Here as there, God's Mercy and Justice are displayed. He has given us more than enough to know that we alone are made in His image, but not so much that Christians who prefer sci-fi deceptions to the love of God's Word will be incapable of pretending to justify their stance.

Just as Eve was seduced by the promise of her eyes being "opened" (Genesis 3:5), so the men and women of the present age—groomed by decades of delusions from a culture long held in the grip of Satan—have been primed to anticipate the promise of "new knowledge" from "first contact" with extraterrestrials upon the "day of disclosure." And whoever is open to dialogue with these demons in disguise is bound to find the allure of their promises irresistible, even once these promises become contingent upon apostasy.

"**Did God *really* say...?**" —Satan.
(Genesis 3:2)

Part Three: Aliens and Demons

10. The False Prophets

"Before [the Devil] showed himself to [Eve], however, he aroused in her many disturbing thoughts or imaginations, in order to approach her in a state of excitement and pre-occupation.... [He then] took the form of a serpent, and thus speaking to Eve drew her into a conversation, which she should not have permitted."
—Revelations of Venerable Mary of Agreda

Having reviewed the science and the theology, we know that extraterrestrials do not exist. Nevertheless, our most important task remains: *discernment,* wherein we apply the methodology insisted upon by our Lord:

> Beware of false prophets, who come to you in sheep's clothing but inwardly are ravenous wolves. You will know them by their fruits. (Matthew 7:15-16)

Only a few sentences later, He repeats: "**Thus you will know them by their fruits**." (v. 20) Elsewhere, Jesus clarifies that "*each* tree is known by its own fruit," (Luke 6:44) therefore this principle has no exceptions. More tellingly still, Jesus connected the most severe rebuke He ever dispensed to one's refusal to judge a tree by its fruits. For immediately after noting that those who blaspheme the Holy Spirit will not be forgiven, Jesus said, "**Either make the tree good, and its fruit good; or make the tree bad, and its fruit bad; for the tree is known by its fruit**." (Matthew 12:33) In sum, the Gospel assures us that one cannot overestimate the importance of judging the tree by its fruits. Failure to do this will ensure that one succumbs to deception.

It is not difficult to appreciate the wisdom in this Divine teaching. Suppose there is some movement or belief that has flourished across the world for decades, producing the fruits of heresy, blasphemy, apostasy, Paganism, murder, suicide, degeneracy, diabolical possession, occultism, fraud, and deceit—all while providing no good fruits to speak of. If a Christian defended this movement by insisting that we "*cannot throw out the baby with the bathwater,*" then how could one so willing to embrace naked delusion be capable of *ever* seeing (much less embracing) the truth? If he can find a way to rationalize defending even that, then he can find a way to rationalize defending anything. He may manage to defend the supremely tragic and final decision to walk straight through the doorway where all are welcome and over which is written, "*abandon all hope, ye who enter here.*" And as we will see in the pages ahead, such fruits define the whole gamut of "the phenomenon" of alien belief, Ufology, and ET-contact-expectation.

This knife, however, cuts both ways. The immediate object of Jesus' rebuke was the Pharisees, who fell into the opposite error: they witnessed the *good* fruits of Jesus' ministry and, instead of acknowledging the finger of God, they claimed these fruits actually came from the Devil. Thus they received Jesus' rebuke quoted above in Matthew 12. Therefore, when the fruits indicate that some phenomenon is in fact from God, all must beware condemning it, for that audacity is the surest path to hell.

This is the path taken by certain putative "good Christians" throughout the ages—spiritual spawn of the Pharisees who are *so very good and smart and pious* that they knew better than almighty God. And they are never in short supply whenever there arises some true messenger of God or vital movement inspired by the Holy Spirit. Thus, they wind up condemning God Himself, and none of their apparent orthodoxy will excuse them from His wrath when, on the Day of Judgment, they stand before the One they had cruelly maligned. Therefore, we will be particularly careful to ensure we are not adopting the methodologies employed by the modern-day Pharisees here.

For one, we will not "nut-pick." We will not select for scrutiny those "fringe" voices associated with Ufology who are making the field's most outlandish claims. Instead, we will assess the movement honestly; focusing on the open and clear statements of the most popular individuals and groups who are most representative of the contemporary ET phenomenon. Next, we will not scandal-monger. This tactic is a favorite of "grifters" who drool over juicy gossip. With all the zeal of a detective, they seek out the "horror stories" associated with their object of derision, no matter how unrepresentative of the overall movement these stories are. By highlighting such stories (and only such stories), these scandal-mongers pretend to have succeeded in painting the movement itself as "demonic," whereas all they have actually done is observed that human beings populate its ranks. So we will not focus on horror-story testimonies of ex-UFO nuts who have become disgruntled—however accurate these testimonies may be. Nor will we mention, much less base arguments on, mere gossip or accusations that are inadequately substantiated. Instead, we will focus on what is most frequently and openly praised and promoted—by those within the movement.

We will not leverage ambiguity or blow out of proportion that which is merely tangential. This underhanded Pharisaical approach consists in applying extreme focus to a select few difficult quotes or events that are surgically extracted from the entire corpus of works produced by their object of scorn. These are selections that could be interpreted in a variety of ways, but the Pharisees neurotically insist that only the most deeply erroneous and damaging of interpretations are what had been "truly" intended by the one they seek to crucify. Ravenous for detecting minute missteps or imagined slights against tradition, they ever *"watch"*—not to learn, but *"so that they might accuse…"* (Mark 3:2)

The Pharisee defends his malice by insisting, *"The Devil will speak 99 truths to promote one lie!"* But in truth, the Devil cannot help but show his teeth, and there is never a need for a militia of self-proclaimed inquisitors sniffing out alleged errors with all the acuity of a bloodhound. Nor is the Devil behind every rock; his power is quite finite, so he must exercise it economically. If he is the one truly behind some individual's message or movement's motivation, then he simply cannot afford to remain content only lurking in its fine print or in select interpretations of some of its claims. He will, instead, act in such a way that straightforward Christian discernment will discover his presence. Jesus assured us as much when He taught that we will *"know"* false prophets by their fruits (Matthew 7:15-16); not that we are stuck "guessing" who they might be, as when one reads some mystery novel. Jesus settled that it is simply not possible for the Devil to fail to reveal his presence clearly if Christian discernment is employed.

Therefore, nothing in what follows will require exaggerated focus on small details to describe "what is *really* transpiring." Instead, we will find the overwhelming stench of Satan permeating all of Ufology's and ET promotion's most representative dimensions, most central claims, and most popular pundits.

We will not waste any breath insisting this-or-that is what is "really meant" by some obscure remark associated with ET promotion, for the astonishing apostasy will be apparent in the open declarations of those we will review: from Ufologists, to "ET Contactees," to Hollywood's own ET promoting films, to UFO Religions, and more.

Our first order of business will be to review the messages of the men who have distinguished themselves as the heralds of the UFOs.

Steven Greer

After less than a decade working as an emergency room physician, 43-year-old Dr. Steven Greer forever left medicine to devote himself full-time to Ufology. This self-billed "Father of the Disclosure Movement" has now spent almost thirty years heralding the arrival of (and facilitating communication with) the "interdimensional UFOs" peppering our skies, while imploring the masses not only to await, but also to work to usher in, a messianic techno-ET-utopia.

The numbers appear to bear out his own claims to prominence. Millions of people across the globe follow him on social media, watch his documentaries, or have joined his "CE5" movement. Dr. Greer is also very strategic. Appealing to both sides of the movement, he insists that most UFO sightings are actually "Project Blue Beam" government psyops designed to portray these "truly benevolent extraterrestrials" as hostile threats. Greer incessantly reminds his followers of the establishment's zeal to prevent the masses from acquiring the Godlike knowledge that ETs want to give, lest the emergence of utopia loosens the elite's grip on power.

The populist message he promotes works wonders in the hearts of his eager and multiplying followers. The gravest danger in Greer's message, however, is found in his "CE5" movement, which instructs everyone to "make contact with the interdimensional ETs and UFOs" and dialogue with them, thus becoming "emissaries to the ETs." Here, Greer is not merely making amusingly irrational promises that will simply disappoint. Here, he is facilitating dialogue with demons.

Entitled *Close Encounters of the Fifth Kind: Contact Has Begun*, Greer's 2020 Documentary lays out his agenda, and it has become a wildly popular film. As of this writing, it has over twelve thousand reviews on Amazon, indicating that perhaps millions have already watched it.[195] In the film, Greer relies heavily on the testimony of Daniel Sheehan, a well-known Catholic Ufologist and a Harvard-educated lawyer who has long collaborated with the Jesuits, even serving as the legal counsel at the legendary Catholic religious order's U.S. headquarters. (Sheehan also founded the "Christic Institute" based on the teachings of Fr. Teilhard de Chardin, whom we will consider later.) Early in the documentary, Sheehan relays his own role in the Disclosure movement: he is working "**to help get the Vatican and the Jesuit order**" involved in heralding the Age of the Extraterrestrials. If he is successful, Sheehan promises, we will finally discard the outdated, traditional Christian idea that the events which have transpired on earth have any sort of special (much less unique) place in God's eyes.

The documentary soon transitions to the narrator dismissing our current "primitive" view of reality and advocating instead for "**God. Prana. Chi. Magic. Consciousness.**" These, we are told, are all the same, and are just synonyms for the "*intangible energy that connects everyone and everything*." Greer diligently runs the gamut of New Age deceptions, but unlike others competing over the same market share, he has shrewdly perceived that all these lies can now be collected under one umbrella-deception: that of the extraterrestrials. Greer then claims: "**The ETs are trying to encourage people to understand higher consciousness and the universal mind, to be able to have a relationship with them.**" The assertion itself is accurate. Demons always desire a "relationship" when initiating contact; so too these "ETs" want a *relationship*. With the help of Greer and other ET promoters, they are more successful today than ever before in history.

> "In all of this, demons are essentially trying to establish an abusive relationship with us as human beings."[196] —Exorcist Fr. Chad Ripperger

The film next declares: "**In our time this search [for ETs] will eventually change our laws, our religions... space, the mirror, waits for life to come look for itself there.**" After ensuring his audience is sufficiently ravenous for this awakening, Greer shares his method for finding it. He details his (in)famous "CE5 Protocols" to dialogue with the aliens:

> Sit in quiet consciousness. ... Connect to that aspect of your mind that is unbounded. And then intend to sense, feel, see, know where the ET craft

> is. It could be in another galaxy. Connect to those beings, politely invite them to come and visit you. … And you show them where you are. ... like a vectoring; a zooming in from wherever they are to where you are. … ZOOM, down to a few square meters around where you are, where you *clearly* show them where you are. And you do this in a coherent thought sequence that is continuous.

Over the years, Greer has remained insistent on this method. It must consist in a careful meditation that is dedicated to a focused visualization (which he also calls "vectoring" or "zooming"). This results in a construction of an image in your sense impressions, which indicates to the "extraterrestrials" exactly where you are. Unfortunately, this approach is not mere nonsense. Greer has taught millions how to invite a demon into their lives.

The Devil is not omniscient, and demons cannot read anyone's mind (only God can do that).[197] They instead must remain content waiting to see who is especially (explicitly or implicitly) inviting their presence, and then "looking" at what a man is physically doing, or at least observing what he is *imagining* sufficiently strongly that the *effects* of his thoughts are found "impressed" in his neurochemistry. The demons do have access to those phenomena, and Greer repeatedly tells his pupils to carefully visualize their physical location. The documentary continues, explaining how this contact is undertaken in a group:

> Sense where [the ETs] are, then invite them... you're actually making contact in consciousness and thought before they arrive in their craft or transdimensionally around you. And when the ETs approach us most profoundly, is when everyone in the team is in a coherent state of consciousness and are clearly doing this with the intent.

Like what Greer describes here, Pagan séances will indeed often produce a diabolical encounter (masquerading as a conversation with the deceased). The success is in part due to this "coherence" of the group, all unified in their intent to contact the deceased, or "spirit guides," or what have you. CE5 is also not unlikely to result in demons manifesting as beings of *light*, as they are prone to do.

The demons, of course, are not stupid. They do not come announcing their true nature, as that would risk spoiling their prospects of continuing the relationship. Instead, *they play along*. They present themselves precisely how these foolish people *want* to see them: including as "UFOs" and "ETs." As the Catholic spiritual warfare researcher, Charles Fraune, observed after reviewing a multitude of occult practices and practitioners:

> ... [D]emons will play along with the big game of the occult that [occultists themselves] have invented. When witches, magicians, and Satanists cast spells and seek to acquire higher and higher magical abilities and ranks, demons will assume their part in the game. They will pretend to ... obey commands, and give every impression that what the occultist believes is true.[198]

One should not expect the demons to act any differently when an extraterrestrial is the most convenient costume for them to wear. If, as Scripture teaches, "**Satan disguises himself as an angel of light**" (2 Corinthians 11:14), then—as we will see in later chapters—it would certainly be easy for him to assume the disguise of an alien or a UFO.

Clearly aware that some of his Christian followers will initially recognize the phenomenon for what it is, Greer seeks to rebut this recognition before it is acted upon. He insists, "**THEY'RE NOT [ANGELS] ... they're ETs with star systems and planets and bodies... if [people] are experiencing fear, or if they feel apprehensive about [making ET contact], make sure you do it in a group.**" Dr. Greer knows there will be a natural instinctual response of fear upon the entry of the demonic—even when disguised. Because of this, he recommends the séance approach—ensuring that contact with the "ETs" is done in a group. This will facilitate the arrival of demons not only due to the greater magnitude of the pull, but also because the mere presence of peer pressure will prevent otherwise prudent souls from backing out.*

One Catholic author with decades of professional experience in the occult recently studied Greer's program extensively. Here is what she concluded:

> ...[It] is exactly like the "bridge" that many Instrumental Transcommunication practitioners have been trying to build between the living and the dead by holding worldwide, synchronized spirit communication sessions on a regular schedule ... the result is a huge and increasing international séance going on right now, just as UFO reports are going through the roof, occult activity (including witchcraft, wicca and magick) is now mainstream, and, exorcists tell us, demonic possessions are off the charts...Unsurprisingly as well, Greer ends these sessions by speaking the Hindu "Namaste"—literally telling the beings, "I bow to you." [199]

---

* Dr. Greer did not share all the details of his protocols within the film. Many are secret; only revealed to his students who pay a massive price and, according to Fr. Spyridon Baily, "are required to sign a non-disclosure agreement." (The UFO Deception, Ch. 12) But some testimonies of those who have participated can still be found, including within a popular book co-authored by an American born university professor now working in Japan. Entitled *A CE-5 Handbook* (2018), its authors first share why they regard their mission as vital, claiming, "**CE-5 contact protocols were co-created by Dr. Steven Greer and several extraterrestrials in 1973. The beings shared with Greer the importance of teaching this protocol to humans**..." (Hatch, Cielia; Koprowski, Mark. *A CE-5 Handbook: An Easy-To-Use Guide to Help You Contact Extraterrestrial Life* (p. 8). Calgary CE-5. Kindle Edition.) Within the book, aspiring contactees are instructed as follows. First, "*connect to one mind consciousness.*" Then, "**detach yourself from your individual consciousness... [and] Name yourself... expand the boundary of who you are so far and wide that your whole body encapsulates the entire universe. You are the universe... Slip into the void**..." In other words, embrace Paganism and Pantheism, while asserting your presence—with your name—to the demons. Then, in order to "*raise your vibration... your energetic frequency,*" you are told to "*...let loose a little. Get sleepy. Enter into a theta brain-wave state.*" That is to say, let go of inhibitions. The less active one's intellect and will, the better. Inhibitions, well, *inhibit* the entry of demons (which is why demons love drug use). Next, "Group Cohesion" is emphasized, as in all séances. Everyone must be of the same mind. Aspiring contactees are advised to "*hold hands to connect your energy together,*" just as in divination.

By popularizing this veritable global conjuring, Steven Greer may well be the single most successful individual on the face of the planet right now by one sorry benchmark. For if there were a Guiness World Record bestowed upon the man who introduced the most unsuspecting souls to direct dialogue with demons, then this prize would likely adorn Greer's mantlepiece.

Whitley Strieber & Jeffrey Kripal

Whitley Strieber was born in Texas in 1945. Educated in Catholic schools, he soon became a bestselling author whose popular '70s and '80s horror novels (about—what else?—elusive werewolves and bisexual vampires) were made into feature films (another of his works was turned into the climate alarmist film, *The Day After Tomorrow*). Strieber's own life, however, quickly became its own horror story. On the second day of Christmas in the year 1985, he became an "alien abductee." These dark and terrifying otherworldly encounters became a cultural sensation upon his recounting the experiences in his 1987 book, *Communion*. Another two years later, the book became a popular Hollywood movie of the same name, starring Christopher Walken as Strieber.

Despite many proofs that the encounters Strieber experienced were demonic, the film concludes not with the protagonists' triumph over these onslaughts, but rather with the opposite: Strieber at last learns to *caress* the "alien" upon asking it what he should call his book (*Communion*). Thus the moral of the story, in all its immorality, is a most germane commentary on the Great Deception now at work in society: literally embracing the demons to enter into "Communion" with them. "**The sacrifices of the pagans are offered to demons, not to God: and I would not that ye should have communion with demons**." (1 Corinthians 10:20)

Although we are discussing a man called Whitley Strieber, this is not about him. It is about the world. Strieber is to 40 years of "alien abductions" what Bill Gates is to the overlapping 40 years of PC operating systems—his work defines the era. And the only difference one can detect between undisguised demons and Strieber's "visitors" (as he calls his abductors) is the name. If a rose by any other name would smell as sweet, then a demon by any other name would inflict as much ruin—or rather, still more. Comparing the open machinations of demons to their disguised tactics, St. Augustine declares that the latter causes "**harm all the greater for the deception**."[200]

Diabolical ruin is seen wherever one looks in Strieber's testimonies about ETs, but our purposes will be served by focusing on his recent efforts with renowned Rice University religion professor Dr. Jeffrey Kripal, another prominent voice in Ufology. Kripal is much more than just another commentator on Strieber—the two enjoy a disturbing bond. Kripal believes he is in occult contact with Strieber's deceased wife. Four years after Anne Strieber died, Dr. Kripal wrote:

> I have interacted with Whitley for about a decade now, often on a daily basis. I believe that I have also interacted with his departed wife, Anne, through a most remarkable drawing channeled by a Canadian medium … I have slept in the same room with Whitley at a symposium (a risky enterprise, I assure you) and experienced my own psyche "split" in two in his sleeping presence.[201]

Kripal concludes this bizarre passage by insisting, "**I do not believe any of the traditional religious ... mythologies**," because religion will only "**take us further and further away... [from] the new world"** that abduction experiences announce. Kripal's parting words lament that he has colleagues who still live in that "old world" of religion, declaring: "**I choose to live toward and for the new one. Which world do you choose?"** Behold the Anti-Fiat, seen everywhere in ET promotion.

Three years before publishing these remarks, Strieber and Kripal together published a book entitled, *The Super Natural: A New Vision of the Unexplained.* The space they placed between the words "Super" and "Natural" in the title was included deliberately. With it, they declare that what used to be understood as simply *supernatural* (or preternatural) is actually just *natural*, only in an anomalous ("super") way; thus atheism triumphs even while embracing Pagan occultism.

Whoever is unconvinced that the diabolical is behind all this need not look far for proof. As with demonic encounters throughout history, so with contemporary "ET contact," including Strieber's, the most grotesque sexual perversion—the details of which I will refrain from quoting here—is pervasive. Throughout *The Super Natural's* ninth chapter, Kripal scrutinizes Strieber's encounters with "extraterrestrials," determining that "**If [Strieber's] *Communion* is not a piece of modern erotic mystical literature, then I do not know what it is.**" He justifies this by noting that Strieber experienced in these "ETs" a "**divine presence that was feminine, that broke [him]... an erotic presence into which he disappeared in waves of terror and pleasure so overwhelming that they literally washed away his personality."** Kripal writes that, in light of Strieber's experiences, "**sex is a 'demon'.**" He observes:

> ... [As Ufologist] John Keel notes in his famous book on the Mothman, *"[The ET] displays an almost pornographic preoccupation with our mating practices"* … many men and women who have encountered these things often find their marriages on the rocks, *"after they begin their liaisons with the space people."*[202]

These degenerate accounts make no sense in the context of space aliens arriving in craft from other planets, galaxies, or so-called "dimensions." They do, however, make perfect sense in light of the traditional Christian understanding of the behavior of a succubus or incubus demon. These demons have always mocked and degraded human sexuality in ways identical to what Strieber and Kripal describe. Prominent exorcist Fr. Chad Ripperger explains that such demons are especially difficult to cast out.*

Kripal continues his exposition of the lesson of Strieber, bemoaning "**the Catholic tradition into which both Whitley and I were born**," which fails to welcome the deep mysticism provided by the aliens. These ETs, Kripal teaches, reveal that "**the demon is God reversed**," (*Demon est Deus Inversus*—a popular occultist slogan) and that appreciating this revelation is "**incredibly important for any remotely adequate understanding of [ET] encounter experiences**." Lest anyone suppose Kripal is cautioning *against* the demonic—by rightly noting it apes the things of God—the professor states the opposite. He *encourages* communion with both God and demons; as he says, "**we no longer need to deny one for the other**." God says otherwise:

> Or what fellowship has light with darkness? What accord has Christ with Belial? ... What agreement has the temple of God with idols? (2 Corinthians 6:14-16)

One who follows Kripal and Strieber supposes he no longer need bother with even the first of the Baptismal Vows, wherein any would-be Christian is first asked—and only permitted to enter the Church if he responds in the affirmative—"**Do you renounce Satan? And all his works? And all his empty show?**" Nevertheless, it is precisely Kripal's (and, in general, the contemporary world's) perverted views on sex that provide an opening the demons use to associate with humans under the guise of "ET Contact."†

Kripal moves on to confess the source of his expertise on Strieber's experiences: his own firsthand knowledge of Pagan eroticism. He writes that he had "seen them before"—that is, the *eyes* of the ET Strieber describes. He states: "**I had seen them many times, countless times... remarkably similar eyes can be seen on thousands of Hindu goddesses throughout India**." Kripal goes on to insist the antics of these "goddesses" are "more or less exactly like" Whitley's "aliens."

---

* He writes: "One kind of spirit that exorcists will commonly come across is called an incubus and succubus. ... once a person allows the demon to act upon him in a sexual manner, unless God intervenes, the person has allowed the demon into that aspect of his or her life, and he or she is not going to stop [the demon]. In effect, the person enters into a stage where he is being raped or molested by the demon, sexually. Cases involving incubi and succubi have been known to be difficult to liberate [i.e., exorcise]. (Fr. Chad Ripperger. *Dominion*. 2022. Page 336)

† The demons know well that the contemporary world is so saturated with sexual degeneracy, which has now reached the very depths through transgenderism, that this portal may be the easiest point of entry for the formal phase of the ET Deception. We have seen that Pope Benedict XVI said, "**Gender ideology ... will be the ultimate rebellion against God the Creator.**" Ultimately, the ET Deception and the transgender deception are both rebellions against God's first revelation about human nature, found in Genesis 1:27, wherein human beings are only said to be male and female and only they are said to be made to bear the Divine Image. Little wonder that this verse is the most maligned one by ET promoters, and that the largest UFO religion today—Raëlianism—traffics especially in transgenderism.

Strieber himself also unwittingly alludes to the apocalyptic nature of the deception under which he labors and which he now promotes. In a 2023 book, *Them*, he issued his own prophecy:

> Our visitors have been ghosts in our world, but now that time seems to be growing short, they may well be bringing themselves more into focus. ... now, as another extinction event shadows the future, the long, ungainly poem of life on Earth comes to the singing of what could be the final stanzas of this latest verse. Many of us are looking up into the stars and out into the shadows that are gathering around us, seeking help from anyone who might linger there. The worse things get, the more will join them ... we have a charge before us: we need them and therefore need to do all we can to make the relationship work.[203]

Strieber later reiterated, "**Our visitors said to me years ago that they would emerge as we faced the peril our planet is in. This is exactly what is happening now**."[204] Although Strieber chose the wrong side, he correctly acknowledges the existence of the present and coming distress. "**Woe to you, O earth and sea, for the devil has come down to you in great wrath, because he knows that his time is short!**" (Revelation 12:12) Because of this shortness of time that Strieber observes, we should anticipate the continued explosion of the UFO and ET Deception.

Indeed, when the world appears to be on the brink—perhaps due to looming (or actual) nuclear war, or a great pandemic or famine, or some great technological breakdown, or the so-called "climate crises"—that moment will be the opportune one for the demons to announce themselves more widely than ever before under the guise of "enlightened extraterrestrials," presenting to mankind a "**deception offering men an apparent solution to their problems at the price of apostasy from the truth**." (*Catechism of the Catholic Church*, §675)

Confirming this assessment, Strieber described his 2019 book, *A New World*, as a "**conjuring [of the] visitors**." He concludes it with an exhortation to give that conjuring our free assent—which is so necessary for the fruition of the ET's plans—by issuing the following plea:

> Will the visitors come closer? ... In other words, will the conjuring that is this book work? ... The visitors must open the doors of their school wide, to us all. We have a planet to lose and our lives along with it, or we have a journey to take. Which shall it be? The decision belongs to all of us and each of us, and to them. Shall we join in what is essentially a new world and a new way of life...? We must decide, and now, and so must they. Surely it is time.[205]

ET promoters always conclude their exhortations with a presentation of "The Choice." (The *choice* Kripal himself presented, quoted earlier, was in the preface to this book.) They are begging us to make a deliberate act of will to open ourselves to, listen to, and embrace, the "extraterrestrials." Nor does the incoherence of this plea bother the ostensibly intelligent scholars who promote it. If these "visitors" in their "UFOs" were really ultra-evolved "aliens," or "interdimensionals," or what have you, then our free assent would be irrelevant. They would simply have their way with us, for better or for worse.

Yet, that free assent *is* always the fundamental first step in enslavement to the Devil. Without it, he is powerless. Hence, Ufology's constant insistence upon it. Without that act of will—that choice, that *Anti-Fiat*—the demonic plan is in peril, just as the Fall of Man itself would not have transpired if Eve rejected communion with the Serpent.

As the greatest prayer man can utter is saying to God, "Thy Will be done," the demons' ultimate desire is to hear that same prayer offered to themselves, and this Anti-Fiat will precipitate the return of Eden's demise. Thus, The First and Last Deception.

Jacques Vallée

A household name among ET believers—and an expert in all things Ufology—the internet pioneer, successful investor, bestselling author, and computer scientist, Dr. Jacques Vallée, enjoys a stature surpassing even Whitley Strieber's. Already in the 1970s, Vallée's fame for Ufology was so great that Hollywood's preeminent alien film, Steven Spielberg's *Close Encounters of the Third Kind,* had a main character modeled after him. In the half-century since, Vallée has remained prolific in advancing the movement (including a 2020 appearance on the world's most popular podcast, and a 2024 History Channel episode dedicated entirely to him).

In a matter as bursting with spiritual undertones (and overtones) as Ufology, knowing the religion of its promoters is essential to understanding their message—and to exposing their goals in promoting it. Like Strieber, Vallée is a close friend of Dr. Jeff Kripal, who in turn writes that Vallée "**sees [religion] primarily as a kind of social control system, certainly not as a deposit of eternal truths**."[206] Dr. Vallée does have his "religion," however. He is a Rosicrucian, and even his promoters concede the importance of this pseudo-religion in his Ufology. D.W. Pasulka wrote that both Allen Hynek—another Father of Ufology—and Vallée were Rosicrucians, and that their "**ufology, furthermore, cannot be separated from their esoteric practices**…"[207] (Vallée himself approved this assertion of Pasulka's, thereby reminding us to examine Rosicrucian principles to grasp his own aspirations in promoting UFOs.)[208]

Rosicrucianism is an occult, cabalistic, theosophic movement that began in the early 17th century, amid the birth of the ET Deception—just several years after Giordano Bruno's execution for promoting it. (Rosicrucianism reveres Bruno.) Esoteric and pantheistic, Rosicrucianism—which is now a branch of Freemasonry—seeks after the so-called "secrets of nature," and it is avowedly messianic. Rosicrucianism's U.S. headquarters is in Vallée's own Silicon Valley, and its 2001 manifesto proclaims:

> The survival of the great religions depends more than ever upon their ability to discard the most dogmatic moral and doctrinal beliefs and positions they have adopted through the centuries… Nonetheless, we presume that their disappearance is inevitable and that, under the influence of a worldwide expansion of consciousness, they will give birth to a universal religion…

There could not be a more fitting religion for a pioneer of Ufology, as these are also the aims of ET promotion. Despite Vallée's cutting critiques of mainline Ufology, he only replaces its space-alien narrative with a still darker one.* Although Vallée often couches this deception in nuance—a candied shell upon a poison pill—no discerning reader will fail to miss the evil of Vallée's thesis. Kripal, endorsing Vallée, describes it as follows:

> [UFOs] are designed or even staged to confuse us, to baffle us, to shock us into another level of consciousness and culture, rather like the mystical paradoxes of Zen Buddhism and Jewish Kabbalah, Vallée suggests.[209]

Commenting on Strieber's *Communion* and Spielberg's *Close Encounters*, Vallée himself declares that "**our religions seem obsolete,**" and excitedly proclaims that encounters with extraterrestrials will be found "**achieving within a single generation behavioral changes that might otherwise take hundreds of years to complete. ... [W]e are in fact dealing with one of history's major transitions**."[210]

Vallée developed the so-called "Interdimensional Hypothesis," which regards UFOs and "ETs" as beings not traversing physical space to arrive on earth from other planets, but rather as "control mechanisms" appearing here from other "dimensions" altogether, with the aim of bringing human consciousness to its next level.†

> "[For Jacques Vallée,] UFOs are some kind of unspecified power or force that permeates the whole universe and directs, at a deep unconscious

---

* Illustrating the deception is the cover of Vallée's "seminal master-work," *Passport to Magonia.* Emblazoned with artwork depicting an alien being holding several masks, the cover presents one mask portraying a demon, another mask of *a different type of alien*, and the third mask of a female face which could represent any number of mythical beings (or perhaps even a real being, the Virgin Mary). Vallée's thesis is thereby made clear: the demons themselves were simply one disguise the "Interdimensionals" once wore, not vice versa. And for him, this diagnosis applies to everything supernatural—it was merely our unenlightened ancestors who believed in the miracles of God and the activities of angels and demons. Vallée, however, has come to the rescue, reminding us that all such talk is identical to talk of fairies, UFOs, and everything else of so-called "high strangeness"—these are all merely "interdimensional" visitors who are truly physical beings, but are using technology so advanced we can scarcely comprehend it. Traditional religious thought, in sum, is to be replaced by this new, superior understanding of reality's highest level and the universe's bedrock.

† Since various aspects of Vallée's critiques against mainline Ufology lend credence to the correct understanding of the *diabolical* nature of ET phenomenon, some Christian authors who have warned against the UFO deception have subscribed to Vallée's "Interdimensional Hypothesis." They reason, "*I suppose demons are themselves from another dimension, therefore I can make this hypothesis work with a Christian worldview.*" Such Christians are only falling for yet another trap. Demons are spiritual beings, not so-called "interdimensional" ones. The very suggestion of "another dimension" or "another universe," much less another class of incarnate intelligent creatures inhabiting them, is itself a sci-fi delusion and a form of neo-Paganism.

level, human destiny. They are "the Force" of Star Wars. This new religion would represent a radical change from the Judeo-Christian view of a personal God who intervenes in history." —Fr. John Saliba[211]

By shifting the language from "demonic" to "interdimensional," Vallée clamors to preserve empirical science as the mediator of the conversation. As the saying goes, "*he who controls the language controls the debate*." This shift ejects the authority of Divine Revelation from the discussion, and the demons attain still more destructive leeway than they enjoyed while being regarded as space aliens. These evil spirits can remain disguised—happily being called "interdimensional"—so long as they are not recognized as the very *demons* who are only accurately described by Christian Revelation.[212]

Whitley Strieber was deeply impacted by Vallée's thesis. Selected by Vallée to write a foreword for his book, Strieber glowingly wrote that, "**[Vallée] demonstrates... that the miracle of Fatima was... a classic UFO encounter**." Strieber continues, declaring—on Vallée's authority—that "**[UFOs] could very well be the single most important influence on our history**."[213] Contemporary Ufology is indeed determined to define Marian apparitions as simply misunderstood UFO phenomena, and this diabolical move derives primarily from Vallée. He claims that, **"[Fatima's early] sightings cast serious doubts on the interpretation of the 'miracle' given by the Catholic Church**,"[214] before declaring that the Blessed Virgin has "intriguing similarities" with "the Egyptian goddess Isis."[215] He goes on to state that "**Several [apparitions], which bear the official stamp of the Catholic Church (such as those in Fatima and Guadalupe), are in fact ... nothing more than UFO phenomena**..."[216] Yet this is just a sampling of Vallée's crusade against the Blessed Virgin.*

Although Dr. Vallée understands nothing about supernatural discernment, his judgment is treated as Gospel by millions. Fixated upon his own puerile observation that the Virgin Mary is often luminous, or in the air during apparitions, along with his jotting down a few other superficial resemblances that chance alone would ensure exist—Vallée regards it as an affirmative "fact" that history's greatest apparitions are just UFOs.

If that is not a Satanic assertion, nothing is.

"The unspiritual man does not receive the gifts of the Spirit of God, for

---

* Vallée also wrote a glowing foreword to another Ufologist's book—*Heavenly Lights*—which argues that Fatima was a UFO. Therein, Vallée denounced those who believe Our Lady of Fatima was, indeed, the Blessed Virgin Mary, saying that this view **"reframes [the events of Fatima] within the structure of dogma and belief. This cosmology assumes that the witnesses were privileged to take part in a miracle... This is the religious path,**" (Foreword. *Heavenly Lights: The Apparitions of Fatima and the UFO Phenomenon*. Anomalist Books Edition, 2007) and he proceeds to mock it as "**cowering in obedience**" before God. The cover of this Vallée -vaunted book, moreover, claims that it successfully "**establishes that the famed Fatima Incident of 1917 did not involve 'Marian Apparitions'—as is conventionally believed—but a series of close encounters with alien beings**." (Rear Cover. *Heavenly Lights: The Apparitions of Fatima and the UFO Phenomenon*. Anomalist Books Edition) As we can see, it is futile to attempt to massage Vallée's thesis into something that is not utterly blasphemous.

they are folly to him, and he is not able to understand them because they are spiritually discerned." (1 Corinthians 2:14)

Concluding his remarks on Vallée in his book quoted earlier, *Authors of the Impossible*, Dr. Kripal praises those who, following Vallée's exhortations, "**somehow manage to shake loose from both their religious and rational chains,**" whereafter "**everything changes, instantly and forever. They have been granted a passport to Magonia. They leave the cave**."[217] That is, thanks to UFOs, they leave the "cave" of both Faith and Reason, and leap into the welcoming arms of demons, to the thunderous applause of Vallée and the many Christian ET promoters who extoll him.

But before exposing the destruction Vallée's teachings are now inflicting upon the Faith through Catholic scholars who promote them, let us consider another scientist (and his protégé) who, though from a different angle than Vallée, also heralded the contemporary world's messianic fixation upon extraterrestrials. Born the same decade as Vallée, Carl Sagan also resembled Vallée by providing cutting critiques of mainline Ufology—only to replace it with his own still more degraded thesis.

Carl Sagan & Neil deGrasse Tyson

Carl Sagan (1934—1996) was an Ivy League astronomy professor, an astrobiologist and astrophysicist, and, more significantly, the chief prophet of 20th-century sci-fi pseudo-religion.

Although famously and astutely critical of reports of UFO sightings, Dr. Sagan was nevertheless among the most prominent ET promoters of the contemporary era. He was religiously dedicated to SETI—the "Search for Extraterrestrial Intelligence" (the effort to contact aliens remotely, using radio waves), and his fame in this arena is no doubt due to his reputation as the late 20th century's most revered pop scientist.

Sagan's popularity especially derived from the 1980 television series, *Cosmos*, which he starred in, wrote, and narrated, and which—over four decades later—remains the most viewed series in the history of PBS. One biography of Sagan reported the show had a half billion global viewers.[218] Today, a significant portion of the world's population has been formed by its teachings, the cultural impact of which is difficult to overestimate. According to *The New York Times*, Cosmos was "**arguably the most successful popularization of science since Albert Einstein ... In 2012 the Library of Congress designated the book version of the show as one of 88 books that shaped America**."[219]

The series began with one blunt assertion from the lips of Sagan that still haunts the memories of millions. Strolling towards imposing cliffs overlooking the sea, Sagan prayed his own false doxology to himself: "**The cosmos is all that is, or ever was, or ever will be**." The thirteen episodes that followed expounded upon this lie, presenting a worldview—a *cosmos*-view—with no place for God.

Sagan's popularity arose not from scientific insight (he had little to contribute there), but from the pseudo-mysticism of the replacement religion that he preached. His messages—including his famous TV show—were peppered with pithy remarks seeking to prick the soul in ways only religion authentically can. Atheists, agnostics, and skeptics of all stripes found in him a corrupt wellspring of that transcendence which they longed for but lacked. One prominent biographer of Dr. Sagan observed:

> In the space age, when "mankind doesn't need gods" (to quote Captain Kirk), science-fiction writers, speculative scientists, and UFO buffs have in fact replaced the old cosmic power structures with scientific ones... Sagan, too, believed in superior beings in space, creatures so intelligent, so powerful as to resemble gods. ... [who] will even share the secrets of the cosmos with us, if we'll simply tune in to their radio transmissions. [220]

Sagan's own advocates concede that his ET beliefs can only be accurately regarded as downright *messianic*. Despite all of Sagan's indictments of religion, man remains—above all else—a religious creature by nature. Dr. Sagan had his religion. The same biography notes:

> Sagan's insistence on the inevitability of cosmic intelligence is important partly because it undergirded his quasi-religious belief in alien super-beings. He believed that these creatures, perhaps dwelling in other galaxies, were benevolent and might help us to solve our terrestrial problems.[221]

Dr. Sagan's overarching life mission was his plea to the world to abandon Jesus Christ for the sake of placing one's faith and hope in alien contact. His aim was to "**replace [conventional religion] with a scientifically based belief system**,"[222] according to *Britannica*, and no observer will fail to note the prophetic relevance of this unrivaled cultural giant seeking above all to *replace* Christianity with belief in aliens.*

***

* The conclusion of *Cosmos* is also worth noting. Its thirteenth and final episode is entitled "*Who Speaks for Earth?*" A veritable secularist manifesto, it is presented as a meditation on what the extraterrestrials will think of us upon "first contact." The massively popular show opens with a solemn pronouncement of the words, "**I call heaven and earth to witness against you this day; that I have set before thee life and death; the blessing and the curse. Therefore choose life, that thou mayest live; thou, and thy seed**." While any Christian knows that the Holy Bible's Deuteronomy Chapter 30 is there being quoted, the show provides no attribution. Thus the anti-Fiat of ET Deception is yet again presented; this time, by employing the very Sacred Text that this show is dedicated to rejecting. Mirroring Sagan's introducing of the "Cosmos" series with his own false doxology spoken in idolatrous adoration of the material universe, he concludes it with his own false Gloria to the praise of the wonders of science. In its final stanza, he claims, "*these are some of the things that hydrogen atoms do after 15 billion years of cosmic evolution,*" degrading humanity as the product of only chance, time, and particles. In a video update about the series recorded ten years after its release and attached to all later showings, Sagan revisits its opening and reiterates the anti-Fiat, saying, "**It is AS IF there were a God who said to us, 'I set before you two ways**.'" The choice? Embrace science and its cosmic orientation. For Sagan and his half-billion viewers, this ET-expectation provides all the mysticism and excitement of religion, but none of its "moral or dogmatic baggage."

If one man holds the distinction of being Carl Sagan's successor, it would certainly be the celebrity astrophysicist, Neil deGrasse Tyson.* Dr. Tyson now enjoys a fame like his mentor's, easily blowing away all other scientists' social media presence. All of his accounts have several million followers—his Twitter account alone numbers over 14 million.

Also like Sagan, ET belief was found in the earliest years of Tyson's professional career. Almost thirty years ago, Dr. Tyson wrote that rejecting belief in aliens is "inexcusably bigheaded," before condemning the generations who did so as being "led astray by anthropocentric assumptions," and declaring that we should "**be guided by the notion that we are not special.**"[223] More significant than this article, however, is Tyson's starring in two reboots of Sagan's "Cosmos" series—one airing in 2016 and another in 2020. His shockingly famous pop-sci preaching is noteworthy not by bringing the ET Deception itself to new levels. Like Sagan, he is rightly critical of UFO claims. Its significance derives from bringing the deception full circle in its historical context and idolizing the ET Deception's originators we considered in Part Two. Thus, those "too mainstream" for Ufology nevertheless find their appetites whetted for another type of ET-contact-salvation, though equally disastrous. Indeed, Sagan and Tyson are no Strieber or Kripal. The Devil is not so shortsighted as to hinge his long-prepared Great Deception on one approach alone.

The new "Cosmos" series begins by playing the old recording of Sagan repeating the lie he announced in the first *Cosmos'* opening moments—"**the cosmos is all there ever will be**"—whereafter Tyson appears, standing on the same seashore beside looming cliffs. Then, to an image of the Milky Way Galaxy, he wonders, "**How many worlds? How many ways of being alive**" are found therein? The view continues zooming out, until the "billions of galaxies" of the Universe are shown in one frame. But then, Tyson says, "many of us" believe that the whole Universe itself *"is but one tiny bubble in an infinite ocean of other universes—A MULTIVERSE! Universe upon Universe."* Then at last, Tyson concludes the same false doxology the show began with, praying to himself, *"**worlds without end!**"*

> "Claiming to be wise, they became fools ... they exchanged the truth about God for a lie and worshiped and served the creature rather than the Creator." (Romans 1:22,25)

With the addition of a single serpentine "s" to one of Christianity's most central prayers, the Glory Be, Tyson knows exactly what he is doing; slyly undermining the Christian worldview—and cosmosview. For he then sarcastically remarks that, "**back in 1599, everyone knew we were the center**

* As a teenager, Tyson received a personal invitation from Sagan to study at Cornell University. A 2014 article in *The New York Times* recounted their interaction, noting that Tyson shares that, "**I learned from Carl the kind of person I wanted to be.**" The article itself remarks that "**[Tyson] could hardly have guessed that he would grow up not only to be like Dr. Sagan but in some sense to be him.**"

of ... a universe made for us." Alas, Tyson's hero arises to denounce this bedrock Christian principle. "**There was only one man on the whole planet who envisioned an infinitely grander cosmos**," Tyson tantalizingly declares. "**His name was Giordano Bruno**." The screen displays a demonic-looking Pope glaring menacingly at Bruno, who, Tyson claims, "**became an evangelist, spreading the Gospel of infinity throughout Europe**." The show then portrays Bruno demanding that a group of scholars at Oxford "**reject Faith and authority**," because, as he tells them, "**Your God is too small!**"

Thus, Tyson celebrates Bruno's introduction of the "limiting God" accusation still hurled against those who disbelieve in ETs. The show later portrays Bruno, in his prison cell about to be executed, finding consolation by mystically soaring through the galaxies, before turning away from the crucifix in anger as the torches light the fire under the stake to which he is tied. Tyson concludes by referring to this not as Bruno's execution, but reverently announces it as his "**martyrdom**." Thus, with seductive dramatization, Tyson presents the first distinguished promoter of the ET Deception in Christendom as an object of contemporary adulation and a call to action. Untold millions have followed Dr. Tyson's call.

Four years after that series was launched, the world watched the first episode of yet another reboot of "Cosmos," wherein Tyson returns to direct the same religious (but religion-hating) reverence he gave Giordano Bruno towards another. Tyson asks, "**Why was there no hint of other worlds and their living things in the sacred books?**" He gives his answer by way of presenting his second great hero, Baruch Spinoza—inventor of the "Principle of Plenitude," which we also discussed earlier. Tyson says:

> Whatever disquiet this contradiction [between Scripture and ET belief] may have stirred in the hearts and minds of the leaders of the Enlightenment, there was only one man who dared to address it head-on… [a] wizard of light, Baruch Spinoza... he began to speak publicly of a new vision of God. Spinoza's God was the physical laws of the universe; his sacred text, the book of nature.

Tyson's error here is not inaccuracy, but hatred of the truth. Unlike the nauseatingly nuanced, he is at least frank enough to observe the contradiction between ET belief and Divine Revelation—but instead of choosing Divine Revelation, he chooses aliens, and he begs his viewers to do likewise. As everywhere, this decision entails redefining God entirely.

The episode concludes with Tyson using his best movie trailer voice to the tune of inspiring orchestral music and a backdrop of flashy CGI visuals depicting what he foresees as imminent human technology. We are shown a massive spacecraft with "SPINOZA" emblazoned upon its hull majestically embarking upon its journey to *Proxima Centauri* to meet the extraterrestrials. Thus Tyson's thoroughly Godless and supposedly "scientific" message is imbued with the same messianic longing seen everywhere in the

ET Deception. It merely uses a different mode of appeal than those we have considered earlier and will consider next.

Another popular PhD scientist and collaborator with Sagan approached the ET Deception from an entirely different angle, and, in fact, with more intellectual rigor, although his discipline was not a "hard" science. While Sagan led his audience to the religion of scientism, the next ET promoter we will examine used his scholarly efforts to lead them to Paganism. The demons remain satisfied with either.

John Mack

The Pulitzer Prize-winning Harvard psychiatrist, Dr. John Mack, was already quite famous when—in the 1990s—he found his life's real passion: aliens. Although he died in 2004, he remains among the biggest names in ET belief. Mack was not himself an "ET experiencer," but he is the foremost scholar of abduction experiences, and his fame rivals that of the most prominent "abductee," Whitley Strieber. Mack is to abductions what Jacques Vallée is to UFOs.

In his bestselling 1994 book, *Abduction: Human Encounters with Aliens,* Mack belittles medieval Catholicism's battles against Paganism, and then declares: "**There can be little place, especially within the Judeo-Christian tradition, for [aliens]**," noting that such beings, when they appear, may be recognized as demons by those faithful who hold to "**the polarizing perception of Christian dualism**." No one—including scholars like Mack—can refute that these "aliens" are demons, therefore they dismiss the view with academically unappealing labels (the bogeyman of "dualism" is among their favorites). Recognizing that ETs are simply demons would ruin the whole glorious plan—for the aliens, Mack assures us in the next paragraph, are benevolent. They are "**concerned primarily with two related projects: changing human consciousness to prevent the destruction of the earth's life, and a joining of two species for the creation of a new evolutionary form**."[224] Nevertheless, not all religions are antagonized by Mack. While Christianity (rightly understood) opposes the evils of ET promotion, the Harvard psychiatrist hastens to assure his readers that Buddhism is just fine. He writes:

> Eastern religious traditions, such as Tibetan Buddhism, which have always recognized a vast range of spirit entities in the cosmos, seem to have less difficulty accepting the actuality of the UFO abduction phenomenon than do the more dualistic monotheisms, which offer powerful resistance to acceptance.[225]

Professor Mack's last book, *Passport to the Cosmos,* published in 1999, lays out his thesis still more clearly.

Early in the book, Mack quotes the theologian Thomas Berry: "*For peoples generally, their story of the universe and the human role in the universe is their primary source of intelligibility and value."* The observation is correct. It summarizes the supreme danger of the ET deception, which, with a single premise, undermines all Biblical and Catholic teaching on man's "role in the Universe," creating a gaping chasm within his existential grasp of reality, thus rendering any coherent and solid Faith impossible. Mack, however, celebrates this chasm instead of perceiving it as a cautionary tale, commenting:

> The UFO abduction phenomenon seems especially well suited to challenge the anthropocentric humanistic worldview. It suggests that we are but one of the higher forms of intelligence that have evolved in the cosmos, and not the brightest one at that. ... Abduction experiencers might be thought of as ontological pioneers who, like [in the Truman Show], are helping us to break out of the bubble of a constricting worldview.[226]

For this author, the aliens come not merely in peace but in magnanimity: to deliver us from Plato's Cave, wherein Christian revelation is the primary source of the lies we are told. Traditional so-called "dualistic" Monotheism casts the proverbial shadows on the wall of the cave, which entrance the shackled prisoners, and from which we need (extraterrestrial) deliverance. Thus is the story of "the phenomenon" of abductions, UFO sightings, and the like, according to its exponents like Mack.

On the other hand, Mack was no Robert Dione—a man we will consider shortly, who promoted aliens by sneeringly denouncing the supernatural as medieval nonsense. Like many other ET promoters, Mack presents his message as an *indictment* of today's secular materialistic rationalism, and he appears to endorse some essential mysticism of Christianity. (Key word: *appears.*) He condemns the "*materialist juggernaut,*"[227] and he emphasizes the importance of believing in the realm of the spirit and in some form of higher power, even rightly lamenting "*the loss of our relationship to nature and the Creator.*"[228]

In all this, however, John Mack is just another wolf in sheep's clothing. Though he cannot spare so much as a nod to the billions of Christians who have never lost that relationship with the Creator he rightly extols, Dr. Mack misses no opportunity to congratulate any indigenous or eastern form of idolatry. He is especially fond of Native American Indian Paganism, explaining its view that:

> "People from other planets," the "invisible beings" who were living here before we were, are now trying to help us heal the Earth, for the male energy has gotten out of balance with the female or goddess energy, and we have turned our backs "on the original teachings of the Creator."

There's a religion Mack can get behind.

The primary strategy of Mack and his ilk is to present traditional miracles, mysticism, and the supernatural in general as "anomalous" or "extraordinary"—just as UFOs, abductions, and the like are! Surely the promotion of one must benefit the other, he leads his naïve readers to conclude. Yet in place of the postmodern West's prevailing secularism, Mack presents nothing but a return to the same demons that populated ancient Paganism. One thus leaps out of the frying pan, and into the fire—yet again, as with Vallée, to the thunderous applause of even today's "Christian" ET promoters. Cheerleading our "evolution" in how we view the Creator, Mack insists that Jesus will not save us, but aliens will. He writes:

> Our notions of the Divine, like everything else, seem to grow along with the evolution of our consciousness. ... Nor is it likely that a messiah/savior will lead us into the Divine Light. For that light, we are learning from phenomena like the one in this book ... The god for this time, we seem to be learning from abductees and others, is more of a partner than anything else... The alien abduction phenomenon is largely an opportunity or a gift, a kind of catalyst for the evolution of consciousness...

For Dr. Mack and for Ufology in general, "The Phenomenon" of the "Non-human Intelligences" now interacting with society on an unprecedented level reveals a new "god," and one totally in contrast with the Almighty God known by our fathers in the Faith. This "god" that the aliens reveal does not save us from sin, much less judge our sins. Rather, it provides a "**catalyst for the evolution of consciousness,**" and we need no longer conform our wills to the Divine Will. We need merely regard this new "god" as a *partner*. Only with this alien-induced evolution, Mack insists, will our planet survive.

> "In their case the god of this world has blinded the minds of the unbelievers, to keep them from seeing the light of the gospel of the glory of Christ, who is the likeness of God." (2 Corinthians 4:4)

For reasons that doubtless escaped Mack himself, he concludes his final book with another frighteningly relevant exhortation. For the Anti-Fiat yet again rears its ugly head, and it is given the position of supreme prominence in Mack's work. In this case, it does so by way of a quote from an abductee he studied extensively, "Karin." This "Karin" delivered a speech in New York City in the year 1997, imploring:

> ... We are once again faced with making a choice. Couldn't we this time, finally, make the right one? Couldn't we simply reach out and embrace this experience and all that it brings with it, including the terror as well as the questions, truth and beauty? Because don't you think in doing so, we would at last learn about who and what we are? And isn't it possible that we just might learn what it is exactly that is the nature of the connection to the Creator as well as the universe? Please, can't we this once choose the path that will finally set us free?

For Karin and Mack, humanity has never before made the right choice; has never before learned "the nature of the connection to the Creator." Not with Abraham, not with Moses, not with Jesus, and not with Christianity. But now, with the aliens, we can—for the first time ever—"be set free" if "**we simply reach out and embrace**" the ETs.

Karin is correct about one thing. As in Genesis 3, so today, we are *once again* confronted with an existential choice of whether to embrace or reject the Nonhuman Intelligence that presents to us the Godlike knowledge. What she does not realize is that it is the same choice Eve was confronted with in the Garden. We know how that went.

Wolves Galore

The men we have discussed thus far are joined by an army of similarly minded ET promoters and Ufologists. Limitations of space prevent including a thorough exposé of each here, so before moving on to exposing a whole new outlet for the ET Deception—that is, the proliferation of the Strong Delusion within the Catholic Church—we will conclude with only brief mentions of several more.

Before Jacques Vallée was publishing his research into UFOs, there was Erich von Däniken's work, including his 1968 book, *Chariots of the Gods?*, and that of Robert Dione, especially his tome published the following year, entitled *God Drives a Flying Saucer*. These works continue to inspire contemporary productions, including the History Channel's "Ancient Aliens" series that arose primarily from Däniken's work, and whose outlandish claims are watched and believed by many millions of credulous viewers to this day. The series so far has completed 20 seasons and 260 episodes.

Within Däniken's and Dione's disastrously formative books, we are told that all the miracles of the Bible (not to mention the world's great ancient monuments) are merely the results of alien technology from ETs visiting earth at various points throughout history. For Dione, one such ET managed to achieve immortality, and through artificial insemination, Jesus Christ was brought about. Dione beat Vallée to the punch in his denigrations of Our Lady of Fatima, writing it was an "illusion" from a flying saucer. Dione was perhaps the first to formulate ET promotion's obsession with denouncing "dualism," writing:

> Philosophers, other intellectuals and clergymen have bumbled, and continued to bumble, with the idea that the essence of man is a strange mixture of physical and spiritual being. ... Such a belief [in an immortal, spiritual soul], when analyzed objectively, falls into the same category as belief in miracles: both are remnants, in an otherwise enlightened culture, of a heritage of superstition...[229]

The message delivered by this molder of contemporary Ufology is only distinct from some other ET promoters by being presented slightly more explicitly, and the message is this: Christianity is an outdated superstition, and it has no place in the modern, scientific world; everything man once regarded as miraculous was and is nothing but the intervention of extraterrestrial technology. The title of Dione's 1976 follow-up work says it all: "*Is God Supernatural?: The 4,000 Year-Old Misunderstanding.*" Despite the blasphemy, one cannot accuse Dione of inconsistency—whoever believes in aliens has forfeited his own ability to convincingly argue for the supernatural origin of the miracles recorded in Scripture and Church history. "Surely," an ET believer would respond, *"an alien civilization billions of years more evolved than we are could easily replicate such feats."*

***

Ufologist John Keel was another pioneer of the Interdimensional Hypothesis, and today—fifteen years after his death—his work remains among the most popular writings in Ufology. He coined the term "Men in Black," an idea which eventually generated major Hollywood films, and he wrote *The Mothman Prophecies*, which also made its way to the silver screen replete with A-list actors in starring roles. Like Vallée and Däniken, Keel collects all so-called "paranormal" occurrences throughout history and subsumes them under the banner of the UFO. Keel, however, argued that Dione's thesis went too far by insisting that the ancients saw flying saucers and simply misinterpreted them as angels or miracles. Keel teaches, rather, that "**the Ultraterrestrials deliberately conveyed [a religious] impression [to the ancients].**" They did this, Keel teaches, because **"the phenomenon is constantly reaching down to us, creating frames of reference that we can understand and accept...[but] the first step to understanding UFOs is to discard all frames of reference...**"[230]

Despite the caveat, Keel nevertheless remains insistent that we regard all the supernatural content of Salvation History and Sacred Tradition as comparatively lowly, unscientific, and archaic ways that the so-called "Ultraterrestrials" *adapted themselves* to our fathers in the Faith. But now that "the phenomenon" has adjusted its tactics, we must "discard all frames of reference"—that is, discard the Deposit of Faith—and embrace the UFOs "as they are." Therefore, with only a minor modification to the thesis of his slightly more blatantly anti-Christian forerunners, Keel dismisses Christianity just as wholly as do Däniken, Dione, and Vallée.

In his most famous work, *Operation Trojan Horse*, Keel wrote that the apparitions of Our Lady of La Salette "**conform to the now-familiar tactics of the ultraterrestrials**,"[231] before arguing that the real phenomenon behind these apparitions was the same one behind the occult séances of Hydesville, New York (by the infamous Fox sisters) the following year. "**The incident at La Salette in 1846 was definitely ufological in nature**," he reiterates. And, needless, to say, Keel resolutely affirms that Our Lady of Fatima was a UFO, lamenting that Fatima "**is already clouded with the distortions of belief... entrusted to the religionists.**"[232] Discontent with only blaspheming the Virgin Mary's miracles in Church history, Keel turns his attention to the Bible, writing that in Scriptural times, UFOs "**were given a religious interpretation in those ancient days when all natural phenomena and all catastrophes were blamed on a Superior Being.**"[233] The Three Kings who came to adore Christ? Keel insists they were actually the same as today's "UFO mystery men"[234]—the "men in black."

So much for the miracles of God recorded in the Bible being... *the miracles of God recorded in the Bible*. Despite his apostate message, Keel is revered even by putatively Christian Ufologists to this day. One popular social media Catholic UFO researcher recently declared to his tens of thousands of followers that Keel's book remains "**the greatest written study**" on UFOs—fifty-four years after its publication. Nevertheless, Keel at times seems unsure as to whether he believes demons are just one disguise UFOs have worn in history, or if UFOs are just one disguise the demons use today. Of course, the latter is true, and Keel had his moments of seeming to prefer it. He certainly deserves credit for moments such as these and other similar ones we will quote in the following chapters: "**The demons, devils and false angels were recognized as liars and plunderers by early man. These same impostors now appear as long-haired Venusians.**"[235]

***

Dr. Abraham "Avi" Loeb, whose antics we considered in Part One, is an Israeli-born, world-renowned Harvard professor who is among the nation's top astrophysicists: serving on presidential advisory councils and lauded by TIME magazine as one of the world's top 25 astronomers. But he is, above all, an impressively prolific ET promoter. Loeb's own tenure as a Harvard Professor overlapped with Dr. John Mack's by a decade, and since at least the year 2020, the passion of Loeb's life has been his search for extraterrestrials and his proselytizing of the message he believes they will bring. The following year, he founded the "Galileo Project" to further his conquest, and dozens of other scholars have since joined its ranks. What Loeb's résumé lacks in the number of years spent traveling amid Ufology, it makes up for in density of line items. Loeb has already published multiple bestselling books arguing for his ET-inspired secular-messianism, his Galileo Project becomes more influential by the day, he has been the subject of documentaries, his efforts are backed by massive funding from billionaires, and he is frequently on the news. Suffice it to say, we will continue hearing from him.

A Netflix film director recently asked Loeb what his life's project was, and he responded without hesitation; saying, "**To acquire a God-like knowledge about the Universe. This is my ultimate goal**," before assuring us that this aspiration is quite humble. He explains:

> Humans emerged on Earth in the last few million years but with artificial intelligence, our information processing capability is growing exponentially ... What does God-like knowledge mean? It means understanding how to create superhuman intelligence and ultimately how to create artificially a baby universe with life in it...The inspiration for an intelligent state of mind may come from encountering a more advanced extraterrestrial civilization.[236]

Like his confreres, Dr. Loeb misses no chance to combine his promotion of the ET Deception with the AI Deception while also ensuring both are understood in pseudo-eschatological terms. Elsewhere, he writes:

> Genesis 1:27 posits that God created mankind in its [sic] own image. Currently, humans are creating AI in their image…Eventually, advanced AI scientists might become God. Instead of following the cycle of life in Genesis 3:19: "from dust to dust" we might fulfil the inspiring transformation: "from God to God."[237]

Loeb cannot resist couching his mission in Biblical terminology—though only to invert the Word of God. His malicious intent, however, does not deprive him of all insight. He accurately views the heralding of the Age of the Extraterrestrial as a fundamentally anti-Christian message. Many news outlets have reported on Loeb's teaching that we saw in this book's opening chapters: his claim that the messiah is coming—a "messiah" that will simply be a highly evolved extraterrestrial, arriving on a spaceship to bring about a "cosmic exodus," and using super-advanced AI systems to propel human understanding to "Godlike" levels.

Of course, no message could be more Antichristic. Yet it is not merely despite these claims, but rather because of them, that Loeb is celebrated by ET promoters; including the most mainstream among their ranks. This is evidenced not only by Harvard's open embrace of Loeb's Galileo Project, but also by major organizations like the SOL Foundation retaining Dr. Loeb for their highly selective advisory boards. Now, SOL—founded in 2023—boasts that it "*marshals intellectual insight and policy expertise to meet the scientific and political challenges*" of UFOs, though in practice it is simply a full-throated apologetic, from top scholars (and beyond), that aliens are here and this is "**world-changing in every sense of the term**" (as stated on the homepage of its website).[238] SOL's opening conference, held at the prestigious Stanford University in November 2023, had in-person talks by Christopher Mellon, the Former Deputy Assistant Secretary of Defense for Intelligence; Karl Nell, the former Deputy Chief of Staff in the US Military; Dr. Paul Thigpen, world-renowned Catholic apologist; as well as Dr. Diana Pasulka, Dr. Jeff Kripal, and other eminent scholars. (Thigpen later contributed a 49-page paper to SOL, arguing for ET belief from a "Catholic" perspective—using arguments refuted in Part Two of this book.) Headline-dominating "whistleblower" David Grusch made a digital appearance, wherein he prophesied a "**new era of spiritual awakening**" thanks to the revelation of the "Nonhuman Intelligences."

***

As is evident, we have now reached a whole new level of the ET Deception: its infiltration into Catholicism. That intrusion is the ultimate aim of the Devil, who seeks to enthrone his superlative minion, the Antichrist, in "the temple of God" (2 Thessalonians 2:4) that is, in the Catholic Church. The infiltration begins with the scholars.

D.W. Pasulka

Joining Dr. Loeb is another new entrant to Ufology—and another University professor—who has in a short time taken the movement by storm. Professor Diana Walsh "D.W." Heath-Pasulka's message resembles Jacques Vallée's, but is still more dangerous due to her unabashed profession of the Catholic Faith. This gives her a warm welcome in the Church (her Ufology books are required reading in some Catholic universities[239]), despite her preaching a message that undermines Catholicism itself. No mere politician wearing a religious label for an image, Pasulka was already a celebrated author on Catholic themes before her foray into Ufology (even authoring a popular 2014 book on Purgatory), and she boasts of her privileged access to the Vatican Secret Archives.

Pasulka's books on Ufology often top best-seller lists. She is a frequent guest on massive platforms promoting the ET Deception, including that of "the most popular podcaster on Earth,"[240] Joe Rogan. She has appeared on countless other large platforms, racking up tens of millions of views, and she featured prominently in a September 2023 Netflix documentary, *Encounters,* which was produced by Steven Spielberg.

Although Pasulka glowingly promotes Jacques Vallée, her real hero may be Simone Planté. Simone (Pasulka only uses her first name) preaches shockingly esoteric ET Deceptions with her own spin. Pasulka praises Simone's view that "**aliens, or extraterrestrials, are now here on Earth...from outside of our dimension of space-time made explicit through AI.**"[241] Pasulka remarks that Simone is "**one of a group of successful technologists who believe that AI will be sentient and that it represents humans from the future,**"[242] and she collaborates extensively with Simone, co-teaching NHI-Deception-promoting courses with her. (Simone is billed as a professor at the phony "Singularity University."[243]) In podcasts, Pasulka regularly speaks of Simone with great reverence, and she extols her supposed ability to, as one Catholic author put it, "**'download' knowledge via telepathy from her Buddhist teachers**."[244] This, of course, is nothing but the very occult "channeling" that Christianity has always condemned as Satanic. As Pasulka's collaboration with Simone primarily concerns the AI Deception, however, we will save further expositions of this for Part Four.

Returning to UFOs, let us consider just one example of multitudes illustrating Pasulka's influence and danger. Appearing on the world's biggest podcast in January 2024, Pasulka listened to Joe Rogan tell her—and his tens of millions of listeners—of his endless stream of blasphemous views. "**The universe itself is God**," Rogan says, proclaiming unreserved Pantheism, and Pasulka's response is "*yeah*!" Rogan echoes Simone and asserts that Artificial Intelligence will soon form life, which will "**become God**." Pasulka's response to this AI-worship of Rogan is, of course, excited agreement; she follows up by saying we now have technology that can replicate sightings of the Virgin Mary. "**There will be intelligent beings that will appear like**

deities to us" thanks to AI, she prophesies—treating this as not only true, but also good. Pasulka then goes on to say that St. Francis of Assisi's stigmata actually came from him being "**irradiated by a UFO**," and that St. Teresa of Avila was actually abducted. But, Pasulka laments, traditional Catholic art has simply "domesticated" these things; doing away with the UFO intrigue they carry. Finally, Pasulka admits that "some people" refer to so-called UFO related events as demonic, only to reject that assessment, saying it consists in "**using the terminology of people from 100 years ago**." Her lesson that emerges is this: we must now embrace both the angelic and the demonic, regarding them as just two previously misunderstood examples of the "interdimensional" UFO phenomenon, which the saints of Church history just couldn't have comprehended.

Pasulka attempted to bolster her point to Rogan by claiming that if we "**put physics on**" the happenings at Marian Apparitions, we can conclude that what the faithful once called Marian miracles are just the result of some advanced technology. (Thus disclosing she knows nothing about physics, Marian apparitions, *or* technology.) We are only "**at the very beginning of learning about this other world**," Pasulka concludes, relaying a dogma among Ufologists which holds that traditional religion is utterly incompetent in dealing with what ET promoters call "The Phenomenon," and the "ontological shock" it brings. "**What do you think the UFO phenomenon is?**" Rogan finally asks Pasulka point-blank. In a word, her response is that it is a new religion. Here, she is correct—a point we will consider more in a forthcoming chapter. Yet instead of warning that this new religion is false and destructive, she rejoices in its growth and eagerly proselytizes it, amassing for it countless converts.

"**It is urgent to get this message out**," she says, declaring, "**there is what I believe to be an awakening happening ... as long as we have some portion of the collective awake, there's hope.**" This is her final message of the podcast. With it, she reminds us that her occasional claims that she is *merely an academic studying these subjects indifferently*, are only misleading preparations for plausible deniability through which she hopes to evade criticisms like those presented here. It reminds us that (as she is still more prone to reiterate) she treats her mission as one of leading the enslaved and deluded masses out of Plato's Cave (or the Matrix, or what have you). In fact, she is only dragging souls that Christ delivered *from* the caves of the Pagans straight back into the clutches of the ancient demons. The only difference is that the trap she lays is more alluring than ever, bedecked as it is with terminology borrowed from sci-fi delusions pretending to constitute real science, and preached in an ostensibly Catholic way that, in fact, undermines the Faith.

Summarizing Pasulka's deeply troubling preaching on other platforms would quickly turn this chapter into a large book of its own, so we must move on. We have here considered her appearance on Rogan's platform not because her teachings there were any more outlandish than elsewhere, but only due to the immense number of people already formed by them. If one were to add up all the people reached by the various evangelization programs put out by the United States Conference of Catholic Bishops itself, the sum would doubtless amount to only a fraction of those Pasulka has reached in the past year with her ungodly UFO message that wears the attire of Catholicism. One can scarcely exaggerate the danger here, yet Pasulka is just one of a growing army of Catholic false prophets heralding the ET Deception.

Professor Pasulka dishes out effusive praise for Dr. James Madden (whom she herself introduced to Ufology). When I wrote my 2023 book, *Only Man Bears His Image*, I had never heard of Dr. Madden, much less of his writings about UFOs. Nevertheless, in that book, I wrote:

> Pasulka is only one member of a steadily growing group of prominent "orthodox Catholic scholars" who are becoming heralds of the "extraterrestrials." There will be more like her soon entering the fray, and whoever wishes to persevere in the Faith must be sure not to be seduced by their works.[245]

Dr. Madden recently published a Ufology book that Pasulka extolled as "**necessary reading… as the contemporary means to free one's mind**." She lauds Madden for doing "*what no philosopher, or author, has attempted yet,*" because he "**argues that the UFO is necessary, now, for humanity's cognitive breakthrough into a freer and more deeply meaningful life experience**."[246] To him, then, we must now turn.

James Madden & Timothy Morton

Dr. James Madden is a Catholic philosophy professor at Benedictine College in Kansas—an institution with a reputation for conservative Catholic orthodoxy. His latest book—*Unidentified Flying Hyperobject: UFOs, Philosophy, and the End of the World*—contains his foray into Ufology. His thesis is that UFOs are so-called "**Hyperobjects**,"[247] which, we will see, also winds up meaning for him that *the UFO is everything*. He teaches that UFOs obliterate our entire prior "mythological religious" view of reality and replace the positioning of Divine Revelation's categories with an entirely new superlative category of being.

Published in 2023, Dr. Madden's book is based on the teachings of Vallée, Pasulka, Strieber, Kripal, and a certain Timothy Morton. (Madden himself takes pains to repeatedly remind his audience that most of his work here is simply a re-presentation of the works of these individuals.)

Yet, the fact that his book is completely formed by authors who, as we have already seen, promote decidedly and thoroughly evil messages, does not merely fail to bother Dr. Madden. These authors, rather, excite Madden to further their same quest. Nor is there any sense of him "baptizing" these ungodly works—he makes no effort to correct their errors, but instead fills his own book with unqualified support for them.* For Dr. Madden, Whitley Strieber even reveals to us "**a fundamental structure of human existence.**"[248] Yes, Strieber reveals the ministrations of demons! But Madden dismisses those who argue UFOs are diabolical, saying this constitutes promoting a "**preferred religious mythology**"[249] and succumbing to the "**tyranny of the past.**"[250] As an alternative to this "mythology"—of Christian Revelation—Madden proposes we adopt the thesis of Kripal,[251] who in turn, recall, insists that religion will only take us further away from the truth about UFOs, and also teaches that his thesis means we must all become occultists.

Clearly, Dr. Madden is not simply drawing from the valid insights against mainline Ufology provided by the "interdimensional" Ufologists—as we have done in this book. Rather, Madden has made his own the same deception they promote. Demons are just another myth, according to this yarn, whereas to grasp UFOs, we must conjure up a whole new ontological category of being that supersedes those of Scripture.

Of Madden's primary influences is a certain Dr. Timothy Morton. An exposition of his work is in order here, as he is now a de-facto molder of contemporary Ufology. Not only the title, but also the subtitle, of Madden's own book is a tribute to Morton's 2013 book, *Hyperobjects: Philosophy and Ecology After the End of the World*. (Pasulka, it is important to note, also draws heavily from Morton.) With that publication, Morton first developed Madden's notion of a "hyperobject."

A cross-dressing, so-called "non-binary"[252] professor at Rice University, Morton—a self-professed Christian—is primarily distinguished by his ferocious advocacy for various far-left causes (including anarchic Marxism) with which his aim is to defeat the Christian belief that humans are uniquely made in God's Image. His latest book, *Hell* (Columbia University Press, 2024), though written by a 56-year-old academic, resembles an angsty teenage rant with content even more blasphemous than one would suppose by reading its shocking description. (For example, Morton therein calls God an "idiot psychopath.") Not one to miss any chance to embrace evil, Ufologist Dr. Jeffrey Kripal heaps praise upon *Hell*, writing, "*This book is classic Tim Morton, and it's more. ... It's religion reloaded.*"[253] (Kripal ringingly endorsed Dr. Madden's UFO book as well.) Morton's *Hell* concludes by casting its lot with the NHI (Nonhuman Intelligence) deception; altering the Glory Be

---

* "I am in the debt of these trailblazers," Madden says, before declaring that Vallée's view is the "most plausible ontological model" of Ufology. And even though Vallée takes the cake, "Kripal's notion of 'Super Nature'", Madden writes, "is a very useful concept for us to have in mind as we go forward." (This is precisely the "useful concept" that Kripal finds especially useful indeed… for radically rejecting the Catholicism of his youth, discarding the supernatural entirely, and embracing degenerate occult practices.) Although Madden is careful to ensure readers do not think he is fully endorsing Aristotle, writing, "The point of this chapter is not to defend Aristotle's natural theology," I saw no such caveats affixed to his repeated effusive praise for Kripal, Strieber, Vallée, and Morton.

prayer the same way Dr. Neil deGrasse Tyson did—writing, "*Worlds* without end, amen."

Similarly, the conclusion of Morton's 2017 book, *Humankind: Solidarity with Nonhuman People*, calls for the "queering" philosophy because of the existence of these "hyperobjects" he imagines. Its opening chapter condemns "**Patriarchal religion,**" because it "**intertwines speciesism and anthropocentrism,**" but Morton is especially furious at such religion because it is "**a device for excising the X-power of things.**" This "X-power," Morton says, consists in the "**dæmons that sit on your shoulder,**" which he insists are quite good. He then issues his own plea to the world: "**For the sake of lifeforms, it is time to release this seemingly evil spectral shimmering from its confinement in the realm of art and start allowing porpoises and humans to have it, too.**"[254]

Morton's teaching about the "hyperobject" is openly presented as an invitation to the "dæmons"—*demons,* that is—to come back onto the shoulders of men. Morton explicitly admits that his "hyperobject" teaching **"has all the hallmarks**" of the "**gods of the Axial Age religions,**"[255] which catechized Christians know are demons (cf. Psalm 95:5, 1 Corinthians 10:20).

Nor are these teachings disseminated only in the echo chambers of academia's ivory towers that are renowned for their self-absorbed and overfunded pretension. A 2021 article in *The New Yorker* commented on Morton's bizarre behavior (such as treating his cat as a person), but more significantly, his insistence that "**humanity had entered a new phase**" because "for the first time," Nonhuman Intelligences ("hyperobjects") have been recognized as being "responsible for the next moment of human history."[256] Similarly, a 2017 article in *The Guardian* reported:

> For Morton, our task is to become ["dæmon"] haunted beings again ... This follows from his adherence to object-oriented ontology, the argument that nothing has privileged status and [humans] exist equally with [excrement] ... The new human [Morton] yearns for passively allows him or herself to be infected by the healing solidarity of non-humans.[257]

Yes, Morton's "hyperobjects" are quite obviously evil spirits, and the message of this so-called Christian aims to undermine Christianity itself.* And Morton's notion of these fables, Madden repeatedly reminds his audience, is the basis for his (Madden's) own UFO thesis. †

---

* Philosophically speaking, however, Morton and Madden's notion of a "hyperobject" is simply a haphazard amalgamation of modified extreme Platonism (that Plato himself would abhor), deified pseudo-neuroscience (a cancer on contemporary philosophy), and occult Jungian psychobabble (the entire purpose of which has always been to overthrow Catholicism)—though all rebranded with terminology that will appeal to sci-fi deluded masses eager to adopt another iteration of Paganism while pretending they are still Christian. There is no point wasting time giving these incoherent ravings serious consideration here.

† Madden may succeed in evading any criticism whatsoever. His teachings on "hyperobjects" are empty; based entirely on distinctions-without-differences describing false middle ways that would, like Morton's teachings, be pointless to examine in detail here. Therefore, no matter

Just as troubling as Dr. Madden's book that furthers the teachings of Timothy Morton—and those of Kripal, Strieber, Pasulka, and Vallée—are his now frequent podcast appearances amid Ufology circles. In an August 2024 "Society for UAP Studies" podcast with Dr. Kimberly Engels,[258] Madden **"agreed"**[259] that Marian apparitions were really UFO sightings. "**Reality is weirder than all of our religions predict**," [260] Madden declared on another YouTube channel—this one, devoted to parapsychology and numbering 169,000 subscribers. Thus, he casts his own proverbial pinch of incense before the NHI Deception's idols—supposing that the "high strangeness" of "The Phenomenon" is so very high indeed that Christianity cannot possibly deal with it. In the same podcast, Madden said that UFOs signal the end of "fundamentalism," since they do not abide by any "orthodoxy," including the Catholic variety. In yet another podcast, Madden appeared with Pasulka herself. There, he maligns "**Catholic nostalgia,**" saying that UFOs will "**de-world us.**" They will "**undo the world**," he repeats, before indicating this is best for us. "**There is no if**," he prophesies, "**there is a radical break coming**." He then ponders, "*can I look hopefully to something that I might not even be able to recognize as a human life?*", before lamenting that in the past his answer was "no." (Thanks to UFOs, however, that has evidently changed.)

Let us reflect on the bald insanity here. Two putatively orthodox Catholic scholars are proclaiming—to massive audiences—that UFOs are wonderfully positive for humanity, absolutely necessary for the world, radically beyond the very categories defined by Divine Revelation, and are, moreover, going to completely change the world so much that once they are through with us, our new lives will—if seen now—not even be recognizable as human life. The faithful should unanimously condemn this message without hesitation. Instead, we only see it being welcomed with open arms by "professional Catholics," while exploding in popularity.

At Benedictine College, Madden now teaches courses on UFOs; courses he claims are among his most "successful" ever, thus a new generation of UFO deluded young Catholics is being formed with the approval of seemingly orthodox Catholicism. And just as Madden singles out Pasulka as his immediate inspiration, Madden himself is sure to inspire other Catholic scholars to follow suit.*

---

what is said about his thesis, he can always respond, "*that is not what I meant.*" Indeed, he means *nothing*, so that response will always be a fair enough one. *Ex falso quodlibet*—by proposing a contradiction, one can claim that anything (or nothing) follows.

* Already, Professor Matthew Ramage, a popular Catholic theologian who is also often described as orthodox-minded, has sung the praises of Madden's book. Ramage wrote (in a review giving the book 5 out of 5 stars) that "*Madden's interdisciplinary paradigm transcends traditional academic boundaries…Inviting readers to reconsider some crucial default assumptions about the nature of reality…with his Uber-Umwelt Terrestrial Hypothesis (UUTH). . [from his] Object-Oriented Ontology (OOO). According to OOO, reality is structured by nested hierarchies of substances that do not fit neatly into our conventional ontological categories…*" Then, Ramage congratulates Madden's ability to "*[incorporate]*

(Note: here, an important addendum must be included. It applies to everyone, but especially to Dr. Madden and Dr. Pasulka. In this book, I am merely addressing the objectively erroneous and gravely destructive content of certain public teachings. I claim no insight into the hearts of those who voice them. They may be good people who simply "know not what they do." (Luke 23:34) Whether they are guided by ignorance, confusion, blindness, malice, self-interest, diabolical influence, or remarkably misguided good intentions, I do not know. Nevertheless, the fact remains that it is both dire and urgent that their errors be strongly condemned; countless souls are at stake. If I had to guess, I would say that Madden and Pasulka's intent is positive, if impressively confused—much like the priests we are about to discuss. Perhaps they are like the 19th-century Catholics who promoted engaging in séances with good intentions, thinking this constituted evangelization because it "displayed the anomalous," when in fact it only opened the doorway to demons.)

***

The Devil's plan often begins with the scholars. But they are only stepping stones. Therefore, we must now turn our attention to a level higher still, and address not only the Catholic scholars promoting the NHI Deception, but the Catholic-clergymen-scholars doing the same.

Perhaps the chief false prophet of the religion of scientism, with its corresponding pseudo-messianic hope in extraterrestrials, was the globally renowned Jesuit theologian, Fr. Pierre Teilhard de Chardin, from whom Pasulka draws heavily, and to whom we now turn, though in doing so we turn our gaze from contemporary talk to its mid-20th-century origins.

Fr. Pierre Teilhard de Chardin

Fr. Pierre Teilhard de Chardin was a French Jesuit priest, theologian, and scientist who became extremely popular in the 20th century for proposing a radical re-writing of Christianity based upon the premises of Darwinism.[261] In his own words, he sought to turn Christianity into "**the very religion of evolution**."[262]

In the eyes of the starstruck, de Chardin became the embodiment of a much longed-for synthesis of the seemingly warring factions of faith and science. In fact—as generally occurs amid such misguided Hegelian efforts—no synthesis was wrought, but only the triumph of the more seductive and politically powerful. What Fr. Pierre orchestrated was no marriage between faith and science. Instead, it was a total reshaping of faith in the image of modern science—or rather, in the image of his own wild conjectures posing as scientific. De Chardin described his mission as follows:

Timothy Morton, Jeffrey Kripal, Whitley Strieber, Diana Pasulka, and Vallée, and more," sharing that reading Madden's book *"helped frame the disparate insights of the above thinkers into a single coherent picture."* The fact that each of these thinkers' "insights" are, in fact, diabolical lies, does not appear to bother Professor Ramage either. Not only did Ramage offer this glowing review, but he also helped enough with the book's development to warrant Madden naming his discussions with Ramage as "indispensable" in the book's acknowledgments.

> What increasingly dominates my interests, is the effort to establish within myself and define around me, a new religion (call it a better Christianity, if you like) where the personal God ceases to be the great monolithic proprietor of the past to become the Soul of the World which the stage we have reached religiously and culturally calls for.[263]

Many of this Jesuit's teachings were condemned by the Congregation for the Doctrine of the Faith in 1962.* Unfortunately, this did little to blunt de Chardin's influence. Instead, his name and his impact have only grown.

One example of this influence was a two-hour PBS documentary, *Teilhard: Visionary Scientist*, released in 2024. In the film's opening remarks, Georgetown University professor John Haught joyfully exclaims, "**Teilhard de Chardin said we need a new God... smart people like Teilhard become aware that the story, the narrative, of Adam and Eve in Genesis can no longer be taken literally**."[264] Surely, "smart people" know better than God. The film's final scene is given to pro-abortion clergyman Michael Curry, the primate of the Episcopal Church. To inspiring epic music and amid the backdrop of stock footage showing all manner of scientific research, advanced technologies, and humanoid robots, Curry extols de Chardin with palpable enthusiasm, saying unironically that he gives "**a vision of hope that *really is* hope, and not pipe dreams. And that is the stuff of prophets**." If that remark was facetious, it would be an accurate description of de Chardin's message, which indeed is only one very large pipe dream that fails to inspire anyone who lacks a "PhD" after his name. Prophets call ordinary people to repentance and holiness, and to hope in Heaven and in the intervention of Jesus Christ. De Chardin robs us of a focus on Heaven, insisting that Christianity becomes preoccupied with temporality, and he dethrones Jesus Christ for the sake of his own "Universal Christ," which—whatever it may be—is anything but Jesus of Nazareth, the Son of Mary. De Chardin's message neither converts nor sanctifies anyone. But it does provide the framework for "Catholic" ET belief.

It is not only American secular public broadcasting companies and their academic contributors who idolize Fr. Teilhard. A major effort is now afoot in the Church to not only officially rehabilitate his legacy, but also urge—with a large and growing petition behind it—that de Chardin be named a *Doctor of the Church* due to his "prophetic vision." In truth, De Chardin only numbered among the very "Pagan Prophets" whom Pope St. Paul VI warned against the rise of in the same decade the Vatican condemned de

* In 2014, the Prefect for the Congregation for the Doctrine of Faith, Cardinal Gerhard Müller, denounced de Chardin's teachings (although without naming him) in the strongest possible terms, noting that they "**lead almost necessarily to fundamental errors**" about God, Jesus Christ, Sin, and Salvation. He even declared that acceptance of de Chardin's teachings "**evidences that a *de facto* movement beyond the Church and sound Christian faith has already occurred.**" (Cardinal Gerhard Müller. *Meeting of the Superiors of the Congregation for the Doctrine of the Faith with the Presidency of the Leadership Conference of Women Religious* (LCWR). April 30, 2014) The trappings of Christianity are maintained, however, therefore the apostasy generally remains unacknowledged; they remain *"holding the form of religion but denying the power of it."* (2 Timothy 3:5)

Chardin. That is, "prophets" who brought the smoke of Satan into the Church while wearing the guise of science. Fr. Pierre made no efforts to preclude this diagnosis; he openly declared:

> The idea of an earth chosen arbitrarily from countless others as the focus of Redemption ... is one that I cannot accept ... There were worlds before our own, and there will be worlds after it. ... Christ has still to be incarnate in some as yet unformed star ... [but] there are times when one almost despairs of being able to disentangle Catholic dogmas from the geocentrism in the framework of which they were born.[265]

De Chardin's teachings are not only a rotten fruit of belief in extraterrestrial intelligence, but a cancerous limb from which innumerable other rotten fruits have emerged.

ET promoters themselves confirm the influence of de Chardin in molding their movement. The *Mammoth Encyclopedia of Extraterrestrials* explicitly credits de Chardin for providing the "**model of human awareness that emphasizes those changes in human consciousness [which] occur [as] a result of UFO encounters**."[266] A recurring and foundational theme in contemporary Ufology is the assertion that ET-contact "raises consciousness," and this "awakening" or "ontological shock" that results is what will, in turn, bring about an extraterrestrial-originated utopia on earth. Fr. Pierre provided the impetus for this expectation, defining it as the "Omega Point" or "noösphere" that has now become a mantra in Ufology and AI-Deception-promotion.

Ufology's current Catholic star, Diana Pasulka, writes that although de Chardin "**is not a saint by official Catholic recognition**," he is nevertheless "**a saint to many, including me**."[267] She then lauds his sci-fi-themed Paganism, explaining: "**Chardin described Earth's soul, or the noösphere, as the culmination of a process of human evolution ...the next stage of the evolution of human intelligence. It could be the next branch of Homo sapiens**."[268] Chardin's teaching looms large not only throughout Pasulka's promotion of the ET Deception and Transhumanist Deception, but also her advocacy for the AI Deception—whose so-called "Singularity" is also presented in terms defined by Teilhard. Many other contemporary Christian ET promoters openly extoll de Chardin's teachings as the motivation for their own, even while conceding that the same teachings overturn the entire basis of Christianity.*

---

* For example, Dr. Robert John Russell is a Christian theologian and physicist who promotes belief in extraterrestrials. After insisting that God has dealt with ET races just as He has with the human race, he reminds his readers that, "**This is, of course, the horizon-breaking image of the 'cosmic Christ,' as Pierre Teilhard de Chardin called it.**" (Robert Hermann. *God, Science, and Humility: Ten Scientists Consider Humility Theology*. Templeton Press. 2000. Page 41-42) Russell then makes a prediction, insisting that we will eventually contact aliens, and only then will we finally arrive at a "**...point where we at last feel truly at home in the cosmos. What a wondrous event that**

Joel Parkyn is a theology professor at a Catholic seminary who also published a book in 2021, *Exotheology*, arguing for belief in aliens. Parkyn extols de Chardin, writing that, in promoting belief in aliens, he **"abandoned the rigidity of the pre-scientific classical solution offered by the likes of Aristotle, Augustine, Aquinas**..."[269] This Catholic theologian ET promoter would have us believe we must move on from the "rigidity" of St. Augustine and St. Thomas Aquinas in order to embrace the "scientifically" informed theology of Fr. Pierre.

One globally famous Catholic priest did just that.

Born in 1943, the Franciscan priest we discussed in Part Two, Fr. Richard Rohr, has been described as a modern-day Teilhard de Chardin. Rohr himself recently confessed that during his college days, he was "**fascinated by the poetic-scientific-mystical-cosmic language**" of de Chardin, which the priest excitedly describes as being "**like spiritual porn!**"[270] He said that it will "**change the foundational evolution of history and spirituality**."[271] We need not guess at what (or who) formed Rohr's deepest aspirations.

Although a prolific author and speaker, Rohr wrote one book that stands out among them all: *The Universal Christ*, wherein he re-packages de Chardin's "Cosmic Christ" in simpler language to promote it to a still larger audience. We have already seen that, in a viral sermon[272] summarizing the views of this book, Rohr teaches that "when" we discover aliens, the type of Christianity that the saints have always lived and died for will prove itself nothing but "**our little story line... shot to hell**." As an alternative—which he promises *will* survive our discovery of extraterrestrial intelligence—he presents his own version of "Christianity" which, even more than de Chardin's version, all but leaves aside Jesus of Nazareth, the Son of the Blessed Virgin.

Despite the outlandishness of their message, however, neither of these priests (de Chardin or Rohr) were active within Ufology itself. The next clergyman we will consider presented similarly deceptive teachings, while also traveling extensively within ET promotion circles and becoming a veritable celebrity of Ufology.

> "The work of the devil will infiltrate even into the Church..." —Our Lady of Akita to Sr. Agnes.

---

would be!" Another ET promoter influenced by de Chardin's teachings is Dr. Steven Dick, an American astronomer and former chief historian at NASA. There is a planet named in his honor, and he is a prolific author who has written extensively on his belief in aliens—publishing many books that have proven quite influential. Dr. Dick uses his "astrotheology" to not only promote a whole retinue of heresies, but also to reject the supernatural entirely. He does this even while admitting that what he rejects is "**the basis for everything they do in the Christian church**." Nevertheless, he insists that "cosmotheology must be open to radically new conceptions of God...a God grounded in cosmic evolution." He defends this exhortation by insisting it "**resonates with the Jesuit Teilhard de Chardin's evolutionary cosmology**." (Steven Dick. *Space, Time, and Aliens*. Springer. 2020. Chapter 13.2.)

Monsignor Corrado Balducci

It is easy to see why Monsignor Corrado Balducci is one of the most frequently cited Catholic names in support of belief in extraterrestrials. Those afflicted by clericalism's primary symptom—the supposition that any priest, and certainly any Vatican-exorcist-priest-theologian, is an infallible mouthpiece for God—regard Balducci as the perfect "proof" that Catholics should believe in aliens. This Italian theologian, who served on Vatican congregations and died in 2008, worked as an exorcist for the Diocese of Rome, where he claims to have counted Pope St. John Paul II as one of his friends. *"If Balducci believed in aliens, we also should,"* today's Catholic ET promoters say. Yet one need only glance at Balducci's own words and actions to see how wrongheaded that advice is, and to discover how loudly the diabolical delusion resounds in the message of Balducci. Fr. Corrado did not merely ponder abstract theological questions related to extraterrestrial intelligence like de Chardin. He unreservedly embraced Ufology—occultism and all—and became a celebrity, commonly appearing in both television episodes and documentary films, while declaring himself the Catholic Church's "**representative to the star peoples**."

An interview with Balducci from the year 2000 was falsely billed by the media as a revelation that, now, *"the Vatican endorses extraterrestrials!"* In it, Balducci spoke with Ufologist Paola Harris, who relayed that she and Balducci "admired each other" and "often dined together," and that **"[Balducci] was optimistic and hopeful that we would have some kind of celestial 'intervention' [i.e., from aliens] and be saved from cataclysmic destruction."**[273] While some Christian ET promoters are more guarded in confessing their messianic views of imminent alien contact, Balducci boasted of it.

He openly and repeatedly collaborated with charlatans, fraudsters, and occultists such as Dr. Steven Greer, Jaime Maussan, and Whitley Strieber. Instead of offering these troubled souls any correction or Catholic guidance, Balducci only confirmed them in their delusions and sent them on their way as they not only continued treading the path to perdition, but also continued seducing millions more into its clutches. To this day, many Ufologists—including these three—defer to Balducci's influence on them to justify the shamelessly corrupt teachings they promote. One researcher noted that Strieber—the "world's most famous alien abductee," with, as we saw, a resolutely anti-Christian message—declared that **"[Balducci's] expertise ... lent 'authority' to [Steiber's] views on the question of whether the Visitors (as Strieber calls the aliens) are demons."**[274] Strieber's "visitors" were, in fact, demons. Due, however, to Balducci's insistence that they were not, Strieber remains under demonic dominion to this day while vigorously proselytizing it. Steven Greer has also remained deeply impacted by Balducci's collaboration with him. When Ms. Harris interviewed Greer, he brought up Balducci's words to justify his own encouragement to become "**an ambassador to non-human life forms...[to] use meditation techniques and the contact techniques to be able to enter and experience Cosmic Mind and then remotely view**." Balducci also gave Greer a personal interview, which the latter recalls frequently and included selections of in his 2013 book, *Disclosure*. In that interview, Balducci shares with Greer that he believes in aliens because "**a human person... [is] the worst.**" The priest continued:

> Lower than [us], there isn't anything... [Extraterrestrials] are probably so evolved that evil doesn't even enter into [them] ... God in his wisdom wouldn't have created only us ... Don't even think about it [ET/UFO sightings] being the devil. ... [Aliens] are going to intervene… they are going to help us...[275]

In the years since this interview, Greer went on to use Balducci's never-relinquished backing to deceive untold millions with a New Age, anti-Christian message based on belief in extraterrestrials.

Claims like those above were not isolated comments. Balducci preached them to a global audience for many years. Elsewhere, he declared that: **"[Aliens] must be much more evolved because the human species is the lowest rung on the ladder of spiritual development.**" [276] Balducci constantly demeaned human nature, indicating a blasphemous disregard for the supreme status of the humanity of Jesus Christ (whose human nature was like ours in all things but sin), the Blessed Virgin Mary, St. Joseph, and the Church in general, but also a capitulation to the apostate Darwinist rewriting of Christianity pioneered by Fr. Teilhard de Chardin.

Balducci also featured prominently in a popular 2006 Ufology documentary entitled *Fastwalkers*. Other major names in this film run the gamut of charlatans and occultists, including known hoaxer George Adamski (via archival media), Steven Greer, and others. The documentary begins, however, with a few pithy statements from some of its stars, including Paola Harris, who declares that the human race is **"the virus of the universe**," lamenting that of all the extraterrestrials in space, "**nobody really wants us out there**." Later in the movie, she sits beside Msgr. Balducci, translating his Italian, as he reiterates his usual arguments in support of ETs and their presence on earth.

The film endorses the most egregious tenets of the ET Deception. It informs us that aliens have been on earth for millions of years, that Sodom and Gomorrah were destroyed not by God but by a UFO, that the US Government is working with aliens to upgrade our bodies to better house our souls, and that we as a species are now in a "conditioning" period before being given the evolutionary step up to our intergalactic-community-status. Eventually, we will get our "life charter" from these aliens—since that is what they have "been doing forever," and this will tell us "**who we are, what we are, how we came to be, and where we might end up going**." Ninety minutes of Antichristic claims such as these bring us to the conclusion of the film, and Msgr. Balducci is given one of its final words.

Moments before the documentary ends, Balducci bids viewers farewell by again declaring that humanity "**is the lowest in the category of the evolutionary scale.**" The directors were strategic. The movie's audience thus leaves with one twisted impression that stands out among all the film's outrageous content: *"the Vatican stands behind these shamelessly Luciferian teachings presented in the preceding hour and a half."*[277] Clearly, this is also what Balducci himself desires to be his celebrity legacy: his dogmatic and repeated insistence—as a "certainty" that "**cannot be doubted**" (as he said on Italian television in the 1990s)—that the entire human race is a pathetically "unevolved" species in desperate need of cosmic intervention from "evolved aliens," to save us from ourselves.

If this is not a diabolical deception, then nothing is.

One Christian scholar—Dr. Michael Heiser—became understandably incensed at Balducci's promotion of the most baldly anti-Christian ideology. He wrote that Balducci is "**disturbingly naive**," and pointed out that if Balducci is going to continue in his Ufology pursuits, "**He could at least be honest and divorce himself from Catholicism and go his own spiritual way, rather than pretend he or his faith is something that it is not.**"[278] Of course, one should never encourage "divorcing from Catholicism," but one can only sympathize with the sentiment that prodded Dr. Heiser to say these words, upon seeing a Catholic priest relentlessly preaching a message which radically undermines Christianity itself.*

Some Catholics may be surprised to hear that an exorcist can be just as deluded as anyone. Yet, occasionally, they are horrifically more deluded, as in Balducci's case. One priest, Fr. Carlos Martins—himself an exorcist—wisely cautions:

> If [an exorcist] is unprepared ... he can become a victim himself. ... [He] may end up as the plaything in the Devil's mind games ... If he is morally impure—if he has sin on his soul—he loses all protection that grace affords him and opens himself to being possessed.[279]

Indeed, whenever seeing an exorcist promoting UFOs, we must remember that some exorcists wind up unknowingly becoming a "**plaything in the**

---

* Three years before *Fastwalkers*, Fr. Balducci appeared in a similarly problematic UFO documentary, the 2003 film *Touched by an Alien*, in which contactees recount, in detail, their encounters with demons (especially sexually abusive succubus and incubus demons), thinking them to be "aliens." Yet another film, *UFO Chronicles: Alien Science and Spirituality*, was released five years after Balducci's death, and it includes more of the footage from his interview for the 2006 *Fastwalkers* movie. There, we see and hear him declare that he is an expert in demonology—very "famous" for his work therein—therefore, no one is allowed to suggest that anything diabolical might be afoot here, since he says otherwise. He moreover states that "It CAN'T be just us on this planet"—that this is simply "NOT POSSIBLE"—and that there are certainly "millions of inhabited worlds." These are the ravings of either a lunatic or a diabolically deceived man. As far as Balducci is concerned, we are not even allowed to *doubt* whether or not ETs are among us.

Devil's mind games," even while they retain all the vocabulary and academic prowess needed to (seemingly convincingly) pretend that could not possibly be the case. And we know that among the Devil's favorite *mind-games* today is precisely that of extraterrestrials.

Catholic priest exorcist-theologians promoting diabolical ET Deceptions did not end with Balducci's death. It continues to this day. Others like Balducci (even if not matching his fame) can easily be found on the internet, and whoever desires to keep his soul safe will be sure to stay far away from an exorcist who promotes Ufology or "exotheology."

(Note: Due to Balducci's uniquely destructive legacy in contemporary "Catholic" ET promotion, a more thorough analysis of his impact is called for than we can fit here; it can be found in *Only Man Bears His Image*, in pages 404-418, and in the Appendices. Balducci's word is also the sole source of the claims made by contemporary ET promoters who say that St. Padre Pio and St. John Paul II believed in aliens. Both claims are outlandish and false. They are also refuted in the Appendices of *Only Man Bears His Image*.)

***

These real-world false prophets heralding the Age of the Extraterrestrial, however, are joined by an army of perhaps even more influential wolves in sheep's clothing on the silver screen. Without Hollywood providing the mythology for a new pseudo-religion, it will likely fail. With that endorsement, its success is a foregone conclusion.

No modern myth that Hollywood has promoted comes close to that of the ET-religion.

11. Hollywood Sci-Fi Grooming

"Have nothing to do with godless and silly myths."
(I Timothy 4:7)

Man lives by the stories he honors. If he reveres true stories—and if he lives in accordance with the greatest story ever told, Salvation History—then he will become a saint. If instead he lives by the "*godless and silly myths*" that Scripture warns against, then he has only invited the lies they contain to form his innermost principles.

Both the world and the Devil understand this well, hence such enormous emphasis long being placed on what messages are conveyed through Hollywood, presently by far the world's preeminent storyteller. Long before Hollywood, however, both the power and the danger of storytelling were known by the wisest. Centuries before Christ, Plato went so far as to have the poets banished from his ideal society, envisioned in the *Republic*, for fear of the corruption they are prone to cause. A century earlier still, Confucius insisted upon his doctrine of "wen;" the regulation of all the arts to ensure they only promoted virtue and truth. Contemporary Ufologist Jacques Vallée also understands; he said, "**To control human imagination is to shape mankind's collective destiny**."[280]

Science fiction, the modern world's prevailing myth, has succeeded in controlling human imagination. Thus it is shaping our collective and tragic destiny. As we will see in the forthcoming pages, sci-fi teachings have, especially for the last 75 years, steadily led the world deeper and deeper into deception while fomenting in the masses an increasing expectation that we will soon either be attacked by extraterrestrials or saved by extraterrestrials (or both). What is telling is not merely that such an outlandish plotline has become so popular. (Even absurd settings can sometimes provide legitimate entertainment; there is certainly nothing intrinsically wrong with presenting a fictional scenario based on aliens.) What concerns us, rather, is the culture-wide grooming process we see meticulously undertaken by the steady descent into apostate darkness of the themes found in ET-based science fiction throughout the last several decades.

It is that darkness, strangling Christian Faith and heralding the Strong Delusion—once subtly, but now openly—which we must expose.

The Fathers: Voltaire and H.G. Wells

Hollywood's formal grooming of society began shortly after the 1947 Roswell event, but two cultural giants—one in the 18th century, and another in the 19th—had already paved the way.

> "Science is our religion!" is a Raelian [UFO Religion] motto. Raelians are true children of the Enlightenment. They verbally assault the Catholic Church, applying the familiar critiques of Rousseau, Marx, and Voltaire."

—Dr. Susan Palmer.[281]

The year 1752 saw the publication of Western history's first alien-themed novel, *Micromégas*. Its author, François-Marie Arouet, is better known by his pen name, Voltaire. This ferociously anti-Catholic Freemason died in 1778, eleven years before the onset of the Pagan atrocities of the French Revolution that his writings helped inspire. Today's secular encyclopedias eulogize Voltaire, with Britannica describing him as a "*courageous crusader against bigotry.*" Reality (and discerning Christian commentary) has other things to say.

Voltaire was "**the most perfect incarnation of Satan that the world ever saw,**"[282] according to Jacques Crétineau-Joly (a devout Catholic French writer who worked with Popes and defended the Faith). Whoever understands Voltaire will find it difficult to disagree. Voltaire was a sexual deviant who openly declared war against Christianity, calling it *l'infâme* ("The Infamous Thing,")[283] and writing, "**The Christian religion is ... an abominable hydra which must be destroyed by a hundred invisible hands. Let us crush the wretch!**"[284] Nor was he speaking hyperbolically. He collaborated closely with Frederick II of Prussia, and to this King, Voltaire wrote affirmatively of the following situation:

> Finally, when conditions were ripe, a reign of terror was to spread over the whole earth to wipe out all Christians, and then establish a Universal Brotherhood without marriage, family, property, law or God.[285]

Of course, a reign of terror did spread, thanks in large part to him, and thousands of Christians were indeed murdered in his own motherland. The French Revolution embodied the Enlightenment—ET belief and all—and Voltaire was that Revolution's chief forerunner and prophet. A million lined the streets of Paris for atheistic pseudo-veneration of his remains as they were returned there in 1791. Within two years, amid *la Révolution's* Reign of Terror, the blood of thousands of innocents—slaughtered in the name of "Enlightenment"—soaked those same streets.

One cannot find a more resolutely evil Enlightenment thinker than Voltaire, therefore it is supremely fitting that the ET Deception in fiction began with this man. *Micromégas* is quite short—a "novella," really—and it is concerned with what happenings on earth would look like to the eponymous massive alien Voltaire envisions. This ET, Micromegas, is many miles tall, and was banished—for heresy—from his home planet orbiting Sirius. In his exile, Micromegas makes it to Saturn, and meets a vastly inferior alien there (only about a mile tall). Both then travel to earth, and by using a magnifying glass and a hearing tube, they manage to interact with the tiny humans they first stumble upon here. These humans happen to be philosophers, all making a journey aboard the same ship. Here, in the final chapter, we have arrived at Voltaire's purpose in writing the story. The aliens pose various questions to the philosophers, finding each response somewhat odd. In the story's final paragraph, however, one philosopher points to the *Summa Theologica* of St. Thomas Aquinas as having the answers these aliens sought. The Thomist then "**looked the two celestial inhabitants up and down. He argued that their people, their worlds, their suns, their stars, had all been made uniquely for mankind**." The aliens' response? They "**nearly fell over with that inextinguishable laughter which, according to Homer, is shared with the gods**." However, Voltaire then describes the nobler of the two aliens, saying that despite its laughter, "**in the depths of his heart he was a little angry that the infinitely small had an almost infinitely great pride**."

Voltaire made no attempt to conceal the purpose of this book, through which he invented the whole genre of ET-based science fiction. It was to mock Christianity, and especially Catholicism. This degenerate anti-Christian crusader paints it as at once hilarious and deeply offensive to believe what both Sacred Scripture and Catholic teaching repeatedly make clear: that God made the material Universe *for* the only race whose nature He assumed into Himself: mankind. "**God created everything for man**..." (*Catechism of the Catholic Church*, §358)

Thus is the story of aliens during the ET Deception's embryonic phase in fiction. What of its infancy?

***

Born in 1866, the revered "Father of Science Fiction," Herbert George Wells, died in his native England in the year 1946—only months before the onslaught of the UFO craze he helped generate.

A fierce Darwinist, atheist, and socialist, Wells also advocated for eugenics and euthanasia—which he hoped would be used for "mercy killings" of the disabled, the elderly, and non-whites. These views he shared with a woman he entered a lifelong adulterous affair with—Margaret Sanger, founder of Planned Parenthood. Wells extolled her as "**the greatest woman in the world**," and applauded her nefarious organization as one which "*will grow to be, a hundred years from now, the most influential of all time in controlling man's destiny on earth.*"[286] Sanger's movement, so championed by Wells, has indeed been influential: one billion children have been murdered through Planned Parenthood's chief activity—abortion—since Wells issued that anti-prophecy.

Wells wrote among the most popular early novels about extraterrestrials; 1898's *The War of the Worlds*. It became an instant best-seller, and even more significantly, major Hollywood movies (including 1953 and 2005 iterations) were later made based on it. The work has become a staple in contemporary public school education, and its impact has been culture-forming.

When the story was read by Orson Welles via radio broadcast on the eve of Halloween in the year 1938, it induced a mass panic. Listeners believed it was actual breaking news describing an in-progress Martian invasion! The next day, *The New York Times* reported:

> A wave of mass hysteria seized thousands of radio listeners throughout the nation … "The War of the Worlds" led thousands to believe that an interplanetary conflict had started with invading Martians spreading wide death and destruction in New Jersey and New York… [the broadcast] disrupted households, interrupted religious services, created traffic jams… at least a score of adults required medical treatment for shock and hysteria.[287]

The lengthy article shares dozens of detailed accounts from the mass panic, thus silencing any suggestion that commentators have exaggerated its significance. "**Mass hysteria mounted so high in some cases that people told the police and newspapers they 'saw' the invasion**,"[288] foreshadowing the deception that would formally begin nine years later, when people would also claim to "see" nonexistent aliens in Roswell and then across the globe. Another report states: "**In Indianapolis a woman ran into a church screaming: 'New York destroyed; it's the end of the world. You might as well go home to die. I just heard it on the radio.' Services were dismissed immediately**."[289]

Remarkable indeed was the panic induced, but that aspect of the broadcast's damage was mostly restricted to one day. Far more problematic than the fake news was the subtler deception contained in the principles the broadcast relayed. It opened with the following somber words:

> We know now that in the early years of the 20th century, this world was being watched closely by intelligences greater than man's and yet as mortal as his own. ... With infinite complacence, people went to and fro over the earth about their little affairs, serene in the assurance of their dominion over this small spinning fragment of solar driftwood which by chance or design man has inherited out of the dark mystery of Time and Space.

Here at the onset of ET-sci-fi's formal phase, we see Christianity being meticulously (if, at first, subtly) deconstructed from the ground up. A skilled author, however, Wells conceals his true intent by using only language he knew would achieve his own ends, while evading detection by his mostly Christian audience.

He begins by assuring us we are being "watched"—i.e., ET promotion's most repeated phrase, "*we are not alone*." He immediately adds that the beings watching us vastly exceed our own intelligence, but are assuredly not angels (or demons), for they are "as mortal" as we are. Wells could not risk allowing any genuinely Christian interpretation of his story to prevail, which would correctly understand any created non-human intelligence (other than angels) to be demonic. He subtly mocks mankind's being "busied" by our "various concerns" and "little affairs"—primarily impugning religion, which he singles out for ridicule later in the story. Then, emphasizing just how ridiculous Wells regards mankind, he labels earth—which in fact is the Theater of Salvation—a mere "*spinning fragment of solar driftwood which by chance or design man has inherited out of the dark mystery of Time and*

Space," not unlike Carl Sagan's later, and still more famous, denigration of earth as a pathetic "*pale blue dot.*" For Wells, earth is not God's *delight* (as in Revelation 21:3, Proverbs 8:31, Psalm 115:16, and so on). It is only an insignificant piece of cosmic debris. The lesson is clear: we are not unique, we are not special, and, in fact, we do not matter at all. It may even be that blind "chance" brought us here. More pointedly still, Wells satirizes mankind's "serene... assurance of their dominion" over earth—rejecting Genesis 1:26, wherein God promises that to man and to man alone.

With H.G. Wells' seminal work, the audience begins pondering if its plotline could really transpire—proven by the fact they thought it *was* happening when a dramatized reading of it was being broadcast on the radio. Readers begin wondering if God's promises are not absolute after all. They question whether earth, and the human race, really have the status that Scripture describes.

The Day the Earth Stood Still

Two years after the Martian invasion panic elicited by the radio broadcast of H.G. Wells' *The War of the Worlds,* American author Harry Bates published a short story entitled *Farewell to the Master*. Eleven years later, it became a major 1951 Hollywood film called *The Day the Earth Stood Still.* Borrowing from the Bible's final entry, the Book of Revelation—which prophesies the great persecution of the Church undertaken by the "Beast" and his "Image" (cf. Revelation 13)—the film has both a humanoid "godlike" alien and a large AI robot that does his bidding and accompanies him. (More on the relation between the ET and AI Deceptions—the Beast and its Image—in Chapter 18.) In a perverse twist, this movie has the aliens arriving with a benevolent intent. The original story describes the alien as "**godlike,**" with "*an expression on his face, which radiated kindness, wisdom, the purest nobility.*"

This "god," however, was shot by—no one will have any trouble guessing—a religious fanatic who declared that the Devil had come. Next, the 1951 film adaptation shows the robot vaporizing all the weaponry of the soldiers gathered there, and its alien master is taken to a hospital and heals. He expresses a desire to address all the world's leaders, but is denied that opportunity; he escapes and eventually shares that his task is to save us from ourselves—for so many alien civilizations are now concerned about the destructive abilities of earthlings since atomic power has been discovered.

Before departing earth, the alien does get an audience; not with world leaders, but with scientists, clergy, and others. He tells them that the extraterrestrial civilizations have built an army of invincible robots like the one alongside him. These robots will—automatically, immediately, and irrevocably—eliminate any source of aggression at the first sign of violence. They were built, the alien explains, because:

> The Universe grows smaller each day, and the threat of aggression by any

> group, anywhere, can no longer be tolerated. There must be security for all, or no one is secure. Now this does not mean giving up any freedom, except the freedom to act irresponsibly ... we of the other planets have long accepted this principle.

Thus, the possibility of war, or of violence of any sort, is eliminated. A utopia is offered. This alien protagonist—and, by extension, the story's creators—exhort us that we need only cast aside the greatest gift God gave us: our free will. The alien continues:

> If you threaten to extend your violence, this earth of yours will be reduced to a burned-out cinder. Your choice is simple. Join us and live in peace, or pursue your present course and face obliteration. We shall be waiting for your answer.

The Choice. The Anti-Fiat.

Many somber but understanding faces there gathered are panned across as they gaze upon the alien in wonder. He promptly departs in his spaceship: a perverse re-enactment of the Ascension of Jesus Christ after His final exhortation, which stands in radical contrast to the one given by this alien.

Now, grooming occurs in phases, with sufficient time allowed for the shock of the earlier assault to wear off before the next surreptitious denigration of the victim's innocence and dignity is undertaken. With this film, Hollywood introduced the next, more dangerous phase of the ET Deception. For while any belief in extraterrestrials is among the most perilous of premises, it is more dangerous still to specifically believe in *benevolent* aliens, as opposed to the hostile ones depicted in the *War of the Worlds*.

I should add that in the 2008 film adaptation of the same story, the alien is not visiting earth in response to aggression and advances in weaponry which "may threaten other planets," but rather in response to the prevailing cultural phobia of the time of its own release: so-called "global warming." Therefore, in the latest movie, the alien has come to "save earth *from humanity*." Here, it is not only the Book of Revelation that is perverted, but also Genesis. The alien is a modern Noah with a wicked twist. He comes to earth to collect living creatures from each species and take them off the planet in his many spaceships—Arks—before the earth is "sterilized" of *all* human beings, who are "destroying" it. Derivative of the same anti-human ideology that is part and parcel to alien belief in general, the modern film promotes the standard Hollywood radical environmentalist message, which, in Pagan fashion, fails to acknowledge that the earth exists for the sake of man. The alien explains he has watched for so long, but that we "refuse to change," and therefore he will "**undo the damage we've done and give the earth a chance to begin again**." (One wonders how many contemporary activists, blocking traffic or gluing their hands to various famous objects, derived their zeal from this film's premises.) A professor, however, convinces the alien that "we will evolve," now that we are "at the precipice."

We simply "needed this crisis," he explains. Persuaded, the alien manages to call off the imminent destruction of humanity, which has "finally" learned its lesson from its extraterrestrial visitation.

2001: A Space Odyssey

The seventeen years that followed its release saw society digesting, tolerating, and finally accepting the teachings relayed in *The Day the Earth Stood Still*. "Aliens will save us" was not an entertaining fictional thought any longer, but a tangible expectation. It was time for another step in the grooming process.

Released in 1968, the film *2001: A Space Odyssey* initiated a new era in popular sci-fi. Viewers everywhere still refer to watching this film as a "religious experience," though failing to grasp that the "religion" they imbibed throughout its two-and-a-half-hour runtime is a neo-Pagan and quasi-atheistic one. *The Encyclopedia of Science Fiction* refers to the power of *2001: A Space Odyssey*, *Close Encounters of the Third Kind*, and *ET: The Extra-Terrestrial* (we will consider the latter two shortly) as promoting "**UFO mysticism.**"[290] *2001*'s director and writer, Stanley Kubrick, insisted this was his intent, saying of it: "**All the essential attributes of such extraterrestrial intelligences are the attributes we give to God. What we're really dealing with here is, in fact, a scientific definition of God**."[291]

To this day, *2001's* haunting images of HAL's red eye camera and mesmerizing voice, and the awe-inspiring black monolith (a large, thin box gifted by ETs) stand as cultural icons. Hearing only a few successive notes of *Also sprach Zarathustra* (German for "Thus Spoke Zarathustra"), building up to its famous crescendo and clashing cymbals, still elicits images of space travel, sentient computers, and alien-guided evolution. It is no accident that the piece itself—composed in 1896 but given a resurgence in popularity by this movie—was inspired by the anti-God existential nihilist, Friedrich Nietzsche; particularly, by his work of the same name. For in this book, Nietzsche declared that "God is dead," and with Him morality has died. That old system must be replaced by the Übermensch—the "Overman" or "superman"—who "rises beyond good and evil," by willing only the "will to power," for he has rejected the "slavery" that is Christianity. These are precisely the aims of the ET Deception.

The inclusion of this piece as *2001: A Space Odyssey's* most memorable score relays a simple message: there is no need for God to explain either our past, present, or future. Our ancestors, according to this message, were irrational apes who were simply granted knowledge of the use of tools by the alien monolith, and from that event human civilization emerged. This "first ontological shock" spurred on our evolution which, long thereafter, assumed its next phase: artificial intelligence becoming sentient. Then, the monolith appears again, with this next threshold having been passed, to bring us to yet another level of space travel and super-evolved "superman."

Thus is the film's plotline in brief, though a few more observations are in order.

The movie begins with a Purgatorial opening scene of preposterous duration depicting the evolution of apes hastened by the appearance of the alien monolith. (Pretending to find such pretentious scenes to be "deep" grants one access to certain clubs.) Fast forwarding to the year 2001 A.D., a discovery is announced. An artificial object was found on the moon, and it had been buried four million years ago. The object? Another mysterious black monolith.

Eighteen months later, a mission to Jupiter is in progress (we later learn that the buried monolith issued the directive to travel there), and we are introduced to the infamous HAL 9000. (This is the AI supercomputer controlling most of the spacecraft's functions—described in the film as a "conscious entity.") During the mission, this computer begins seeking its own ends. Sensing the craft's crew is a risk to its self-interest, it manages to kill all but one of them. This remaining pilot does succeed in disabling HAL, at which point a pre-recorded message is broadcast to him, revealing the truth about the monolith found on the moon issuing directions. The craft arrives at Jupiter only to find there floating in space beside the gas giant... another black monolith. This monolith is heralding neither the birth of tool use (as with the apes), nor the dawn of space travel and AI (as with the astronauts on the moon), but rather another level still. To provide the revelation, it immediately brings the pilot through a wormhole.

What follows is intentionally obscure. The pilot eventually enters a bedroom, and successively finds himself progressing through all the ages of life in this small room. On his deathbed as a very old man, in this same room, he again sees a black monolith appearing at the foot of his bed. Reaching up with his right hand toward the object, he is suddenly transformed into an unborn child floating in a sphere (on that same bed), only to, immediately thereafter, be found (still as an unborn child), looking down upon earth from orbit. The movie then ends to the tune of the same notes inspired by Nietzsche's work declaring God is dead, and man is now the super-being. Stanley Kubrick confirmed that this choice of concluding music was intentionally symbolic. He said:

> [At the film's ending,] as happens in so many myths of all cultures in the world, [the pilot] is transformed into some kind of super being and sent back to Earth ... We have to only guess what happens when he goes back.[292]

Kubrick here refutes any suggestion that we are *not* here confronted with a return of the Pagan demons. The movie's basic message—even when the details are presented so incoherently as to appear derivative of psychedelic drug use—is straightforward in its perversity: the genesis of man has nothing to do with God, but only with evolution aided by the intercession of technology from aliens; moreover, the future enlightenment we should long for and seek out will come from the intercession of these same extraterrestrials and AI.

Again clearing up any confusion, Kubrick gave an interview specifically about this movie. In it, he rejects "**Earth's monotheistic religions**," replacing them with an ET-belief inspired religion of scientism replete with its own apocalyptic pseudo-messianism that will bring about "**immortal machine entities**." The famed director stated:

> I don't believe in any of Earth's monotheistic religions, but... there must be, in fact, countless billions of such planets where biological life has arisen ... [including many] where it is hundreds of thousands of millions of years in advance of us. ... They may have progressed from biological species ... into immortal machine entities—and then, over innumerable eons, they could emerge from the chrysalis of matter transformed into beings of pure energy ... Their potentialities would be limitless and their intelligence ungraspable by humans.[293]

Kubrick, who did not believe in demons, nevertheless laid out the red carpet for them. How convenient a disguise would be provided for demons if only mankind would anticipate instruction from "benevolent extraterrestrials" who are so astonishingly evolved that they have become "**beings of pure energy**"—extraterrestrials, moreover, whose activities not only replace the Book of Revelation, but also the Book of Genesis. Extraterrestrials who explain both our origins and our destiny.

One may as well beg the Antichrist to arise. Yet these are not the words of some crazed fringe-blogger; these, rather, are the open convictions of men who have crafted modern culture into what it is. And they confess with a smile precisely what they were doing with their world-dominating works of science fiction the whole time.

***

Kubrick did not direct the film's sequel, whose own title also contains an amusingly failed prophecy—*2010: The Year We Make Contact*. The new director, however, ensured the same infectious principles formed its own message. The 1984 sequel's opening scene takes place in the "very large array" in New Mexico. (This is the same setting Carl Sagan used as the point of "first contact" with ETs in his book, "Contact," released the following year and later made into a major Hollywood film.) In *2010*, the *2001* pilot has returned to earth as a godlike entity who can influence electronics and manifest in apparitions to console earthlings and to announce impending great events. He has become a Vallée-esque "interdimensional" of sorts—the monolith's alien curators gave him this "permission."

One of the protagonists in the film, defending the HAL AI supercomputer (which, it turns out in this film, is actually quite benevolent and was simply misunderstood in the first movie), declares the essence of the AI Deception, saying: "**Whether we are based on carbon or silicon makes no difference. We should each be treated with appropriate respect**." The film then has the godlike former pilot soothingly talking to HAL (which "knows" it is about to be destroyed once its ship explodes), telling it not to be afraid, since "**We'll be together... where I am now**." The film impressively combines the AI and ET deceptions into one Antichristic manifesto.

The world—on the brink of annihilating itself through nuclear war—is eventually saved by the ET monolith's work to turn Jupiter into a new star (and its moons into new planets, which then resume the process of evolving life through the intervention of the ETs), and the film concludes with a monologue by the main character, who announces in a letter to his own son:

> We have seen the process of life take place. Maybe this is the way it happened on earth millions of years ago. Your children will be born in a world with two suns. You can tell your children that you remember when there was a pitch black sky, with no bright star, and people feared the night. You can tell them when we were alone; when we couldn't point to the light and say to ourselves, 'there is life out there.' Some day the children of the new sun will meet the children of the old. I think they will be our friends.

In a word: in the "old days," before we made "contact" with the ETs, things were dark. "People feared the night." But now, for the first time in human history, there is hope. We now know that "we are not alone," and that we will soon be embraced by a civilization of aliens. Whoever wonders whether *2001* presents an ET-messianism need wonder no longer; its sequel renders it explicit.

Arthur Clarke, the science fiction author of the novel on which the film was based (and who collaborated with Kubrick), made even fewer attempts to conceal the story's motive. In his book, people name this new star "Lucifer." To reward HAL, Clarke has the aliens who control the monoliths grace HAL by making it just like the Godlike pilot himself; some sort of immortal, immaterial being. Thus—all through the ministrations of extraterrestrials—mere animals become men, mere men become gods, and mere computers likewise become omniscient. In an Epilogue, Clarke has the primitive peoples that evolved on Jupiter's moon, Europa, worship fragments of *human* spacecraft that fall down there, relaying his view that all religion is simply the muddled worship of space aliens.

It was in Clarke's most famous work, *Childhood's End*, (published in 1953) that he rejected Christianity most vociferously. There, the "Overlords" (aliens) give the world a technology through which all past events can be seen perfectly, at which point, Clarke writes:

> ... [I]t had always been obvious to any rational mind that all the world's religious writings could not be true... [but] here, seen by some unknown magic of Overlord science, were the true beginnings of all the world's great faiths. ... Within a few days, all mankind's multitudinous messiahs had lost their divinity. Beneath the fierce and passionless light of truth, faiths that had sustained millions for twice a thousand years vanished

like the morning dew.

Behold the premises and aspirations of ET-science fiction.

Close Encounters of the Third Kind

If 1968 was the most pivotal year of sci-fi cinema, 1977 was a close second. By fast-forwarding one decade, we now find a society adequately primed for the climax of the first phase of Hollywood's sci-fi deception agenda. That year saw the release of what is considered the "greatest alien movie of all time," *Close Encounters of the Third Kind.* It was directed by Steven Spielberg—and was itself a remake of his first-ever film, *Firelight* (1964).

Any ambiguity regarding the benevolence of the aliens in *2001: A Space Odyssey, The Day the Earth Stood Still,* or similar films, was absent from this blockbuster. While the dark deception of *Close Encounter's* plotline is unmistakable, an entirely new realm of danger is also evident in the film. The entire buildup of the movie is itself a reenactment of the thirty years of news reports leading up to its own release (reports which, in the almost fifty years since, have followed similar, but heightened, patterns).

In brief, the protagonist, Roy (played by a young Richard Dreyfuss), is "fortunate" to be among those intimately brushed by a UFO making various displays of itself in Muncie, Indiana. He becomes obsessed with discovering more. In the various challenges he faces during that quest of discovery, he perseveres despite receiving many of the dismissals one often hears of "ET contactees." Viewers are thus induced to identify with him and to derive inspiration to likewise fight through any such "obstacles" they might encounter. Roy becomes increasingly eccentric in his obsession, and eventually his wife leaves him, taking their children with her. Although briefly distraught, Roy quickly forgets his now lost family; the pursuit of alien contact more than fills that hole for him.

The parallel plotline follows a character, Lacombe, who is openly modeled after Ufologist Dr. Jacques Vallée. Immersed in a crusade to discover more about the ETs, he globe-trots with his colleagues in the U.S. military. They are especially zealous to learn the origin of those famous five musical notes which, to this day, people around the real world will immediately associate with aliens thanks to this movie (not unlike the similarly famous notes of *2001: A Space Odyssey*). This effort brings them to northern India, where they come across a large group of Hindus chanting these very notes in unison. (The words they are chanting to the tune—not translated in the film's subtitles—are those of the messianic prayer, **"Aaya Re! Aaya,"** which means, "**He has come!**") The scientists send the same notes out to space, and in return they receive a set of coordinates leading to the infamous and appropriately named Devil's Tower in Wyoming.

Roy and several other contactees, however, had been receiving those same signals more directly—*telepathically*. Thus, the film simultaneously promotes both the (pseudo)scientific and the (pseudo)mystical search for extraterrestrials. The contactees' obsession with images of Devil's Tower and the five notes leads them to the same place, where Roy finds companionship with a woman who is not his wife, and whom he eventually romantically embraces. When all the obstacles and government conspiracies are finally triumphed over, both Roy and the woman (whose young son was also abducted the same night as Roy's first encounter) find their way to the secret military site constructed to welcome the ETs at Devil's Tower.

The aliens do not disappoint. In a scene no 1977 theatergoer will ever forget, their majestic mothership draws nigh, "coming with the clouds." (Revelation 1:7) It descends triumphantly, clothed in luminous splendor, and after a friendly musical exchange with the scientists and military men gathered, its doors open. First, the unharmed abductees of the past few decades emerge; then, the aliens themselves. An ecstatic Roy is soon after selected by them to board the craft. Enraptured onlookers watch him enter the spaceship with outstretched arms, bathed in refulgent light, and surrounded by a cohort of cute little extraterrestrials as if carried aloft by a cloud, reenacting the genre's recurring obsession with perverting the Ascension of Christ. (It is no surprise to discover that this scene has Roy finally reveal that he was born in 1944, making him 33 years old at the time of the film's setting and release.) The ship departs, implying Roy has willingly and permanently abandoned his family,[294] and the movie's viewer is left with a veritable *religious* longing to himself meet the same extraterrestrials.

Steven Spielberg not only directed this film but also wrote it, and his motive was not to provide escapist fiction, or an engaging thought experiment, or a setting for a meaningful moral of the story. Rather, his intent was clear: to merge Ufology "reality" (both in its personal and mystical aspects, and in its governmental and scientific aspects) with those themes that had hitherto been considered in science fiction. Spielberg aimed to immortalize and mythologize the "heroic conquest against government coverups for UFO and ET truth;" to summarize a movement and inspire its continued pursuit; to elicit a pseudo-religious faith and hope whose object is not God, but benevolent extraterrestrials.

All of these goals are made clear in the film's plot, but they are rendered undeniable in Spielberg's choice to openly model the movie's chief Ufologist character (Lacombe) after the real-life Ufologist Jacques Vallée. More significant still, yet another renowned Ufologist, J. Allen Hynek, even makes a cameo in the film's climax with his trademark pipe in hand. Throughout the movie, the tension and the excitement of UFO encounters, unexplainable phenomena, abductions, debates with stubborn skeptics, and so on, were portrayed as if by an author of political fiction, in order to make the viewer say to himself, "*This resonates so well with what I see in the political world today; this could very well be real!*"

Close Encounters' immense cultural significance is as sure as its corrupt message. Prolific film critic David Thomson's respected tome, *The New Biographical Dictionary of Film*, lauds the movie, writing that "**Close Encounters**

had a flawless wonder, such that it might be the first film ever made. ... [it] is as close to a mystical experience as a major film has come..."[295] *Close Encounters* has won numerous awards, and is consistently ranked among the top movies ever made. Whoever wishes to accurately chart out the most significant cultural trends will be sure to focus on this movie. Spielberg himself alludes to it being his defining legacy. One biographer relayed:

> When asked in 1991 to select a single "master image" that sums up his work, Spielberg chose one with powerful echoes of his first childhood memory: the shot of the little boy in *Close Encounters of the Third Kind* opening his living room door to see the blazing orange light from the UFO, "*that beautiful but awful light, just like fire coming through the doorway. ... there's a lot of promise or danger outside that door.*"[296]

Much promise and danger indeed. The "promise" of this film even magnified a deliberate and occult global conjuring of ETs in the year of its release.

This conjuring was orchestrated by a 1977 hit single, "Calling Occupants of Interplanetary Craft," which rose to chart-topping levels (*including the top 15 in the US and Australia, the top 10 in Canada and the UK, and the number one position in Ireland*). It was originally released by the Canadian rock band, "Klaatu"—named after the alien in *The Day the Earth Stood Still*—but a popular American duo, The Carpenters, released the version which dominated charts. The song is an explicit attempt to "telepathically" contact UFOs and implore them to come and save the world. Written in honor of "World Contact Day," which had been called for in 1953 by the "International Flying Saucer Bureau," the song's lyrics established that it was no joke. They begin with an instruction to "**close your eyes and concentrate**" so as to "**telepath messages through the vast unknown.**" The message thus announced to the cosmos tells the "ETs" that "**we'd like to make a contact with you. We are your friends... Our Earth may never survive. So do come, we beg you**."

After the release of this song, reports of alien abductions began to rise dramatically. Demons do come when conjured, no matter which one of their disguises is beckoned.

***

Spielberg's next alien-themed movie, released five years after *Close Encounters*, became just as famous, and it did so for clear reasons. In order for a societal grooming process to be particularly effective and unshakeable, it must seduce more than just the adults presently in power. It must also entrap the children who, decades later, will lead the world. What had been lacking thus far were films of a similar caliber dedicated to ensuring that boys and girls in their most impressionable and formative years grew ravenous for contact with aliens as a solution to their own ills and as the object of their own dreams. Spielberg, abhorring a vacuum, obliged. *E.T. the Extra-Terrestrial* was born.

E.T. the Extra-Terrestrial

1982's top movie release remained Hollywood's highest *ever* box-office hit until *Star Wars* in 1997, and it was another creation of Steven Spielberg. *E.T. the Extra-Terrestrial* stands in a class of its own as a cultural behemoth which refuses to diminish in influence over forty years later. The movie's own promotional artwork is not coy in illustrating its message. Depicting yet another perversion of history's most famous fresco—Michelangelo's painting of the Creation of Adam in the Sistine Chapel—*ET's* cover shows the finger of an alien reaching out to a child's finger on the backdrop of a view of earth from space.

This "children's movie" tugs on every heartstring imaginable in order to ensure that no child (or adult) who watches it leaves the experience without intimately identifying with the protagonists' fanatical love and longing for "E.T.," and Spielberg knew exactly what he was doing.* With serpentine cunning, the film expertly captures the hearts and minds of children while placing its prey firmly in the hands of the ET Deception. One of Spielberg's biographers describes the result: a "**universal pronouncement**" that would "**send audiences weeping from the cinema**." He then quoted novelist Martin Amis, who said:

> Towards the end of E.T., barely able to support my own grief and bewilderment, I turned and looked down the aisle at my fellow sufferers... Each face was a mask of tears... We were crying for our lost selves.[297]

E.T. was created both to seduce its young viewers and to induce its adult audience to weep "**for their lost selves**," while leaving them with the hope of finally finding themselves in an abiding "childlike" longing for extraterrestrials, who—they now hope—can provide what humans (and religion) cannot. As Spielberg said of the movie, it "**transforms the father-son relationship into something much more cosmic.**"[298] This transformation is noble when applied to God the Father, but is only a diabolical ruse when applied to science fiction.

The film's opening shows E.T.'s craft having landed in a wooded area of California, with innocent and happy aliens appreciating and studying the foliage. Sinister music begins playing as humans promptly show up, causing the craft to flee—though inadvertently leaving behind one alien (the eponymous "E.T."). The left-behind creature finds its way to the home of

---

* One of his biographers described the film's strategy, "**Aware that children are most impressed by what happens to children on screen, Spielberg concentrated on them to the virtual exclusion of adults, who are shot almost entirely from a child's eye view. ... Learning a lesson from Walt Disney's animators, who'd discovered that the most loveable faces were those closest to a baby's, with wide eyes and a high forehead, [to make E.T.] he pasted onto the photograph of a baby's unjudgemental face the eyes of poet Carl Sandburg clipped from another photo, and the forehead and nose of Ernest Hemingway and Albert Einstein. The result was a one-size-fits-all father figure without the censoriousness that comes with age**." (John Baxter. *Steven Spielberg: The Unauthorised Biography*. Harper Collins. 1998.)

the protagonist, a 10-year-old boy named Elliott Taylor (a name with unsubtle initials). As usual in such stories, we see a broken home (the father left the family for another woman) with no little discord dominating domestic life. Elliott, when he eventually discovers E.T., finds in him an escape, a friend, and even a savior.

E.T. forms a relationship with Elliott and soon they find themselves *mystically* linked—whatever one feels, so does the other. Soon, Elliott begins referring to himself as "we" (meaning both he and E.T.), signaling the deepest possible bond with the creature. In one of the film's most iconic scenes, E.T. touches a bleeding cut on Elliott's finger, instantly healing it just as, in the background, we hear *Peter Pan* being read aloud to his young sister by their mother (who is oblivious to the presence of the alien in the closet). The wound heals as she soothingly recites the famous seductive words: "*She says she thinks she could get well again if children believed in fairies. Do you believe in fairies? Say it. 'I do, I do, I do!' If you believe, clap your hands [clap, clap, clap].*" The message is not missed by any child who watches: it is a moral imperative that they believe in E.T.; so much depends upon it. As the scene fades, Elliott puts his arm around E.T.

Later on, the children do their best to help "ET phone home." On the night of Halloween, both E.T. and Elliott become quite sick (E.T., due to being on earth; Elliott, due to his mystical bond with the alien). Soon, the government discovers that E.T. is being hidden in Elliott's house. They descend with an army of men in space suits and quarantine the home to study the alien—who, alongside Elliott, is receiving intensive care in the quarantined home. One of the government men, however, reveals himself a secret ally of Elliott. He had waited for this moment since his own childhood, and the movie's adult viewers are induced to identify with him. He assures Elliott that it is a "miracle" that E.T. is here.

E.T. and Elliott exchange a glance and the boy tells the creature, who now appears to be dying, "*ET stay with me, please ... I'll be right here...,*" with E.T. responding, "*Elliot ... stay, stay, stay...*" before flatlining. At this moment, Elliott begins "separating" from the alien, and no longer mystically bound to its condition, he begins to feel better. After, E.T. is declared dead, and Elliott is given time alone with the corpse. With tears filling his eyes, Elliott says to E.T., "*I'm so sorry. You must be dead, because... I don't know how to feel. I can't feel anything anymore. You've gone someplace else now. I'll believe in you all my life. Every day. ET. I love you.*" At that, the alien resurrects, and communicates to Elliott that the spaceship is coming back to bring E.T. home. A daring escape and a lengthy chase follow as Elliott, along with his siblings and friends, take E.T. to the place from where it "phoned home," so that the spaceship may come and bring it back to its planet.

An exceedingly emotional farewell follows. Before E.T. boards his craft, he says "*ouch*" and points to Elliot's heart, acknowledging the heartbreak of the goodbye. This "tragedy" is rendered all the more bizarre since

E.T. had only been with Elliott for a few days, though the farewell is portrayed as one of brothers who had never left each other's side. They hug warmly to rising orchestral music as E.T. says, *"I'll be right here,"* while touching Elliot's head with his finger (now to triumphant music), and as Elliott's mother, tears streaming down her understanding face, looks on. After waddling up the craft's ramp, E.T. is enclosed by its doors while engulfed in light. The ship ascends and creates a rainbow as it departs, everyone looking on with the same religious awe and astonishment that Spielberg depicted so skillfully five years earlier in *Close Encounters of the Third Kind*. Henry Thomas—the actor who played Elliot—said of this scene, "**When I did the good-bye scene, I couldn't stop crying because I worked with E.T. every day and he was real to me**."[299]

It is not difficult to see what effect such a movie has on any children who watch it, not only on the actors within it. Just as a boy who watches *Batman* finds himself, the next day, certain that he will himself end all crime in his own city, so too a child who watches the present film gazes out his window each night with mystical longing for an extraterrestrial encounter of his own. This often proves to be no passing fad, since—unlike *Batman*—it is relentlessly fed by a steady stream of alluring UFO related claims in the real world. Elliott's connection to the alien most closely resembles Voodoo, and capping it all off is the call to the viewer to identify with Elliott; for in the days that would follow the story's conclusion, Elliott would have to continue his life without E.T. there physically. Nevertheless, his love and longing for aliens would remain "right here"—in Elliott's head—where E.T. last touched the boy before spectacularly ascending into the heavens—just as Christ assured His disciples, before ascending to Heaven, "*Behold, I am with you always.*" (Matthew 28:20) The viewers can be just like Elliott: they can keep the resolute fixation upon aliens ever in their minds, and the flame of hope for ET contact alive in their hearts.

Those who were Elliott's age when the *E.T.* was released are now middle-aged, and millions of them have since then only nurtured the same desires this film implanted. They are our leaders in government, commerce, media, and, yes, even the Church (one theology professor at a Catholic seminary in fact dedicated his lengthy 2021 book promoting alien belief, *Exotheology,* to Steven Spielberg). With each passing day, one learns of more and more men and women in positions of power and influence openly professing their deep longing for extraterrestrial contact, as well as their conviction that the time for this messianic contact is finally upon us. This phenomenon is not accidental; it is the fruition of a plan that has long been in place. The grooming process has worked exceedingly well.

> "In the inner heart of every person the voice of God and the insidious voice of the Evil One can be heard. The latter seeks to deceive the human person, seducing him with the prospect of false goods, to lead him away from the real good that consists precisely in fulfilling the Divine Will." — Pope St. John Paul II

***

Thus far we have focused on only the most indisputably influential films. Had we scrutinized slightly less popular ones, we would immediately see still more outlandish and explicitly diabolical themes. Each movie discussed above, however, has proven so impactful that it has been selected for preservation in the United States National Film Registry due to its great "historical, cultural, and aesthetic contributions." This Registry was established by an act of Congress in 1988, and it is no small matter for a movie to be included therein. Therefore, not only the constituents of culture, but also its curators and chroniclers, have declared these films to be defining stories of the modern era.

One can also "follow the money" to find evidence of the dominance of these trends of ET expectation in secular culture. For example, two men—George Lucas and Steven Spielberg—who both directed ET-based sci-fi movies, remain, by far, the richest movie directors in history. The multi-billion-dollar fortune of each blows away third place (Peter Jackson) by a factor of five. Steven Spielberg—who, of all directors, has proven most devoted to furthering both the ET and AI Deceptions—is the most successful director in history. He alone created six separate films that were each the highest-grossing in the years of their release.

One additional film should be considered, however, before we move on. For although it did not generate as much revenue in the box office or make its way into the National Film Registry, its impact is nevertheless massive and even more explicitly anti-Christian than what we have reviewed thus far.

Contact

The celebrity astronomer, Carl Sagan, is unique; his contributions remain culture-forming in the realms of both science and science fiction. Sagan's ET-derived pseudo-mysticism is illustrated most sharply in his 1985 novel, *Contact*, which became a major Hollywood film by the same name in 1997.

The hero of this story, Dr. Ellie Arroway, is an avowed atheist who believes science has disproved God and who has been dedicated since her childhood to discovering radio signals from aliens. The opening scene shows her at her amateur radio, with her father walking into the room to encourage her efforts. That night, after learning from him that radio waves can reach other planets, little Ellie asks her father: with these same signals, "*can we talk to mom?*" (The implication is clear.) Soon thereafter, her father also dies. A priest attempts to console Ellie, but she rejects his suggestion to trust in God's Will, and instead runs to her ham radio to try to speak with her late father.

As an adult, Arroway joins SETI (The Search for Extraterrestrial Intelligence), a program seeking to contact aliens via radio waves. With a zeal and conviction which can only be described as religious, she triumphs over many obstacles in that pursuit. A climactic scene then depicts the titular first "Contact," when her long-suffering dedication finally reaps its reward. An entrancing but simple melodic signal (ET-sci-fi's recurring theme) begins arising mysteriously as the shot pans across a mesmerizing view of earth from space. Interspersed is a backdrop scene of a famous Christian author appearing on Larry King Live (played by the real Mr. King), expounding upon the need to fill the "holes in our lives" that exist due to the modern world feeling *so alone*. We are then shown Arroway herself, reclining with closed eyes upon the hood of her car in the evening twilight amid the massive telescope array in New Mexico, serenaded by crickets, and listening intently to her headphones for the ultimate revelation from on high.

Suddenly, her eyes open and she sits upright. It has arrived. The signal reaches her ears immediately after the same Christian preacher, while lamenting modern life's holes, asks rhetorically, "*is it any wonder that we've lost our sense of direction?*" (Thus Sagan declares we need no longer fear such encroaching nihilism—now that we have Prophet Arroway's message from the supremely evolved beings to fill that hole.) We soon discover that this signal contains massive amounts of encoded information: blueprints for constructing a machine to enable one person to travel through space via a wormhole. The machine is built, but—harkening back to *The Day the Earth Stood Still* released half a century before—an apocalypse-minded Christian zealot destroys it.

Thankfully for Arroway—who had not been selected as the traveler on the first machine because she was an atheist—a second machine had been secretly built, and she is covertly selected to travel in this one. The contraption brings her, through various wormholes, to the star Vega's solar system. In her journey, she sees technological alien civilizations whizzing by, only to suddenly appear on a beach. There, she is gently approached by an apparition of her late father, whom she embraces emotionally but also realizes is actually an extraterrestrial. This ET had been reached by SETI, and it found humanity worthy of enlightenment (and constructed an image of Arroway's father by downloading her memories).

The extraterrestrial explains that there are "many of us" (aliens), and that humans are an "*interesting species ... capable of such beautiful dreams, and such horrible nightmares. You feel so lost, so cut off, so alone, only you're not. See, in all our searching, the only thing we've found that makes the emptiness bearable, is each other.*" The wormhole then brings her back to earth, with all those who had observed the event thinking no time had passed and that she went nowhere. Naturally, observers suspect a hoax once Arroway recounts her experience. She is grilled during a Congressional hearing, berated for her inability to provide any evidence that she had traveled through space and had seen alien civilizations. She admits she lacks proof, but insists she knew what she experienced was real, an "**awesome, wonderful vision of the uni-**

verse... [showing] that we are not alone," and exhorts her audience to accept her testimony "**on Faith**." We are then shown hidden details which confirm the veracity of her account. These details, however, are kept confidential, and the Search for Extraterrestrial Intelligence continues. The audience thus receives its call to action and, even to "faith"—though not to faith that saves. The end screen says, simply, "**For Carl.**" Sagan had died just seven months before its release.

This is the movie *Contact* in a nutshell. Although I saw the film soon after it came out 28 years ago, I had not seen it since. Nevertheless, the beach scene described above eerily haunted me. I no longer wonder why, for I now see that was exactly Sagan's intent. He knew that to promote his replacement for religion, he must tug the same heartstrings that would naturally lead to religion's authentic expression, while at the same time directing his audience squarely away from God. *Contact* at once provides a God-substitute that is transcendent, childhood-memory-centered, focused on a father figure, mysterious, current-events-relevant but future-oriented, and heroic. It does all this while pretending to remain entirely within the domain of science.

The beach scene, a perverse twist on the Apostles' encounter with Jesus by the Sea of Tiberias after His Resurrection, provides the climax of Sagan's antithesis to religion: an encounter with "The Other" that both consoles and inspires, while offering a tantalizingly partial explanation of reality and a mission moving forward. To further tug at those same heartstrings, Sagan has the alien appear in the form of the person the film's viewers have been vicariously longing for the whole time: Arroway's father. It is tranquilizing to hear his soothing voice—even if it was really an ET—assuring us that "we are not alone" because "we" (the many civilizations in the Universe) make the (apparent) "emptiness" of space "bearable." And it is pseudo-religious as it parodies the same assurance of Christ that is used manipulatively in *E.T.* and other such films; namely, His promise that, indeed, we are *not* alone, because *He* is with us. And just as the Apostles proceeded to proclaim their Faith in the Risen Christ to propagate Christianity, so too Arroway goes on to beseech her listeners to take her testimony—"*on Faith*"—as she propagates SETI with missionary zeal.

No longer did ETs need to be promoted only in the dull and uninspiring terms of electromagnetic waves and astrobiology and the like. It could now be presented as the false faith it was always intended to be—a faith which, like the true one, is propagated not by data discovered in empirical studies, but by the bold and heroic witness of individuals who *know what they saw*, and are willing to suffer in proclaiming it.

Strengthened by Sagan's efforts, ET promotion everywhere is permeated by an aping of the excitement one finds in the Bible's Book of Acts. There, early Christians proclaim Christ to the world because *they know what they saw*, and they know they must tell the world that *we are not alone—God is with us*. ET promoters likewise insist "we are not alone," though they do so Godlessly. In the Appendices, we review what even contemporary mainstream secular scholars of the phenomenon of Ufology note: namely, that ET-belief is now presented as *the* new Faith.

When seeking to make it into a religion, science's strength is, in fact, only its weakness. No one sheds his blood to proclaim Newton's Laws. There is no thrill, much less mystical attraction, in declaring such things. But it is entirely different to say, *"I know what I saw, though I have no physical evidence. It is awe-inspiring; it answers our heart's deepest longings. I will suffer for it and dedicate myself to it. Will you?"*

That proclamation has the capacity to inspire.

This is exactly why Sagan worked his way up to *Contact's* dénouement throughout the whole plotline, knowing he could only succeed in presenting ET-based Scientism as a substitute religion by leaving aside the very thing that supposedly renders science so supreme. Little wonder that Sagan was so dedicated to the project of this movie. It summarized decades of his career which, as we saw in the last chapter, was above all intent on replacing Christianity's faith and hope in Jesus Christ with a "scientific" faith and hope in extraterrestrial contact.

The Grooming Continues

The dawn of the third millennium saw no reduction of efforts to prime society for the Age of the Extraterrestrial. Too many such films were released to continue reviewing their synopses, but we can mention 2005's *War of the Worlds*—yet another Spielberg ET film—which brought H.G. Wells' novel that birthed the era of alien sci-fi to prominence again. Three years later another remake emerged: 2008's release of an updated version of *The Day the Earth Stood Still*, discussed earlier, wherein the "benevolent" extraterrestrial arrives to save earth... *from* mankind. Also in 2008, the "Marvel Cinematic Universe" was born, and it quickly became, by far, history's most popular and highest-grossing film franchise. Although it traffics in all the most absurd sci-fi themes, extraterrestrials are certainly pervasive in this cultural behemoth. A new era of *Star Wars* films began in 1999 and continued until 2019. More problematically, a popular *Star Trek* reboot also saw 2009, 2013, and 2016 iterations (continuing a slew of such films that did not pause through the 1980s and 1990s). Unlike *Star Wars, Star Trek* has always been not just an entertaining "space opera," but a pointed sci-fi deception; set, as it is, not *"a long time ago in a galaxy far, far away,"* but rather in mankind's own future (and often in its past or present on earth, as *Star Trek* frequently employs the time travel deception). Today, many Christians derive their theology not from Scripture and their science not from reason, but rather seek both from *Star Trek*. Since the series is replete with every sci-fi deception one can name, it is singularly effective in presenting a pseudoscientific toolkit with which every element of the real-world demonic ruses can masquerade as technologically possible and therefore potentially beneficent. This is to say that any evidence which a sincere Christian would use to discern the presence of the diabolical in some phenomenon has a Star-Trek-derived "technology" to explain it.

Moving forward, the film *Prometheus* (2012) had scientists in 2089 A.D. discover that aliens were mankind's true creators. 2014's *Interstellar* depicts a "**civilization that's evolved past the four dimensions we know**"—in the words of the film's chief protagonist—to rescue mankind from earth, now dying due to environmental catastrophe. Although this film inclines towards the time travel deception, that distinction does not divorce it from the ET Deception—many ET promoters insist that some or all "aliens" are actually time travelers (perhaps time-traveling humans). Similarly, 2016's *Arrival* has the ETs coming to save mankind by revealing the secrets of time travel. About this movie, Ufologist Dr. Jeffrey Kripal wrote, "**we should be approaching the total UFO phenomenon much more like *Arrival* and much less like *The War of the Worlds*.**"[300]

Perhaps most significantly, Steven Spielberg is slated to release a 2026 UFO themed film entitled "The Dish"—and, like *Close Encounters of the Third Kind*, the plot will be of his own creation. The film is now heralded by Hollywood-watchers as his "*big return to blockbuster territory,*"[301] and by ET-inclined commentators as his push to make the ultimate case for the "Day of Disclosure." Their hope is not unfounded, since Spielberg surpasses all other culture-curators in heralding the Nonhuman Intelligence Deception (both in its ET variety and, as we will see in Part Four, in its AI "Singularity" form). Whatever the movie's thesis, it promises to make a splash, for it will follow the legendary director's longest-ever pause in releasing feature films. He will have had ample time to craft it into what may be his magnum opus. On the forefront of Spielberg's own mind is doubtless the first movie he ever made, *Firelight*. This amateur film which he made at age 18 (in the year 1964) was also about Ufology, and its protagonist fought to convince the U.S. government of the reality of aliens. Sixty-two years later, will Spielberg strive to define himself as his first film's protagonist?

Irrespective of Mr. Spielberg, one thing is clear: the grooming of society has been taking place for many decades. It is malicious—expressly Antichristic—and it will not stop unless it has achieved its objectives.

Serious Christians will not forget that they were warned.

> "For the time is coming when people will not endure sound teaching, but having itching ears they will accumulate for themselves teachers to suit their own likings, and will turn away from listening to the truth and wander into myths." (2 Timothy 4:3-4)

***

Although the impact of fiction—particularly in its most powerful vessel, Hollywood—cannot be overstated, what distinguishes the ET Deception from similar nonsense is that it is not just another channel of sin and error in popular entertainment; rather, it is (like nothing else in history) a single movement whose founding "myth" in fiction is joined by allusions to the same in (false) mysticism and concurrent (deceiving) developments regularly reported even in mainstream news. Each of these prongs of attack is gargantuan and "of one mind" with "the beast" to "make war on the Lamb." (Revelation 17:13-14) What the demons have laid the groundwork for since the Enlightenment, what they have surreptitiously ensured adorns the silver screen, and what they have nudged the false prophets of the previous chapter to proclaim, they have openly spoken to the ears of the false mystics we must cover in the next chapter.

12. False Revelations

"What folly, therefore, or rather what madness, to submit ourselves through any sentiment of religion to demons, when it belongs to the true religion to deliver us from that depravity which makes us like to them!" —St. Augustine

Announcing the existence of extraterrestrials, or heralding their imminent earthly arrival, or insisting they are already here, or even facilitating occult communication with them, is such a common theme in false "revelations" and non-Christian religions that it borders on ubiquity. We see this theme in every outlet of deception we look towards—from ancient Pagan religions founded thousands of years ago, to the world's most popular and destructive contemporary cults, and everywhere in between.

Islam, Buddhism, and Hinduism

Major world religions that have existed for centuries or millennia—with the exception of Christianity and ancient Judaism—explicitly teach that extraterrestrials exist.

Beginning our considerations with the newest major religion, Islam, we see that it was welcoming, from the start, of the idea of non-human, non-angelic intelligences, which can easily be understood in the present day as aliens and UFOs. Islam's basic text, the Quran, repeatedly teaches the existence of aliens by its reference to many inhabited worlds. As one author explained:

> Even strictly literal readings of the [Quran] seem to support the idea of the plurality of worlds, which has been the basis of much science fiction. Qur'an 1:2 translates as '*praise to God, lord of the worlds*'. In total, the expression 'lord of the worlds' occurs forty-two times in the scripture. ... Qur'an 42:29 means: '*And among His Signs is the creation of the heavens and the earth, and the living creatures that He has scattered through them.*' ... [a footnote on a famous Quran translation reads:] '*Life is not confined to our one little Planet. It is a very old speculation to imagine some life like human life on the planet Mars.*'[302]

The great-great grandson of Muhammad himself, M. Ali al-Baquir, was born in the 7th century, and is a revered Islamic scholar. Addressing those who believe "*that God created only this single world,*" he declared: "**I swear by God that God created thousands and thousands of worlds and thousands and thousands of humankind**."[303] Islam, we see, has always unreservedly promoted the ET Deception as practically an article of faith. And it is noteworthy that for the first 900 years of the religion's existence, none of its many military incursions into Christendom succeeded in bringing with them the theological infiltration of this heresy.

Buddhists are even more prepared to welcome an announcement that we are in contact with extraterrestrials. Dr. David Weintraub summarized the Buddhist approach by noting that "**Buddhists view the universe as unimaginably large, ancient, and filled with living beings everywhere. ... The very existence of ET is built into the Buddhist worldview**..."[304] Buddhism's most popular sect believes that *"the universe is [like] a thousand-petaled lotus flower with ten billion world systems on every petal, for a total of ten trillion worlds..."*[305]

Presenting any comprehensive overview of extraterrestrials in Buddhism would quickly turn this section into its own book. Suffice it to say that Buddhism is not merely open to aliens, nor does it—like Islam—merely teach that they exist. Rather, Buddhism is itself a systematic ET deception. Of all major world religions, it is the most philosophically and theologically opposite of Christianity. It is also the one that most strongly favors alien belief.

Buddhism, however, is indeed aligned with the views of those Christians who believe in the possibility of multiple incarnations of God. Replace "Vairocana" with the "Cosmic Christ" or with "other Incarnations of God" in the Buddhist teaching that follows, and you are left with exactly what today's Catholic ET promoters propose: **"Vairocana represents the cosmic possibility of Buddhahood...The omnipresent Vairocana incarnates into countless buddhas... one for every world... all of the life forms in the universe have access to nirvana by way of their own particular manifestations of Vairocana.**"[306]

Hinduism—from which Buddhism took its own origin 2,500 years ago—today numbers over a billion adherents. It, too, is a fundamentally ET-welcoming religion.

> [In Hinduism,] a living being from Earth might be reincarnated into another kind of living being on another planet anywhere in the universe, and those beings in turn could be reincarnated here on Earth. ... Swami Prabhupada [founder of Hare Krishnas] wrote ... *"In that ocean [of space] there are innumerable planets, and each planet is called a dvipa, or island..."* According to the Vedic tradition there are 400,000 species of living beings in the universe with humanlike forms, many of them advanced beyond us. ... In keeping with the Hindu belief that the entire universe goes through cycles of creation and destruction and rebirth, Prabhupada identifies the lifespan for the universe as about 300,000 billion years...[307]

Other Pagan world religions, though not counting as many adherents as those described above, also commend belief in the existence of aliens. Basic texts of the religion of Sikhism hold that *"God created living beings throughout the universe and not just on Earth."*[308] Jainism—another Pagan religion which is about 2,500 years old—is in agreement:

> In Sutra 4.14 of the *Tattvārtha Sutra*, we learn that *"... The number of suns in the entire region inhabited by humans is [132] ... Each moon has an entourage*

> *of 28 constellations, 88 planets and 66,975 × 1014 stars."*... [Nothing] about Jainist concepts confines life to the single planet Earth, as souls could bond with any non-living matter and the matter containing souls could evolve in a biological sense anywhere in the physical universe.[309]

***

While Enlightenment-era Western scientists who first began believing in aliens congratulated themselves for having *"evolved beyond medieval anthropocentrism,"* they in fact only devolved back to an especially outdated thesis—one asserted for thousands of years in various Pagan traditions. The only difference was the new jargon. A scholar putting that feat on his résumé might also boast of believing in a flat earth. That view would entail "boldly rejecting the shackles of Christian tradition," which has taught a spherical earth from the beginning, and instead embracing Pagan Chinese tradition, which always held that earth was flat (until Catholic missionaries eventually showed them it was not).

A Christian ET believer finds himself forced to concede that, while almost all the world's Pagan religions have, for thousands of years, confirmed his belief in aliens, it is only his *own* religion—the One True Faith—that does not. Yet there is but one reasonable explanation for why belief in extraterrestrials permeates Paganism and is absent from (not to mention, condemned by) Christianity.

Even the circumstances that cause this contrast to arise display an exquisite exercise of Divine Providence. If indeed the alien and UFO deception is none other than *the* Great Deception itself—*the* penultimate eschatological test, as we will explore further in Chapter 15—then Christianity is perfectly poised to be the only religion which survives that test. All others will be strained out by the Great Sifting, and this principle will be the sieve.

Swedenborgianism, Theosophy, Adventism, and Mormonism

So-called "New Religious Movements" of the modern era—relatively recent alternative spiritual organizations coalescing around one heresy or another—do not merely draw heavily from the ET Deception. They emerged from it.

The earliest of these movements, Swedenborgianism, preceded the more famous Millerism by several decades. Beginning in 18th-century Europe, the Church today numbers thousands of members dedicated to the teachings of Swedish scientist and theologian Emmanuel Swedenborg (†1772).

It all began after Emmanuel himself—long tormented by strange visions and mental crises*—eventually beheld a vision of one he claimed was Jesus Christ. After this encounter, Swedenborg endeavored to establish the doctrines of "**the true Christian religion**." His religion, however, is most famous for asserting that creatures live on the moon and other places, and that they may be contacted through "astral travel" (a precursor to today's "remote viewing"). Dr. Walter Martin explains that this Church was "*...based largely upon space travel and contact with extraterrestrial beings.*"[310] Importantly, this first major "Christian" ET movement was not just an *example* of occultism infiltrating Christianity—it was the very *origin* of that disastrous infiltration. Dr. Martin also notes that, "**The door opened perceptibly for popularizing occultism in the Western world and merging it with Christianity in 1743...[through] Swedenborgianism**." In 1782, the first official Swedenborgian congregation was established in Maryland, and after that point, many American thinkers began distinguishing themselves as renowned promoters of belief in aliens. Dr. Martin continues his exposé:

> [Swedenborg's] revelations and visions of the deceased marked a new trend ... Many occult, esoteric, mystic, mind science, and New Age groups relied upon [his] groundwork for occultic contact with spirits. Swedenborg's metaphysics shaped the thinking of nineteenth-century New England transcendentalists such as Emerson, Thoreau, Ripley, Holmes, and Alcott, who later influenced Madam Blavatsky of Theosophy, Mary Baker Eddy of Christian Science, the early twentieth-century metaphysical schools, [and more]..."[311]

Indeed, this unholy mixture of belief in extraterrestrials with Christianity—conceived by Giordano Bruno but pioneered in its formally devotional aspect by Emmanuel Swedenborg—gave birth to other movements far more

* Some scholarly tomes refer to Swedenborg as simply having psychological problems, but in fact he proved he was not merely a run-of-the-mill charlatan or mental patient. He had access to a spiritual realm—but not Heaven. For example, reports circulated about his "mystical" abilities, with which he could "see" current events he was not present for and recount them accurately. He did this well before the first telegraph machines were built. In an academic tome published by Routledge, today's leading scholar on Swedenborg—Dr. Jane Williams-Hogan—noted that Swedenborg was interviewed by a Governor who posed many questions to him about a large fire that had happened the preceding day, though hundreds of miles away. The result? "**When reports came from Stockholm, they confirmed Swedenborg's description in every detail. Needless to say, people were amazed**." (*The Occult World,* Routledge Worlds). Routledge. 2016. Page 150) A supernatural origin to Mr. Swedenborg's insights is ruled out by the repeated blatant heresies he claimed to have received, so we can easily discern what remains as the explanation for his "revelations" that extraterrestrials exist. Another hallmark of the diabolical displayed here is that a *current* event was revealed to Swedenborg, through "remote viewing." This type of sight is easy for demons to bestow, since they can be anywhere they please and take note of what is happening in the moment. (The future, on the other hand, is not directly known by demons.) To this day, "remote viewing" remains a practice commonly associated with Ufology. This, for good reason: demons often grant this supposed privilege to people who bow down to them (even if this idolatry is undertaken without the knowledge that its object is a demon).

famous than his. Theosophy, Christian Science, the New Age, and many other esoteric or occult movements, are today rightly denounced by all faithful Christians. Yet they all ultimately derive from alien belief.

Before proceeding to the next century of developments, a word is in order on another "mystic," Helena Blavatsky, and the anti-Christian movement of "Theosophy" which she spawned. This, too, is central to the formation of contemporary ET belief. Religion scholar Dr. J. Gordon Melton describes Swedenborgianism as "**the first of the modern occult religions**," but he notes that Theosophy is "**the single most influential occult body of the modern era**," saying:

> Blavatsky created a new occult system which included a hierarchy of "ascended masters." ... Blavatsky included among the hierarchy masters who dwelt on Venus and with whom she was in contact. These masters she termed the Lords of the Flame and the Lord of This World, the head of the hierarchy for humanity.[312]

Blavatsky's "ascended masters" on Venus became a prototype for much ET "mysticism" of the centuries that followed. (Although later "revelations" quietly migrated the locations of these "masters" to planets outside of our own solar system, once technological progress showed earth to be the only inhabited planet orbiting the sun.) Evidence of Blavatsky's influence can be seen in two influential "ET Contactees" of the 20th century, George Williamson and George Adamski. Catholic researcher Ursula Bielski notes:

> Williamson was primarily contacting his aliens through a method all too familiar ... A Ouija board. ... He traveled to the compound of George Adamski ... Williamson's channeling experiences were truly inexplicable. ... His books, steeped in Theosophy and informed by the incessant stream of voices from "space," would prove extremely influential on future UFO believers. Williamson had been told by his space friends that they had founded Judaism and Christianity, and that extraterrestrials had disguised themselves in the personages of pretty much every key player in the Bible.[313]

We will see such views as these saturating the UFO cults of the 20th century, but we first turn our attention to noting the "denominations" of Christianity that formed in the prior century, also based on ET belief.

Seventh Day Adventism is a Millerite sect founded in 1868, based on the "Divine revelations" of Ellen White. Admired not only by Adventists, Mrs. White was named by the *Smithsonian* magazine as one of the "*100 Most Significant Americans of All Time*," and her religion still counts over 22 million members worldwide. This church surely teaches that extraterrestrials exist, but even more, this belief was also part and parcel to the sect's formation. Without belief in aliens, there would be no Seventh-day Adventism. Dr. Weintraub notes:

> Seventh-day Adventism emerged in the 19th century in part as a solution to theological problems stemming from the possibility of extraterrestrial

> life. The prophetess, Ellen White, described visions of extraterrestrial beings in different worlds that were "tall, majestic people" and entirely without sin.[314]

Among other heresies it promotes, this ferociously anti-Catholic religion teaches that the Roman Church is the "Whore of Babylon," that the pope is the Antichrist, that worshipping on Sunday—the Lord's Day—is the Mark of the Beast, that the elect cannot arrive at Heaven until the Last Judgment, and that hell is not eternal. These teachings prove that White's messages are either man-made or demon-made—and these messages speak in detail of aliens based on her "view of other worlds," and their inhabitants who "were so much more lovely than those on the earth."[315] White's false revelations not only mandate belief in extraterrestrials, but also promise that sufficiently faithful Adventists get to visit them.

A similarly massive modern sect, the Church of Latter-day Saints (LDS—or Mormonism), came into being not long after White's writings. Joseph Smith's own infamous "revelations" on the golden plates constitute the Book of Mormon, regarded by all its devotees as Divinely inspired. It goes without saying that no authentic Christian can put any credence in Smith's messages; they deny the Trinity Itself. Yet these "revelations," too, are replete with extraterrestrial belief, and the insistence that the Universe is teeming with planets filled with aliens. Even God Himself, the Mormons assert, is actually a certain being who lives on the planet "Kolob." The foremost chronicler of the ET debate throughout the centuries, Dr. Michael Crowe, notes:

> [ET belief] was not only propounded in [Mormon] scriptures; it was also advocated in statements by some of Smith's most important disciples... Brigham Young asserted in one of his discourses that '*[God] presides over the world on worlds that illuminate this little planet, and millions on millions of worlds that we cannot see...*'[316]

In its less public texts, Mormonism proposes doctrines that delve into the same radically esoteric content for which all UFO cults are known. (Although official LDS statements often seek to downplay or dismiss these teachings.) God, Mormons believe, was once an ordinary man who attained "exaltation" in order to become "the God of this universe" and beget his son, Jesus. (How it is logically possible that the Creator was first a mortal man is not explained.) Mormons who themselves attain "exaltation" will also become "Gods;" ruling over their own planets filled with aliens just as God rules over ours.*

---

* The other similarly influential quasi-Christian sect that arose in the 19th century is the Jehovah's Witnesses. Today they have almost nine million members. Although they are considered another "Adventist" movement, the writings that spurred it on (those of Charles Taze Russel) did not claim the status of *revelations*, but only commentaries on Public Revelation. Interestingly, Jehovah's Witnesses emphatically reject belief in extraterrestrials. Obviously, this does not vindicate the

***

Although the religions above involve belief in aliens, when we shift our focus to consider just a few of the avalanche of explicit "UFO religions" that have arisen in the last 78 years since Roswell, we see the expressly diabolical nature of what is transpiring even less veiled.

Nowhere else in history does one discover a similar degree of sudden proliferation of *religions* dedicated to a prevailing cultural theme as one finds in ET belief. For contemporary society, which has again become Pagan, UFOs might as well be prophets wielding thunderbolts. Whatever these phenomena consist in, one conclusion is beyond doubt: we are dealing with an intrinsically religious happening, not merely a cultural trend. (And all scholars who study the matter assert as much, as we see in the Appendices.) Confronted as we are with decidedly religious phenomena, however, we must now do what those scholars do not: *discern* them—spiritually.

UFO Cults

> "A California venture capitalist friend tells me that everyone he knows in Silicon Valley holds regular rituals to summon the "aliens" to give them technological wisdom."[317] —Rod Dreher

The post-Roswell era has witnessed the birth of so many UFO "religions" that providing a substantial summary of them here would be impossible. Therefore, we will cover only a small number of these cults, beginning with brief reviews of a few of them before moving on to more thorough explanations of a handful of such cults that are especially relevant.

Few by now will be unaware of the "Church of Scientology," considering it counts some of the most well-known celebrities alive—Tom Cruise, John Travolta, Elisabeth Moss, and others—as zealous adherents. The actual number of members is uncertain; scholars have pegged it at around 25,000, whereas the Church itself claims up to fifteen million members globally. Whatever the actual number, this cult is very rich, powerful, and influential.

Scientology was founded in 1954 by the most prolific science fiction author in history, L. Ron Hubbard. As we saw in Part One, Hubbard was primarily inspired by the uniquely perverse Satanist, Aleister Crowley—even being referred to as Crowley's heir. Hubbard denigrated Jesus Christ in terms too blasphemous to re-print here, but we should briefly review some of his claims, as recounted in the Fishman Affidavit submitted in a

---

movement—it is manifestly heretical. But it is quite significant that, among the three most noteworthy of the new 19th-century "Christian" religions, this is the one that does not claim a new revelation, and likewise is the one that does not believe in aliens. Instead, Jehovah's Witnesses—rightly revering the clear sense of the Christian Bible on this point—insist that in the whole physical universe, earth alone is the object of God's concern. It is, therefore, precisely because Jehovah's Witnesses *do not* claim a new Public Revelation that they have been protected from succumbing to the ET Deception.

1993 federal case. The affidavit recounts Hubbard teaching that every religion except Buddhism is ordered towards "the eventual enslavement of mankind," however he does find some truth in the Book of Revelation:

> No doubt you are familiar with the Revelations section of the Bible … there is truth in it. [The] anti-Christ represents the forces of Lucifer … [which is] a mythical representation of the forces of enlightenment, the Galactic Confederacy. My mission could be said to fulfill the Biblical promise represented by this brief anti-Christ period.[318]

Hubbard's own son—L. Ron Hubbard Jr.—confirmed that his father viewed himself as the Antichrist and as Satan. The younger Hubbard said:

> [My father] thought of himself as the Beast 666 incarnate. … The Antichrist. Aleister Crowley [also] thought of himself as such. And when Crowley died in 1947, my father then decided he should wear the cloak of the beast and become the most powerful being in the universe … you've got to realize that my father did not worship Satan. He thought he *was* Satan.[319]

Hubbard combined his unique talents in authoring works of science fiction with the occultism of Crowley to create a religious harbinger and herald of the Great Deception—hence its impressive degree of success, particularly among the very elite who guide contemporary secular culture.

The theology of his "religion" reads like the worst of science fiction films, with mankind's origin, destiny, morality, and spirituality entirely dominated by ideas related to extraterrestrials. Thankfully, however, Scientology has already been rightly condemned in many books, documentaries, articles, and so on, therefore we needn't here spend time reviewing its thoroughly evil antics. For the sake of our discernment, we need only recall that one of the most expressly Antichristic, large, and powerful organizations on earth today does not merely promote the ET Deception, but came entirely from that Deception.

One year after Scientology's formation, the "Aetherius Society" came into being. Founded in 1955 by George King, this "religion" does not simply claim to derive from some existing corpus of ET-themed revelations. It is actually dedicated to conversing with aliens, whom followers refer to as "Cosmic Masters." Mr. King claimed that, in 1954, after he had long been practicing Yoga, he heard a voice saying to him, **"...prepare yourself. You are to become the voice of the interplanetary parliament**." The Aetherius Society still has thousands of members across the world. The home page of its professionally designed website presents instructions on how to "open the door of enlightenment" through "King Yoga," in order to "**spread and act upon unique cosmic teachings given to Earth by highly-evolved alien intelligences**."[320] Aetherians claim to possess audio recordings of "the master Jesus," received through channeling, and these teachings parrot the standard New Age narrative. Aetherius remains prominent enough that *The New York Times* deemed the group worthy of focus. It published interviews with devotees in an August 2020 article, *Talking to Aliens is Their Religion*, and described the group as an "**international community bonded by the belief that the earth needs extraterrestrial support to survive**."[321] The video of Aetherius presented by the *Times* shows one of their altars, with a cross in the middle, but straddled by Pagan idols. The segment concludes with a quote, from their audio recording of "Jesus," atheistically exhorting: "**Even if you do not believe in God, then believe in good, please. It's the same thing**."

Nor has this society missed the opportunity to capitalize on the latest UFO frenzies. In a press release from July 2021, Aetherius gloated:

> For over 60 years The Aetherius Society has insisted that UFOs are real, friendly and alien, and that governments know it. ... [but] it is now clear that the Pentagon knows that UFOs are real and physical. ... While governments take faltering steps towards the truth, there is an explosion among the public of what might be described as UFORIA. ... Most people in [the US & UK] now believe in UFOs and intelligent alien life ... Disbelievers in UFOs are now in a minority ... In time ... they might be seen as somewhat akin to those who believe in a flat Earth.[322]

Yet another UFO Religion is the still-popular "Unarius Academy of Science," also founded in 1954. Its San Diego headquarters—just twenty miles south of the Heaven's Gate ET-suicides—can still be visited.

Appropriating standard pseudoscientific terms to explain its aims, the group promises that "fourth dimensional" physics enables its contact with advanced aliens on "higher frequency planes." Dr. Kenneth Samples relays that Ernest Norman, the group's founder, is a "**a spiritualist medium with prior involvement in flying saucer groups, [who] identified himself as the reincarnation of Egyptian pharaoh Amenhotep IV, as Jesus Christ, and as a space traveler in ancient Atlantis**." [323] Norman and his wife Ruth's efforts have managed to convince a large following to expect extraterrestrials to imminently come to earth with a "**monumental landing [to] usher in the cosmic golden age of logic and reason**."[324] This landing was first predicted for 1975, and then for 2001, but it is still awaited with conviction by Unarians. Sociologists Dr. Diana Tumminia and George Kirkpatrick wrote of this cult:

> According to Unarius, [Nikola] Tesla was the reincarnation of the Archangel Michiel and a sympathetic Space Brother. ... [who also] facilitated the transmissions of messages from his fellow scientists, Albert Einstein, Louis Pasteur, William Crookes, and Robert Oppenheimer who now reside on the planet Eros...[325]

They go on to recount the many revolutionary inventions the Unarians expect these aliens to bring—free energy, hospitals that heal all diseases, and so on—before relaying:

> The most important gift of the starships are the Space Brothers themselves. Powerful healers, they will eagerly train Earth people to change

their own societies. The collective consciousness of the Space Brothers will "overshadow" the peace efforts of our planet...[326]

Christians will remember to beware a man who claims to descend from above to grant a utopia and peace on the earth.

Another group with a disastrous legacy inspired by ET-belief is the so-called "Nation of Islam;" an eclectic American organization based on ideologies of black nationalism and Islam (recall that Islam's founding text, the Koran, explicitly teaches that aliens exist). Belief in aliens is not a mere tangent to the mission of this group, but a part of its basic motivation. The Nation of Islam's 50,000 followers are led by the black supremacist, Louis Farrakhan, who has announced the imminent arrival of UFOs as "divine spaceships" —which will descend to punish the nations for the sufferings of Africans. He claims to have met "Master Elijah Muhammad" aboard a UFO. One author summarized this group's teaching as follows:

> White devils enslaved the original black race, [but] they would eventually face annihilation brought about by bombs from the Mother Plane… [which] is able to defy Earth's gravity, [and] lift up mountains. ... Farrakhan claimed to have been abducted by [a] craft while on a visit to Mexico in 1985. On board, he encountered Elijah Muhammad… The science fiction elements of the Nation of Islam even facilitated an otherwise unlikely alliance with the Church of Scientology in the twenty-first century ... Like the leaders of the Nation of Islam, L. Ron Hubbard, the founder of Scientology, promised his followers enhancement and salvation through science and technology.[327]

Whoever is acquainted with any of the antics of the Nation of Islam knows of its thoroughly rotten fruits, which we need not review in detail here. The group's essence is captured well by its leader's insistence, on national television, that Adolf Hitler was "a very great man."[328] Here it is worth noting that Hitler was deeply formed by the "Thule Society," an early 20th-century group of German occultists that sponsored the rise of the Nazis. Thule's inspiration? It purported to channel messages from extraterrestrials near Orion.

***

One would be forgiven for supposing a UFO religion cannot get more degenerate than those described above. Yet we have scarcely begun to review the evil generated in the last several decades by combining belief in aliens with religion. The next ET religion we will review is not only openly Antichristic in its aspirations, but also boasts 100,000 members.

Raëlianism

> "[For] Raelians, the UFO enters this landscape as a new iconoclastic symbol in stark opposition to the crucifix." —Dr. Susan Palmer[329]

Some fifty years ago, a man now known across the globe as "Raël"—the eponymous founder and leader of Raëlianism—was just an ordinary twenty-seven-year-old aspiring race car driver named Claude Vorilhon. But one commute to work in the final days of 1973 changed everything. Claude felt a curious urge to depart from his ordinary daily drive in central France and instead trek across a nearby extinct volcano. There at the Puy de Lassolas, the young man described an encounter: amidst shining red light, a spacecraft descended. His own words recount what supposedly transpired next:

> I was petrified and remained absolutely still... Then the incredible happened. A trap door opened beneath the machine... First two feet appeared then two legs... I could see around [his head] a strange sort of halo.[330]

This 4-foot tall being with almond-shaped eyes who emerged from the craft proceeded to bestow revelations upon Claude. From that moment until the present time, Mr. Vorilhon has been "Raël" (the name given to him by the "alien"), and he has devoted his life to the mission of teaching the world about the *Elohim*—"those who came from the sky." These extraterrestrials are our real creators, he teaches, and they will soon usher in their reign on earth. That reign will be initiated through an intergalactic embassy in the Holy Land. The embassy will be the Third Temple—once the Pope and the Catholic Church are out of the way, that is. The foremost scholar of Raëlianism, Dr. Susan Palmer, noted:

> ...[T]he Raelians' anticlerical, pope-bashing, Vatican-closing crusades are in fact a necessary step [for them] ... [The] demise of the Catholic Church is necessary before the Raelians' message can be established on a global scale and the embassy can be made ready for the advent of aliens ... [Rael announced] that he was the true "pope" and "*not that usurper in the Vatican!*"... Raël continues to pit himself against the pope [saying:] "*Everything he is against, we are for ... contraception, homosexuality, divorce. All the values we espouse, he opposes.*" [331]

Like another ET-believing movement, Buddhism, Raëlianism is an atheistic "religion." It replaces any traditional belief in God with scientism, and its ethical teachings are purely hedonistic. The cult superficially praises Jesus and the Bible, yet only as a footnote to declaring that now, with enlightenment from evolved aliens, can we finally grasp the true meaning of what is contained in His words and in those ancient texts. Here as everywhere in ET promotion, the single premise of belief in aliens is leveraged to feign respect for Divine Revelation, while enabling all of its content to be "re-read"—that is, emptied of meaning. For "in the last days," men will hold "*the form of religion [while] denying the power of it.*" (2 Timothy 3). The prolific scholar of new religious movements, Professor George Chryssides, explains that, to Vorilhon, "**The alien claimed to be able to shed proper light on the meaning of scripture, and the name by which Vorilhon is better known to his**

followers, 'Raël', is said to mean 'the one who brings the light'" [332] Satan, indeed, is still fond of his original name—*Lucifer*—meaning, "light bearer."

As is also standard in ET-contact messages, a false eschatology is presented. Dr. Chryssides explains their beliefs: **"[Raël is] the messiah ...[and] time is getting short. The Elohim are expected to arrive around the year 2035, and already the Raëlians have attempted to make preparations**." What is being prepared? *"A new world, if you can take it,"* as ET promoters have long been prone to religiously repeat as one of their mantras. Chryssides summarizes it as follows:

> [The Elohim] now want to see the results of their genetic experiment. They come in peace, not to conquer the earth, but to establish a system of world government... The country that the Elohim have selected for the embassy is Israel. This is because the Jewish people play a particular significance in Raëlian thought. The Jews are believed to be descendants of the 'Nephelin' (i.e. Nephilim: Genesis 6:4), a special race of people who came into existence from sexual relationships in ancient times between some of the Elohim and humans.[333]

This one-world government that Raëlianism teaches the Elohim are bringing will usher in a Utopia that includes a corresponding one-world currency, the end of all war and crime through genetic editing, the colonization of other planets, and AI robots performing all manual labor. Discontent with the mere anticipation of these changes, however, Raëlianism works to bring them about.

In 1997, Raël founded an organization called "Clonaid," dedicated to achieving human cloning, as it sees this initiative as a step towards technological immortality heralded by the "aliens." (Remember Frank Drake?) With symbolic and undoubtedly deliberate timing, Clonaid claimed—on the third day of Christmas in the year 2002—that its technology enabled the first-ever human clone to be born.

In the intervening decades, Raëlianism has remained active in promoting its own political and philosophical ideology; the tenets of which include gender ideology (i.e., sex being a fluid and artificial concept), radical feminism, relativism, nudism, sexual hedonism, abortion, homosexuality, a rejection of marriage, childlessness (or, at most, two children), and other doctrines a Christian should expect from those heralding the Antichrist.

Raëlianism numbers tens of thousands of members—perhaps as many as one hundred thousand—who constitute an army dedicated to enacting the directives received by a demon wearing the disguise of an alien.

Heaven's Gate

> "Not only is suicide a sin, it is the sin. It is the ultimate and absolute evil... The man who kills a man, kills a man. The man who kills himself, kills all men; as far as he is concerned he wipes out the world." —G.K. Chesterton

On one spring day in the year 1997—the UFO Era's Golden Jubilee—thirty-nine members of a group calling itself "Heaven's Gate" were found dead inside a Southern California home. Of the dozens of corpses, each one (save the two most recently deceased) was ceremonially dressed and found lying tidily upon a separate bed. It would soon be discovered that this mass suicide had been meticulously planned and methodically undertaken in the time leading up to the grim discovery that fateful 26th of March.

All the deceased had freely and willingly committed suicide together, confessing their certainty that this act would bring them into union with the "extraterrestrials piloting the UFO trailing the Hale-Bopp comet." That comet's closest approach to earth was surely the right time to—in this UFO Religion's words—"reach the Next Level." This necessitated separating the soul from its "container," the body, so that the ultra-evolved and technologically-advanced aliens could access it.

It remains the largest mass suicide in American history.

"Do's Final Exit," the video filmed by the cult explaining what they were about to carry out, is still readily viewable online.[334] In this 90-minute sermon, we see that the group's founder and leader, Marshall Applewhite ("Do"), is a man who appears in full possession of his faculties while calmly and (apparently) logically explaining the importance of what the group was about to accomplish. Indeed, the horror of demonic possession is no match for the havoc that can be wrought by a diabolical premise freely embraced by a rational mind.

Such a tragedy as this should surely prompt national reflection. The scholar quoted earlier, Dr. Chryssides, also commented on this event, and he accordingly pondered how such a situation was possible. He wrote:

> ... [U]nlike the Peoples Temple in Jonestown, Guyana, in 1978, there were no concerned parents pursuing them, no congressmen investigating, and there was no siege situation, as there had been with the Branch Davidians in Waco, Texas, in 1993. The group was non-violent... [so] the obvious question was how so many people could have been persuaded to end their lives in such a strange way... [335]

This "obvious question" has an equally obvious answer. The Heaven's Gate members who made the supremely tragic decision both repeatedly and clearly explained, beforehand, why they were about to do what they did. Anyone is free to watch their lengthy explanations. Every aspect of their undertaking was justified by the ET Deception; particularly the supposedly "Christian" variety thereof.

> "The final objective is death, because God is the god of life, while Satan is the god of death. How many suicides are inspired by the devil! Even mass suicides are inspired by Satan." —Exorcist Fr. Gabriele Amorth

Those who make a pact with demons will often preserve the narrative from below, come what may. Thinking the pact is with "aliens" does not weaken its death grip.

Evidencing this preservation of demonic influence is a man named Rio DiAngelo. He was a Heaven's Gate member, though absent for the suicides. Nevertheless, he remained convinced of the cult's authenticity, unshaken by the tragedy. Still more evidence is found in two other former members of the cult who regretted leaving it—*after* learning of the suicides. Accordingly, they undertook their own "Final Exit" to meet the aliens, attempting suicide just two months later (one succeeded at that time; the other succeeded in yet another attempt the following year). And even more former members of the cult who survived continued speaking highly of it and remained hopeful that they too could at some point "Level Up" to embrace the extraterrestrials.

Heaven's Gate was not simply another cult with a charismatic founder who latched on to one more or less arbitrary theme he could leverage for personal power. Scholars of the movement have noted that the leaders had no particular "magnetism." Their power, instead, was due to great success in convincing followers that they truly were "evolved" aliens who could teach the means required for others to arrive at a similar state of being. They simply fed off the ET belief of their audience.

Contemporary ET believers may smirk at the "uneducated gullibility" of the Heaven's Gate members for believing its leaders were extraterrestrials. In fact, many members were quite educated and intelligent, and were apparently serious Christians. The smirkers, therefore, should be reminded that pride cometh before the fall, and they should ponder, *"How exactly would one who believes in aliens go about proving that some individual cult leader is not the alien he claims to be?"* The cult leader himself could easily offer a refutation of any such attempted proof—using just those premises which are inextricable from ET belief itself.

Aside from the suicides themselves, therefore, the greatest tragedy of Heaven's Gate is found in the intervening decades of failure to learn any lessons. The UFO rabbit hole's destination is death: whether physical or spiritual (or both). One cannot hope to navigate it in such a way as to find a hitherto elusive happy ending. One can only come to his senses—and escape it—or he can follow it to its logical conclusion. The latter was the choice of "Do" and his devotees—including his "devout Christian" disciples, which Heaven's Gate seduced with a message permeated by apparent Christianity.* Explaining the nature of Heaven's Gate (two years before the suicides), Jesuit priest Fr. John Saliba explained:

---

* The leaders of Heaven's Gate bedecked their claims in Biblical terminology so thoroughly that some scholars of the phenomenon have struggled to determine whether the cult is more accurately called a Christian sect or a UFO religion. Its founders, Bonnie Nettles and Marshall Applewhite ("Ti" and "Do") called themselves "The UFO Two" at times, but also presented themselves as the "Two Witnesses" of the Apocalypse (cf. Revelation 11:1-14). They described their mission as one almost identical to that of Jesus Christ. Heaven's Gate viewed the Incarnation, Death, and Resurrection of Christ in terms borrowed from ET belief, and they deftly applied those same exotheologies to their own group. It was precisely the willingness of Christians to believe in aliens that provided the principles needed to rationalize and proselytize this supreme tragedy.

> Christian UFO cults come into being when the presence and activity of flying saucers are accepted as central to the Christian message. ... Their theology is based on the biblical theme of the imminent coming of God's Kingdom, which they see heralded in the advent of flying saucers. They are so convinced that UFOs will come to rescue the human race in the near future that they travel with their followers from hilltop to hilltop in expectation of their advent.[336]

There are no major American cults seducing Christians into the worshipping of Hindu deities, much less the sacrificing of oneself to these same idols. Christians have always been on guard against such deceptions, so when demons assume the guise of Krishna, Christians will generally have nothing to do with it. But when the same demons assume the guise of a UFO or an extraterrestrial, many Christians will open wide the door and roll out the red carpet.

Dr. Chryssides explains how the Heaven's Gate suicides were prompted by the atmosphere of ET belief which saturated the times (and which, I add, saturate the present times still more thoroughly). "**Many people**," he notes, "**had become disenchanted with conventional Christianity, and were therefore seeking alternatives**." Ever since, belief in aliens has provided that alternative, for the groundwork had already been established. Chryssides continues:

> Belief in space travel and extraterrestrials was part of the *zeitgeist* of the 1970s ... Erich von Daniken's bestselling book *Chariots of the Gods?* Offered an alternative interpretation of Christian and other scriptures, claiming that extraterrestrials had already landed on earth many centuries previously ... The Roswell incident of 1947 had convinced many that a spaceship had crashed on to the planet Earth ... In popular culture, space travel was exploited in films and television programmes, such as *Star Trek, Star Wars, E.T. the Extraterrestrial*, and many more. ... Claims of alien abduction were beginning; space exploration, still in its infancy, coupled with the proliferation of science fiction, raised the question of whether planet Earth was the only location of intelligent life, and belief in extraterrestrial visitation was [widespread].[337]

He concludes by accurately observing, "**Nettles and Applewhite's initial publicity ... was a teaser.**" Nevertheless, ET promoters today desperately argue that this is all unproblematic for their mission. They dismiss Heaven's Gate as being merely an example of the damage brainwashing can cause. Dr. Benjamin Zeller, another expert in Heaven's Gate, easily refutes this dismissal: "**Members said they joined of their own free will. They recorded exit statements. They wrote autobiographies that explained why they joined this movement**."[338] Then, he aptly observes that, "**If we're going to call that brainwashing, then I don't know what we're going to call free will.**"[339]

Other commentators on Heaven's Gate—rightly outraged at this horror led by Applewhite, but lacking Zeller's competence and insight—are known to shout that there were *"38 murders, and only one suicide."* Understandable as is the emotion that leads to such outbursts, it simply does not represent reality. Did Applewhite also "murder" the other Heaven's Gate members who took their lives—for identical motives—well after his own death on that infamous day in March?

No. These were suicides if any deaths ever were. Those who deny that fact do so because they fail to appreciate the supreme danger here, which is found in the degeneracy of the ET- belief principles upon which the cult was built. It is these principles which the facts of history unambiguously instruct us to reject. This is evident not only in the case of Heaven's Gate, but also everywhere we look—even if we merely turn our attention to what was transpiring the very same year, one country to the north. For, also in 1997 (but in Canada), another UFO cult was inducing its members to commit mass-suicide to rendezvous with space aliens.

***

The Solar Temple was formed in 1984 in Geneva, Switzerland. Founded by a homeopathic "physician," Luc Jouret (who would later decide he was not a physician at all, but rather Jesus Christ) along with a New Age yoga instructor, Joseph Di Mambro, this "Temple" claims to trace its origins to the Knights Templar revival that took place after the French Revolution. The group's philosophy is borrowed from Theosophy, the New Age Movement, and Rosicrucianism (the spirituality practiced by Ufologist Jacques Vallée). Unlike Heaven's Gate (which required celibacy)—the Solar Temple broke up existing marriages only to re-pair the separated spouses with other members of the cult. These arrangements were executed at the orders of the "Cosmic Masters." Di Mambro's role in this regard was most important of all, however. At the age of 58, he took a young drug-addicted woman as his concubine, and when she was found with child, this pregnancy was declared a "theogamic conception" which would bring about the great "Child King." But their plans for this "King" did not pan out, so they turned to other methods.

Although equally desirous of union with extraterrestrials as were members of Heaven's Gate, the devotees of this group caused the deaths of almost twice as many people; using both suicide and homicide. This cult's members also believed that ritualistic suicide was their gateway to union with ETs—not by transporting them to a UFO trailing a comet, but by enabling them to be born again on the Sirius solar system[340] (which has long been an object of fixation in various ET "revelations").

The deaths caused by the Solar Temple cult include the events of 1994, when fifteen committed suicide and thirty were murdered, including many children as well as an infant; the events of 1995, when sixteen more suicide and homicide victims of this cult were discovered in France, including an Olympic athlete, Edith Bonlieu; and the events of 1997, when five more members committed suicide in Quebec, Canada. During this period, other attempted Solar Temple mass suicides were thwarted by authorities, preventing the death-toll from climbing even higher. Despite these interventions, seventy-four people would ultimately die at the hands of the group.

We must distinguish the nature of these heinous acts from other mass "suicides," for extraterrestrial-belief inspired mass suicides stand in a historical class of their own.* In both the Heaven's Gate and the Solar Temple cases, we see modern Western history's largest mass suicides, which were undertaken not out of the desire to avert some undeniably impending horrendous fate, but rather out of the diabolically deluded conviction that suicide enabled communion with aliens. Just as significantly, we see the strength of the delusion persisting after the initial tragedies, with multiple rounds of suicides taking place (even with months or years in between), and surviving members continuing to praise the cult's principles. The Jonestown deaths (mostly murders) on the other hand, did not persist beyond November 18th, 1978. There was no lingering diabolical deception. Everyone realized how evil it was, and everyone who had been connected with Jim Jones' "People's Temple" raced to denounce this group and its founder.

With UFO religions, however, we see a foretaste of the "**strong delusion**" (2 Thessalonians 2:11) that Scripture prophesies for the times of the Antichrist. Not only the suicides, but also the infiltration of deception into Catholicism gives evidence of this foretaste. The ET "revelations" we will review next have not remained at a "safe distance" from the Church.

Urantia

Thousands of pages long and incredibly comprehensive, *The Urantia Book* is said to have simply materialized on paper by the ministrations of extraterrestrials in prohibition-era Chicago. Later, in 1955, the text of these "revelations" was published, and they have since been massively influential. Also referred to as "**The Fifth Epochal Revelation,**" *Urantia* is far greater, its proponents insist, than the comparatively lowly fourth such revelation (Christianity). Today, untold millions treat this book as an authentic work of mystical literature, it has spawned religions of its own, and it remains a best-seller in many languages.

---

* For example, the Jonestown Massacre, from which derives the expression "drinking the Kool-Aid," remains the most well-known "mass-suicide" (and with good reason, considering its enormous death toll). Many papers have been published about it, and the consensus appears to be that this event was primarily enabled by the machinations of one particularly evil and charismatic man, Jim Jones, who managed to elicit a fanatical degree of blind obedience from the followers of his cult. Unlike with Heaven's Gate, this event truly *was* much more of a mass-murder than a mass suicide, hence its rightful title as a "massacre." The victims had most likely already been secretly drugged in a meal they were given earlier, and they were surrounded by armed guards who they knew would shoot them if they did not drink Jones' concoction. Most of the other historically largest mass-suicide events were efforts to avert clearly impending horrendous fates. Generally, they preceded the arrival of incoming armies known for atrocities. Clearly, such mass-suicides as these are not remotely comparable to what we are now considering

> "Our Father in whom consist the universe realms ... give us this day the vivifying forces of light" (144:5.2,6) "Our creative Parent, who is in the center of the universe ... deliver us from inertia, evil..." (144:5.12,19) —*Alternate versions of the Lord's Prayer suggested by Urantia*

The "celestial beings" who "revealed" this text teach that earth—which they refer to as "Urantia"—is just one of innumerable inhabited worlds in one particular "local universe." That local universe itself is only a part of "superuniverse number seven," and each of the many "superuniverses" contain tens of thousands of "local universes." All together, we are told that seven trillion inhabited planets exist (cf. *Urantia*, 15:2.17). As everywhere else in ET promotion, the "revelations" claim that earthlings are unfortunately quite "unevolved" in comparison to the beings on other planets, hence our dire need to make contact with the extraterrestrials and learn from their superior ways.

Also resembling other New Age "revelations" that seduce many Christians into their clutches, this one speaks extremely highly of Jesus—even appearing to acknowledge His Divinity—before going on to later deny the Virgin Birth, and, finally, to teach that Jesus is only one of almost a million "Creator Sons" (one, moreover, who has 10 million other inhabited worlds to minister to in our particular "local universe"). For *Urantia*, Jesus is "Divine," sure, and He is the second person of "a" Trinity, sure. But once you hear what the "extraterrestrials" have to say, you realize just how much grander the big picture really is. This, too, is a ubiquitous strategy in ET belief (not only in full-blown "ET revelations")—superficially conceding the truths of the Faith, while aiming to make them seem so *small*.

UFO religions have been spawned by this text, including one that became so influential as to prompt Dateline NBC to go undercover and eventually air an exposé highlighting its nefarious ET-inspired antics on primetime television in 1998. The focus was "Gabriel of Urantia," or "TaliasVan," whose real name is Tony Delevin. A Pittsburgh native, he started the Urantian cult called "Global Community Communications Alliance," then referred to as the "Aquarians." "Gabriel" (Delevin) assures us that "there are trillions of inhabited worlds,"[341] and that this recognition changes everything about life on earth—especially religion.

Delevin's legal threats have made this Dateline episode difficult to find, but those with some web-searching skill can still access it. Therein, NBC's John Larson uncovered Gabriel's "New Garden of Eden," where cult members awaited the arrival of aliens to save believers from a doomed earth. Jesus, the Angels, Adam and Eve, and virtually all Scripturally significant individuals were all extraterrestrials, according to Gabriel. Indeed, most of his preaching was—and still is—couched in Christian terminology. "Almost all of" traditional Christianity could be preserved, but with a sci-fi twist, as is seen throughout today's "exotheologies."

NBC interviewed ex-members who shared that the cult turned their lives into a "living hell." Gabriel demanded everything from his followers, and he did so without remorse, because in return, as he unabashedly said in an interview, "**we give them the bread of life, which is Jesus**." In accordance with his "higher knowledge" of God's Will, derived from the extraterrestrials, Gabriel would break up marriages and direct his followers to the spouses they were "supposed" to marry. (Polygamy was generally not allowed—except of course for Gabriel himself.) Children, too, would be separated from their parents and placed with those whom the ETs, through Gabriel, deemed fit.

The intervening 26 years since the NBC episode's release have not seen the cult's demise. Its leader remains as active and influential as ever.* Located fifteen miles north of the Mexican border in Arizona, Delevin and his followers dutifully await the imminent arrival of Jesus Christ—on a spaceship. But he who is coming, "**The Promised One,**" Gabriel teaches, **"is also known as Kalki Avatar, Maitreya, Krishna, and Imam Mahdi**."[342] Amusingly, this false prophet—sporting extravagant robes and facial hair reminiscent of an Old Testament patriarch—also published a 2022 video,[343] sternly warning against false prophets. He then calls on his listeners to quit their jobs and "*come under the authority of true Divine Elders*." Failing to do this could "*lead you to your destruction,*" because everything "*in the third dimension, which you are a part of, will fall apart.*" He solemnly declared that this dimensional disintegration would occur by August 2024. "**The New Jerusalem is coming,**" are his parting words, before the screen directs viewers to the website of the "University of Ascension Science," which in turn boasts—not inaccurately—of having "**touched thousands of lives for over 30 years.**"[344] Whose lives? Christians, predominantly. And many Catholic Christians at that. Yet this is scarcely a fraction of the faithful infected by *Urantia*.

The breathtaking insidiousness of this "ET revelation," therefore, is best illustrated by a less newsworthy individual than Tony "Gabriel" Delevin. For "revelations from extraterrestrials" do not only result in blatant cults that destroy the lives of dozens, hundreds, or thousands. Even worse, their teachings infiltrate mainstream Catholicism—unopposed—and thereby ruin the faith of millions.

In 2023, author Gregg Tomusko published a book entitled *The Catholic Church and the Urantia Book*, dedicated to extolling the supposed harmony of both. Mr. Tomusko describes himself as a daily Mass-attending devout

* The cult's associated grade school boasts that it acknowledges each child's "*latent astral talents and abilities gained in his or her past repersonalizations (incarnations)*." (gssu.org) One hopes these poor children will soon be rescued. *Urantia's* attacks on youth, however, begin well before childhood. In one popular 2024 video, Delevin quotes *Urantia*, stating "*for the majority of the 8 billion people on our world, the soul begins to be gradually developed after birth, and [is] not considered a soul until a person is around 5 or 6 years old*." Delevin himself is also adamant, declaring, "**Abortion is not murder... it is a woman's right to choose.**" Instead of regarding abortion as murder, he concludes, "**we must be open to the Fifth Epochal Revelation written in the Urantia book**." That revelation will surely deliver us from such "backwards" beliefs as regarding an unborn baby, or a four-year-old child, as a human being.

Catholic, a former seminarian, and a current confirmation class teacher. He presents Catholicism and the *Urantia Book* as being "**spiritually the same,**" while issuing lengthy descriptions of why he regards Biblical and Urantian teachings as mutually supportive. Tomusko writes that "**the Father's will was for Jesus to live life as a man, to gain the experience of being his lowest free-will creation**..."[345] We saw this insidious premise earlier, in the case of Monsignor Balducci. Like most ET promoters, these two believe that the human race is on the "bottom rung of spiritual evolution" among the many civilizations of the galaxies or so-called "dimensions." Nevertheless, we cannot fault Tomusko for one thing. He does maintain internal consistency. Jesus had a human nature like ours, so it follows that whoever dismisses human nature as holding a rank immeasurably below the "more evolved aliens," cannot avoid succumbing to blasphemies against Jesus Christ Himself.

Despite these heresies, Tomusko usually comes across like many other Catholic *Urantia* followers: orthodox-minded. He shares that his Latin pocket prayer book is always kept handy, he draws from St. Thomas Aquinas, and he speaks highly of the Divine Office in its traditional naming: Lauds, Vespers, and so on. He then says that, "**The Catholic Church and the Urantia Book introduces us to the Big View**." Explaining why, he writes:

> They're always talking about eternal life ... The Catholic Church has fought the devil through the centuries. She reminds us that we are in a war against '*Satan, and all the evil spirits, who prowl through the world seeking the ruin of souls.*'[346]

Immediately afterward, he laments: *"Like a light bulb that continues to get dimmer, we've become complacent to evil's entrenchment and steady victories on our world. Christians need a wake-up call..."*[347] Proclamations such as this sound wonderful. Their presence in the works of ET promoters, including theologians, exorcists, and experts in demonology, beguiles their readers into supposing they need not be concerned; that they need not wonder if this scholar himself may have become just another victim of his own field of study's concerns, like a psychiatrist developing schizophrenia and proceeding to write books extolling its virtues.

Tomusko discloses how he teaches his Catholic confirmation students, in the 7th through 9th grades, about *Urantia*. His method consists in sharing teachings from *Urantia* while modifying individual words therein to make it sound less esoteric. For example, as a dedicated devotee of *Urantia*, he "knows" that the "real" name of Jesus is actually "Christ Michael," but when reading passages from *Urantia* to his confirmation students, he replaces that with the "mere earthly" name for Him, "Jesus." This is the Name, mind you, which Sacred Scripture teaches is "**the name above every other name**," so much so that:

> ...[A]t Jesus' name every knee must bend in the heavens, on the earth,

> and under the earth, and every tongue proclaim, to the glory of God the Father: JESUS CHRIST IS LORD! (Philippians 2:9-11)

It is no mystery why ET revelations show a proclivity for drawing their devotees away from the Holy Name of *Jesus*, and towards more vague references to a "cosmic Christ." The name of Jesus burns the demons.

But Tomusko does not stop there; he provides an intricate guide on how to incorporate *Urantia* into everything Catechetical, going so far as to present a selection of readings from it for each day of Holy Week and various hours of the Passion.

Highlighting Mr. Tomusko's case is important here to convey a broader paradigm. For although he is not exactly a *New York Times* bestseller, there are many Catholics like him who reach much larger audiences. (Tomusko's slip was that he revealed the strategy of those who overshadow him.) They operate on the same principles as Tomusko while pretending it is "orthodox" to advocate for a radical re-thinking of Catholicism in accordance with their beliefs about aliens. One would not necessarily discover, upon meeting them—or even being one of their devoted followers, students, or parishioners—that they hold wildly esoteric Faith-undermining views derived from revelations given by "extraterrestrials" (demons). For their disguises are excellent. They cover their tracks well, they blunt the impact of their stealth Paganism with careful word choices, and they have a quiver full of rationalizations—including endless appeals to their résumés—to pretend it could not possibly be demons that are the real origins of their neo-Gnostic "exotheologies."

> "What shocks and amazes me is the perfidy, cunning, and intelligence of how the new postulates are implemented and linguistically disguised. I have always considered theologians to be the greatest and most gifted of the sophists. The phenomenon always makes me think of the Antichrist ... [who] guts the faith from the claim of absoluteness and exclusiveness of Jesus Christ." —Bishop Marian Eleganti[348]

Those Catholics who promote the ET Deception with such seductive cunning—and who are accurately described by Bishop Eleganti's words above, as they *"gut the faith"* of the *"absoluteness and exclusiveness of Jesus Christ"*—haunt the ranks of not just Catechists but also clergy, (putative) mystics, apologists, professors, theologians, exorcists, and, increasingly—as the climax of the Great Deception draws near—they will likely be found among the ranks of Bishops; even those working in the Vatican.

The Divine Messengers

The Voice and Echo of the Divine Messengers were commissioned, we are told, by the Virgin Mary—and the "regents of Andromeda"—as prophets in 2011 to announce to the world a new stage in spiritual evolution by the revelation of the extraterrestrials that populate the galaxies. This, indeed, is what these wildly popular "seers" have convinced their massive and global audience is the case. In fact—as we will see—the group was born from the fraudulent seers' collaboration with a famous Brazilian Ufologist and his outlandish prognostications.

The group has assumed several names in recent decades. Alternatively referred to as the "Community-Light of Figueira," the "Association Mary, Mother of the Divine Conception," or the "Grace Mercy Order," they are now most known as the "Voice and Echo of the Divine Messengers"—or simply—*The Divine Messengers*. Three hundred thousand follow this group on just one of its social media pages (Facebook), and videos of the "apparitions" circulate on platforms such as YouTube, where they likewise rack up hundreds of thousands of views. Their Catholic-themed devotional music videos garner millions.

Whoever attends their events, peruses their website, or watches their videos is immediately confronted with an appearance of pious, traditional, and orthodox Catholicism. Like Tomislav Vlašić, whom we will consider next, these "Divine Messengers" take every measure to pose as a faithful apostolate and an authentic religious order. Their website is replete with traditional images of Jesus and Mary, the Eucharist, St. Joseph, and the Divine Mercy. They extoll repentance, Confession, the Rosary, and *most* everything one would hope to see promoted in an authentic apparition. The leader calls himself Friar Elias, giving the impression he is a Catholic monk. Two of the female "seers" call themselves Mother Maria Shimani and Sister Lucia de Jesus. Elias and the others host large events in tandem with the reception of their messages, and they dress and comport themselves as if they were Carmelite monks and nuns. (In fact, they have no recognition from the Catholic Church.)

Their messages are replete with teachings about the extraterrestrials that they claim populate the planets, blending their own brand of Exotheology with Catholicism. An "*apparition of the glorified Christ Jesus*" in the year 2020 revealed to Friar Elias that "**Andromeda is the regent of your Solar System and others**," wherein "**the Creator Fathers once came to sow the Plan of God in the material universe**." The same messages declare that the "**Celestial Church is showing you... the real science of your origin**," before continuing:

> In the same way that God thought of His human Project, He also thought of other Projects that continue to evolve, similar to this one. With other degrees of consciousness and evolution, [there are] different creatures of the Father...[349]

Here as everywhere in the ET Deception, we see not only a false revelation of origins but also a false eschatology. A message from the year before states:

> ... movement in all living beings of this and other planets [will be affected]. As from the year 2020 on, the Solar System will enter into the last

> and most acute spiritual, cosmic, mental and material stage of its transition.[350]

A new era is upon us, these messages insist; however, the transition consists not of intervention from Almighty God, but rather in an evolution of our solar system thanks to the beneficence of the extraterrestrials.

The "Divine Messengers" also promote religious syncretism (insisting, for example, that "some will be more open to following" Buddha, some Muhammad, others Christ—any of which is fine),[351] New Age spirituality, and animal-personhood (along with a condemnation of meat-eating). They heretically speak of a time "before God existed,"[352] refer to the "human person of Christ,"[353] and embrace numerous other doctrinal errors. All this being said, most missives from the "Divine Messengers" do not directly pertain to extraterrestrials. Most appear quite pious. Thus, they continue to seduce multitudes of unsuspecting pew sitters. A Catholic ET promoter, therefore, might insist ET belief is not especially significant in defining this heretical group of charlatans' motivations and aims. But such an assessment would be ignorant of the group's history and basic principles.

An expert in new religious movements, Dr. Vitor de Lima Campanha, traced out the Divine Messengers' genesis carefully, explaining that the group's initial impetus derived from meetings with the prominent Brazilian Ufologist, Jose Trigueirinho Netto. Mr. Netto teaches that alien spaceships are among us, sent from an "intergalactic confederation" tasked with humanity's "transition" in order to avert an "extermination war." Their "rescue operation," he insisted, is "ready to start at any time," and will create a "**New Man**."[354]

In 1987, Netto founded a community in Brazil with New Age, Theosophical ideals. What happened twenty years later led directly to the "Divine Messengers." Dr. Companha explains that "**In 2008, Trigueirinho [Netto] received two Uruguayans who would later be known as Mother Shimani and Friar Elias**." He continues, noting that these two presented themselves to Netto as coming to him upon a Heavenly mandate "**to look for him**":

> Both said they were seers in frequent contact with the Virgin Mary... [who] guided them to look for Trigueirinho ... As a result, Trigueirinho, Mother Shimani and Friar Elias created a [New Religious Movement], based in the Community-Light of Figueira. This site undergoes a Catholic-inspired reorientation with the creation of monasteries devoted to Mary... There is the ordination of mothers and friars, including the use of habits similar to those of Catholic orders; they take vows like those of chastity, obedience, silence, etc. and participate in reframed rites of Communion, Worship and Foot Wash.

Lest anyone contend that these two were simply authentic seers whom Heaven sent to convert a wayward Netto, Dr. Companha points out that only the opposite is true. These seemingly Catholic trappings "**do not contradict the work of Trigueirinho Netto. Rather, they reinforce the contents, blending the New Age and UFO elements with the incorporations of the Catholic imagery**." As elsewhere in ET belief, the deception is treated as if it were actually a catalyst for the Faith's flourishing.

Nor are these "Divine Messengers" the only false mystics working overtime to seduce Catholics with the allure of the extraterrestrial combined with a superficial guise of piety.

***

In the year 1989, a young man named Giorgio Bongiovanni visited the Shrine of Our Lady of Fatima in Portugal. He emerged from it a self-declared "stigmatist," and proceeded to travel the world announcing his mystical messages, while amassing many followers. But these revelations were not from the Virgin Mary. They were from "the extraterrestrials" who, he insists, had chosen him to help usher in their era on earth. Mr. Bongiovanni's messages, like those of the "Divine Messengers," *appear* completely Catholic (despite occasionally slipping into heresies like reincarnation)—only to, beneath the surface, undermine the Faith. Yet by the time the apostasy is revealed, his followers are too addicted to quit. Bongiovanni continues to be extremely popular to this day. We address him in the Appendices.

Presently, we must move on to yet another level of nefarious machinations in the ET Deception's infiltration of Catholicism. For the deception runs deeper still: we have even pseudo-mystical Catholic priests now promoting it.

The Church of Jesus Christ of the Whole Universe

A man named Tomislav Vlašić was once a Franciscan priest in Croatia. The use of past tense is not due to his death. He is alive and well, but he has been laicized and excommunicated, and is now the founder and head of the "Church of Jesus Christ of the Whole Universe," which claims a mandate from on high to herald the age of the extraterrestrial and demand that the Catholic Church announce it.

> "I strongly recall all the ecclesiastical hierarchy, beginning with the Catholic ones, to an honest awareness about the presence of life in the universe, and to give to the faithful likewise clear information. This requires a radical rethinking of the universal mission of the Church which cannot be delayed any further." —"St. Michael" to Tomislav Vlašić[355]

Vlašić first became well known as an initial spiritual director to the Medjugorje visionaries, a group of Bosnian children who reported apparitions from the Blessed Virgin Mary beginning in 1981. Six years after those apparitions began, Fr. Tomislav moved to Italy and founded a New Age community, which one of the Medjugorje seers, Marija Pavlovic, disavowed after initially mistakenly endorsing (not having understood its true nature). At this point, it was discovered that Vlašić had covered up an affair (with a nun, no less), from which he fathered a child. Fr. Tomislav then publicly went off the rails and abandoned his vocation.

Priests in positions such as his—theologians working closely with *truly* holy seers—are favorite targets of the Devil, who understands well that such priests can devastate his plans if only they abide in orthodoxy and holiness. What better line of attack to derail such a priest than introducing him to diabolical "revelations" from extraterrestrials?

A theologian himself, Vlašić knows how to give the appearance of orthodoxy while spouting ET-messianism. While that defense fools many, it did not fool the Church. In 2008, the Vatican's Congregation for the Doctrine of the Faith began investigating Fr. Tomislav for the "**diffusion of dubious doctrine**," and "**suspect mysticism**." Accordingly, Vlašić was laicized (stripped of priestly authority) in 2009. In 2018, he founded the "Church" he now leads, built upon both his own "revelations" and those of a certain Stefania Caterina. Two years later, he was excommunicated.

Vlašić's promotion of belief in aliens, however, began long before the official 2018 formation of his "Church." In 2012, he had produced a video with Ms. Caterina (the two had already been collaborating for decades) insisting they were both members of a group called the "central nucleus." This nucleus consists of 49 Divinely-chosen individuals, dwelling throughout the galaxies, whom God has appointed to save humanity "in all universes." Today on his website, Tomislav states "**The fundamental points of the revelations we have received**,"[356] which include the pseudo-doctrines that:

> God has not created only the humans of the Earth ... The people of God are present on all inhabited planets and constitute the "Church of Jesus Christ of the Universe" ... Without the work of this Church [*i.e., his "Church of Jesus Christ of the Whole Universe"*], the Christians and the humanity of the Earth will not be able to overcome the trials that are awaiting this planet, nor will they reach the fullness promised by Jesus.[357]

Elsewhere, these revelations promise that extraterrestrials will bring us "health" and "peace," along with "**technology from the power of the Holy Spirit**," that will provide "enough to eat for everyone." These aliens, moreover, have thus far "prevented the Third World War," yet might not be able to prevent it any longer if we do not now begin listening to and obeying them.

Despite the pious attire, these "revelations" are motivated by the same promises which invigorate secular Ufologists. They feed off an insatiable desire for extraterrestrial technology to usher in a false utopia on earth, in a Satanic perversion of the Era of Peace that Jesus and Mary have long promised in their authentic apparitions. Parasitic on those true private revelations (for the Devil always apes the things of God), Vlašić's messages speak of a "glorious" coming of Jesus—but not in a manner compatible with what approved revelations have taught is at hand.[358] Instead, these messages promise a coming of Jesus to earth with a cohort of extraterrestrials to establish the "new Church."

> "When the time is right, the aliens will manifest openly and visibly to everyone. It is God's plan that this happens. Then the great interplanetary fleet will be positioned above the Earth and will be visible from every corner of the world." —Stefania Caterina[359]

The material generated by this galactical "Church" is even more effective in seducing Catholics than the last slew of texts we considered. It avoids the extreme esotericism of *Urantia*, it is better at avoiding obvious heresy than the *Divine Messengers*, and it is filled with traditionally Catholic piety and genuinely holy exhortations. It exalts Our Lady of Fatima, Divine Mercy, and other clearly authentic apparitions to give the appearance of standing shoulder to shoulder with God's true interventions into the world. The messages repeatedly praise and promote traditional Catholic devotions, the greatest of canonized Saints, the most important Liturgical feasts, and so on. Yet it leads its readers to above all affix the desires of their heart of hearts in the expectation of imminent saving intervention from extraterrestrials, insisting that "openness" to them is needed—the very openness that will indeed invite in otherworldly entities, though not "aliens":

> "... [The] universal mission has begun... Beginning from this year, 2013 ... [the faithful] will be sent on a mission in the entire universe ... Alpha Centauri, the humanity faithful to God, has been placed at the head of this mission of the people of God, instead of the humanity of the Earth as God had originally planned ... we are entering more and more into the cosmic Pentecost ... The faithful brothers and sisters [Alpha Centaurians] will allow you to feel their presence and, when God permits it, they will introduce themselves to the open souls..."[360] —January 2013 "revelations" to Tomislav Vlašić

To recap: Tomislav was a priest-theologian who became famous for his association with a pious and fruitful private revelation; a position he then abused—using it for the dissemination of a theology based on extraterrestrials. He is not the only one in such a position, with such credentials, who has done and will do similar things. But the Devil will not use the exact same strategy twice; he is always fine-tuning it.

For now, a sincere Catholic—even if he does not yet understand there are certainly no aliens—can be confident that an ex-priest who radically rejected the vows of his vocation, and was so unrepentant as to incur excommunication, would not suddenly be the singular instrument chosen by God to save the world and usher in a new era. Nevertheless, when addressing Vlašić, we are still dealing with a more subtle exercise of discernment than elsewhere. And the next time, the deception will be still more difficult to detect.

Perhaps the next major infiltration of the ET Deception into the Church will come from another man like Vlašić—a priest-theologian closely associated with an authentic private revelation. But the next time, perhaps, he will *not* be censured by the Church, and will be more successful than Vlašić in keeping up the false appearance of orthodoxy and holiness. Whatever the details, one warning emerges: whoever among the faithful is open to the possibility of contact with extraterrestrials has primed himself to be ensnared in the next trap.

This trap is not a simple mistake or innocent error that will at worst result in wasted time. It is an opening for the entry of the same demons who have always been our chief enemies, who are only interested in the damnation of our souls, and who will rest content with superficial piety and "nuanced" orthodoxy so long as their prey embraces ET-derived pseudo-messianism. To the ancient antics of these demons, we now turn.

13. Aliens, Demons, and Paganism

"Many of the UFO reports now being published in the popular press recount alleged Incidents that are strikingly similar to demoniac possession and psychic phenomena which have long been known to theologians..."[361]

—U.S. Air Force Office of Scientific Research Report

The Devil is a "liar and a murderer from the beginning" (John 8:44) and it is only fitting that as society abandons belief in his existence, an unprecedented degree of his deceptive and destructive activity rises in proportion.

If the greatest trick the Devil ever pulled is convincing the world he didn't exist, then a close second is what he replaced it with: fictional entities he could use as masks and which would equip him to continue—or rather, increase—all of his diabolical ploys even while his victims remain oblivious to the true source. Thus ancient Paganism returns with a vengeance, and the same legion of demons Christ had cast out on the Holy Cross return with "seven other" legions, "more evil than itself." (Matthew 12:45)

> "Uncounted numbers of well-documented demonic events are readily available to every researcher. The manifestations and occurrences described in this imposing literature are similar, if not entirely identical, to the UFO phenomenon itself. Victims of demonomania (possession) suffer the very same medical and emotional symptoms as the UFO contactees."
> —Pioneering UFO Researcher, John Keel.[362]

Demons are more prevalent in the contemporary age of Satan's prophesied "75 to 100-year" greater power we discussed in Part One, wherein he is given greater leeway than ever before. We are seeing their machinations now at unprecedented levels amidst the UFO Deception which is a preparation for the "Strong Delusion." (2 Thessalonians 2:11) The machinations *themselves*, however, are anything but new.

> "The gods, taking the disguise of strangers from other lands, walk up and down cities in all sorts of forms ... by witchcraft and deception they may make us think that they appear in various forms..." –Plato. ~400 B.C.

Satan's minions have always attacked mankind, and they make use of whatever disguise is most fitting in the cultural context in which they appear. In Plato's time, it was the "gods." Today, it is sci-fi. UFO-like encounters are not merely unoriginal, they are ubiquitous in history. Jacques Vallée noted that **"it is difficult to find a culture on earth that does not have an ancient tradition of little people that fly through the sky and abduct humans."**[363] He continued:

> These were the facts we have missed, without which we could never piece the UFO jigsaw together. Priests and scholars left books about the

> … these beings. … [which were] exactly that of today's UFO pilots. The methods were the same. There was the sudden vision of brilliant "houses" at night, houses that could fly, that contained peculiar lamps, radiant lights that needed no fuel. The creatures could paralyze their witnesses and translate them through time. They hunted animals and took away people."[364]

Another author—commenting directly on Vallée's findings here—provides more examples of the Pagan myths of earlier times which behaved exactly as today's "ETs" and UFOs do:

> These "little people" went under a variety of names: the Scandinavian *trolls*, the *ihkals* of Mexico, and the Malaysian *bunians*. The British Isles provide a wealth of such legends, from the Celtic stories of babies abducted by elves to Irish *leprechauns* and Scottish *sleagh maith* ... In France, these legends concern *fees*, or fairies, and a diminutive race known as the *fions*.[365]

Yes, ancient Paganism is certainly reenacted through contemporary Ufology and ET belief. Just as tellingly, one of today's most popular brands of Paganism is itself a product of Ufology: the New Age Movement. The Vatican's 2003 document, *Jesus Christ, the Bearer of the Water of Life*, repudiates the movement categorically. Therein, the customarily non-hyperbolic Pontifical Council for Culture described the New Age Movement as seeking the total eradication of the Faith: "**The New Age which is dawning will be peopled by perfect, androgynous beings who are totally in command of the cosmic laws of nature. In this scenario, Christianity has to be eliminated and give way to a global religion and a new world order**."[366] The document's admonition, moreover, is simple:

> The gnostic nature of this movement calls us to judge it in its entirety. From the point of view of Christian faith, it is not possible to isolate some elements of New Age religiosity as acceptable to Christians...[367]

The New Age Movement is indeed an attack on Christianity itself and human nature itself. However, few serious Christians today are unaware of its dangers. But many of them will be surprised to discover that this movement is not an umbrella term counting Ufology among its disciplines, nor is it a separate or related trap to which most ET promoters happen to succumb.

Quite the contrary, the New Age Movement *itself* is one offspring of expectation of contact with extraterrestrials. Thus, this entire destructive movement is just one of the innumerable rotten fruits generated by alien belief, which is the more fundamental deception. As the Macmillan *Encyclopedia of Religion* explains, in its entry scrutinizing the origins of the New Age Movement:

> The immediate roots of the New Age movement may seem surprising at first. Shortly after World War II, popular curiosity was attracted by unexplained phenomena in the sky referred to as unidentified flying objects

(UFOs). …Study groups were formed by people who wanted to investigate these phenomena…[they] believed that UFOs were in fact spaceships inhabited by intelligent beings from other planets or other dimensions of outer space. Representing a superior level of cultural, technological, and spiritual evolution, they now made their appearance to herald the coming of a New Age. ... In due time they would become the vanguard of the New Age, or Age of Aquarius.[368]

Little wonder that the messages delivered by the "ETs" invariably confirm New Age teachings. Let us now consider the "encounters" themselves that deliver these messages.

The Nature of "ET Encounters"

"If civilizations really existed on other planets, our Holy Scripture, the Bible, would definitely say something about that ... people who have allegedly seen aliens really saw demons, whose existence many people now refuse to believe in." —Russian Orthodox Metropolitan Hilarion[369]

Although the UFO era began in 1947 with Kenneth Arnold and Roswell, it took another five years before direct encounters with living "extraterrestrials" would become a similarly massive cultural phenomenon. Enter George Adamski (†1965) who, beginning November 20th, 1952, rose to prominence for his dialogues on that day with an "alien from Venus." He thus became "**the first contactee of the modern era who elicited veneration from his followers,**"[370] and Fr. John Saliba notes that Adamski claimed to provide revelations from aliens that gave "**a Christian interpretation of UFOs and their occupants**."[371] In truth, they were directed at the *overthrow* of Christianity's essence. Fr. Saliba continues: "**In fact [Adamski] veers from traditional Christian doctrine when he holds that Jesus was, like many other extraterrestrials, a visitor from another planet who came on Earth as a missionary.**"[372] From its very onset in contemporary Ufology, "ET contact" has always been an assault on the foundations of the Faith.

The phenomenon of "alien abduction," however, propelled Adamski-like claims to an even higher degree of popularity. In 1961, Betty and Barney Hill claimed to have been abducted by aliens—and their story became public several years later, being made into a bestselling book (*The Interrupted Journey*, 1966), a movie (*The UFO Incident*, 1975), and then permeating many other productions of pop culture. The following decade, Steven Spielberg's blockbuster film, *Close Encounters of the Third Kind* (1977), was released and alien abduction became a cultural mainstay. Whitley Strieber's release of *Communion*, ten years later still, sealed the deal. In one scholarly tome assessing the phenomenon, John Whitmore noted:

The number of people who claim to have been abducted by occupants of UFOs has been rising almost exponentially since the early 1970s when the

> subject was first granted acceptance by the media and the ufological community. With the publication in 1987 of Whitley Strieber's *Communion*, interest in abductions and abductees exploded.[373]

Throughout the intervening decades, Ufologists claim that millions of people have been abducted by aliens. And although the actual number of alien abductions is zero, what is popularly called "alien abduction" does appear to truly happen not infrequently. But what exactly are we talking about in those cases?

The first thing to know is that most such "abductions" are "recalled through hypnosis." Seated comfortably in a leather-backed sofa and listening to a hypnotherapist give relaxing commands to close one's eyes and imagine something while meditating for a few moments as he counts to ten—the "abductee" is thereby subtly coaxed to suddenly recall a whole volume of objective and accurate information about events that literally happened in the real-world's past. So we are told.

As this idea is utterly ludicrous, we are hereby reminded of the most basic fact about "alien abductions:" they are pure deception. The most competent scholars have pointed out* that such hypnosis is a vehicle for the suggestions of the hypnotist to take on a life of their own in the patient's (that is, *victim's*) now malleable mind. Hypnosis generates fantasies.

While this explanation may account for most abduction reports, it does not explain *all* of them. Here too, however—just as no UFO sightings are from "extraterrestrials," even though many non-debunkable UFO sightings exist—some accounts of abductions cannot be explained only by psychological missteps. Mentally stable and apparently trustworthy adults would not be so prone to psychological manipulation (even amid "hypnosis") as to *all* emerge from their sessions categorically affirming—year after year, and "on the record"—the reality of the abductions they describe, if there were not *something more* to it than mere suggestion. That is, if there were not something more than mere *human-originated* suggestion.

Here lies the crux of the matter. Approaching a hypnotist with the mindset that this silly person will magically (for that is exactly what we are speaking of here—there is no credible empirical explanation for what transpires in this type of "hypnosis") reveal to a person the hitherto entirely secret and enigmatic details of his past has done something akin to using a Ouija Board to consult spirits or approaching a palm-reader to peer into the future. He has chosen to wander outside of the safety afforded to him by a

* For example, Fr. John Saliba writes: "**Because suggestion is the process of inducing an uncritical acceptance of knowledge, emotion, or behavior, the hypnotized UFO contactee is open to accept without much evaluation any ideas about flying saucers that are brought up in the hypnotic session.**" (*The Gods have Landed*. State University of New York Press. 1995. Chapter 9) Thereafter relaying the findings of psychologist Dr. Robert Baker, Fr. Saliba also explains: "**He stresses the unreliability of hypnosis and hypnotic regression and adduces evidence to show how both do not lead to an accurate remembrance of events. Confabulation, inadvertent cueing by researchers, psychological needs, and fantasy-proneness all play a part in the recalling of a UFO experience.**" (*The Gods have Landed*. State University of New York Press. 1995. Chapter 3)

Christian worldview, and he has strayed directly into the radius of the Devil's chain. He has practically begged for what exorcists refer to as "diabolical obsession." Exorcist Fr. Gabriel Amorth relays what happens in these cases:

> Diabolical obsessions are disturbances or extremely strong hallucinations that the demon imposes, often invincibly, on the mind of the victim. ... he is subjected to a powerful force that creates mental activity in him that is repetitive, obsessive, and irresistible. Such representations of reality, even if foreign to his manner of thinking, become profoundly fixed in his psyche. The objects of these hallucinations can be manifested as visions, as voices, or as rustlings; they can also appear as monstrous figures, horrifying animals, or devils. In other cases it can be an impulse to commit suicide or to do evil to others; and, particularly in the young, it can lead to confusion about one's gender. ... This usually does not completely deactivate the mind and the will of a person, who remains conscious and alert. It does, however, heavily condition him in his relationship with the world.[374]

Given what is true of diabolical obsession, abduction reports should not be surprising for a Christian who knows his demonology.

This is not to say there are not also real diabolical encounters that masquerade as alien abductions—these events need no hypnosis to "recall." Those encounters exist too, and there is plenty of physical "evidence" that is easy for demons to leave behind, such as marks on the body. This deceives those who know nothing of the demons' power into supposing that such evidence is proof of contact with "ETs." In fact, everything we have any evidence of transpiring with respect to "abductions" is within the power of demons to accomplish. Most frequently, burns and other injuries are reported after "abductions." Yet these are also just as frequently associated with another class of demonic activity, "diabolical vexation." Fr. Amorth explains:

> Diabolical vexations are the second type of the demon's extraordinary spiritual aggression and are far and away the most numerous... Vexations are true and actual aggressions, physical or psychological attacks that the demon works against a person. At times they result in scratches, burns, bruises, or, in the most serious cases, broken bones. At times the victim is the target of stones or other objects. Typical cases of vexation are illnesses without any apparent cause that affect the internal organs or the limbs or pathologies that provoke pain in a part of the body without visible signs. Vexations can involve health, affections, or work.[375]

In the same passage, Fr. Amorth notes that these vexations can result from a person's dabbling in the occult—which one sees in the lives of almost all "abductees."

"Abducted" women will sometimes claim to have been impregnated by the extraterrestrials. Yet, no child has ever been shown to have been conceived or born from these "alien-impregnated" women who is anything but the offspring of one (human) mother and one (human) father. Demonologist Fr. Jean Vinchon notes that "**Possession upsets feminine functions, causes false impressions of pregnancy by distending the stomach**."[376] Indeed, no "alien abduction" ever has, or ever will, result in an actual pregnancy. Every human being[377]—including the Antichrist himself—is 100% human; the offspring of a human mother and human father.*

So-called "implants" received from "alien abductions" are similarly without a shred of evidence of being anything but natural, earthly objects. Even here, however, it is possible for demons to materialize some objects by borrowing from, and rearranging, material that is somehow already present. Fr. Amorth also notes:

> ...[T]he person exorcized can spit nails, glass, or hair. These objects do not come from the esophagus, in which case they would seriously wound the internal organs; rather, they materialize in the vomit coming out of the mouth. Twisted pieces of iron, woven thread, knotted twine, and similar objects can also be found in the cushions and mattresses of the possessed.[378]

Perhaps more likely than this, however, is the case of a demon-afflicted person himself either placing the so-called alien implant—which is only some ordinary piece of technology—within his own body, and then either lying about it or entirely forgetting about it due to demonic interference.

Most importantly, in addressing the subjective experiential portion of "non-debunkable" abductions, we see the hallmarks of the diabolical everywhere we look. Relatedly, one Catholic demonologist noted, "**The demon almost never appears in ... true form,**" for they "**lie and deceive constantly**." That "abductors" present themselves as "aliens" and not as the demons they really are, therefore, does not differentiate these experiences from other demonic ruses. The same demonologist notes that some free choice allows a demon to enter, and "**permission might be given in response to a deception, but the door is opened nonetheless...**"[379] In other words, it matters not if a demon is disguised—as a fairy, a UFO, a deceased relative, an alien, or what have you. If the disguise is embraced, so is the being wearing it.

---

* Sadly, some conspiracy-minded Christians succumb to diabolical deception here in their very supposition that they are opposing the Devil's agenda. They realize that the demonic is behind these "abductions," yet they radically stray from Christian orthodoxy by supposing that demons can truly become incarnate and impregnate women. Some claim the Devil is hereby amassing a "hybrid army." Often, they base their views on apocryphal texts that contain much heresy, such as the Book of Enoch, which falsely claims that the "Nephilim" of Genesis were literal offspring of women and demons. (For the correct—Catholic—understanding of that passage from Genesis, see *Only Man Bears His Image*, pages 679-684.)

The best place to look in ascertaining the nature of "alien abductions" is surely the preeminent scholar of these events we discussed earlier—Pulitzer prize-winning Harvard psychiatrist, Dr. John Mack.

Dr. Mack's Abductions and the Name of Jesus

Recall that John Mack is an unabashed ET promoter: if his own testimonies of abductions reveal the cloven hoof, then there is little use objecting to them on account of "bias." His bias *supports* the ET Deception.

Over the course of many years, Dr. Mack studied hundreds of "abductees," and undertook a study of the topic that was likely more rigorous than any other. Here is what he found.*

> Some abductees feel that certain beings seem to want to take their souls from them. ... [one abductee said] 'I cannot describe to you the scream, my own scream, being ripped away out of my body.'[380]

There is only one type of entity that would conceivably be interested in "ripping" one's soul away. Mack also wrote:

> [Another abductee said]: "I knew instinctively that whatever that thing was that was next to me wanted to enter me. ..." She felt certain that what these beings were after is the human soul... But "they can't just take our souls," she said, "because they cannot come into this world physically. But they can fool you into handing it over."

This is precisely what the Catholic demonologist quoted above, Adam Blai, explained is the hallmark of demonic activity: *fooling* us into handing over our souls to them. Discussing the "ET" encounters of a man named Credo, Dr. Mack writes:

> "These beings scar you for the rest of your life. You're not able to relate properly to ordinary human beings after they have had a session with you." ... White people in South Africa call these beings "aliens," Credo said. "We don't call them that. They are part of us, part of our lives." ... [they are] "solvers of great problems," he concedes, "in whatever bloody world they live in." Their technology may be "several million years ahead of us."[381]

It is only madness to claim that an alien with technology "several million years" beyond our own would, upon meeting a man here, only result in scarring him so severely that he cannot "relate properly to ordinary human beings" thereafter, while providing no benefits to speak of. Yet this is surely what a demonic encounter could do.

---

* For the sake of decency, I will—here as elsewhere—exclude from our consideration the grotesque sexual perversion that permeates alien "abductions," including those relayed by Dr. Mack. Suffice it to say, those abundant testimonies give clear evidence of "incubus" or "succubus" demonic activity.

Another abductee, Jim, relayed to Dr. Mack his "abduction," wherein he "**all at once he felt as if his body were surrounded by static electricity, and then he sensed a foul, sulfurous smell... he was 'frozen with fear**.'" In another experience, Jim says, "**I could feel him probing through to the deepest fiber of my mind**." Mack summarizes this experience, saying that it was as if...

> ... [the alien] were searching for a weakness. Apparently an area of vulnerability was found, as the image appeared before Jim of his brother in a hospital room clutching his heart and apparently in the throes of dying... The experience left Jim feeling badly shaken and demoralized.[382]

That image itself was a lie; his brother was not dying. But these are the tactics demons use (not to mention, the smells they exude) to elicit the desired response of capitulating to fear. Mack also relays the accounts of another "abductee" he studied, Sara, as follows:

> [It was] "reptilian," almost "snakelike, serpentlike" and quite elongated. "Red vein-things" made the head appear like "a body turned inside out." The creature was not "bad. It's nice enough"... It seemed vulnerable, in need of "understanding" and "cooperation" from her.[383]

Little commentary is required about a "serpentlike" creature who presents himself as "nice" and in need of our "cooperation." Those interested in reviewing that creature's first entrance onto the scene may wish to read the third chapter of Genesis. Mack's own take on Sara's experience? He enthusiastically agrees with her insistence that it appeared in order to spur on "evolution," writing:

> Sara and other human beings are forced to acknowledge the reality of entities and domains that our Western acculturation have taught us in recent centuries cannot, even must not, exist. Yet this kind of ontological, physical shock may, in my view, be an essential initial step in the process of human consciousness evolution that seems to lie at the core of the alien abduction phenomenon.[384]

Any Christian who does not see the demonic agenda in this might as well hang up his hat.

In the conclusion of his last book, *Passport to the Cosmos*, Mack again describes the overall meaning of the abduction phenomenon as he came to understand it better than anyone else upon completing his years of study:

> [Abduction] experiences seem to be created, as if by design, to shatter (the word virtually all abductees use) the previously held idea of reality (which usually had no place for such entities) and topple the experiencer from the sense of being a member of a uniquely intelligent life-form at the peak of the Great Chain of Being... [385]

Here, Mack concedes that the abductors are above all interested in "shattering" precisely what the book now in your hands is dedicated to solidifying: the Christian view of the *Great Chain of Being*, wherein man alone bears the

Divine Image. Only if this worldview is "shattered" through the "ontological shock" of ETs can the Devil succeed in his most apocalyptic of plans.*

***

Perhaps most telling of all is the discovery made after a lengthy study of "abductions" by one "CE4 Research Group."

This group was established in 1997 to study abduction experiences. They quickly discovered over four hundred individual accounts of abduction—including hundreds of accounts they received directly from the mouths of "abductees" themselves—wherein the "alien abduction" immediately stopped, or the "ET" even fled, upon the would-be abductees' invocation of the name of Jesus.[386]

Over a hundred testimonies they have collected are available to be read—including many audio interviews that can be listened to—on their website.[387] Obviously, there is no conceivable reason for a being to flee at the name of Jesus unless it is a demon. This is clear from both Scripture and Tradition. As Jesus said, "**these signs will accompany those who believe: in my name they will cast out demons**" (Mark 16:17), and the Father of the Church, St. Irenaeus, confirmed that: "**By invoking the name of Jesus Christ... Satan is driven out of men**." Yes, "aliens" are demons. And even some ET promoters have conceded that. As we have seen, John Keel promoted diabolical deception as any Ufologist does, but he also had his moments of honesty, as when he wrote:

> Dabbling with UFOs can be as dangerous as dabbling with black magic. The phenomenon preys upon the neurotic, the gullible and the immature. Paranoid-schizophrenia, demonomania and even suicide can result-and has resulted in a number of cases. A mild curiosity about UFOs can turn into a destructive obsession. For this reason, I strongly recommend that parents forbid their children from becoming involved. Schoolteachers and other adults should not encourage teenagers to take an interest in the subject.[388]

Despite the pervasive and obvious darkness, we should add that the emotions arising from "ET encounters" do not always appear negative. Nonetheless, positive emotions reported in relation to "alien contact" experiences are also not difficult for the demons to arouse. If a simple chemical injected into one's veins can accomplish some effect, so can a demon. We must, therefore, not be fooled into supposing a demon, disguised as an alien, is a benevolent entity merely because some who experience contact with them report various sensations of elation or euphoria, as Mack sometimes relayed about abductees he studied. Fr. Ripperger explains:

* Dr. Mack also writes: "…[these alien] beings [possess] technologies and other powers far in advance of our own...[ET contactees] speak of it as an "awakening," … [ETs] have access to what in Western societies is called nonordinary states of consciousness, similar to the symbolic worlds of the shamans of indigenous cultures…" No faithful soul could for a moment find palatable the notion of God sending "extraterrestrials" to give mankind an "awakening" through the same methodology employed by "shamans of indigenous cultures."

> [Demons] can cause elation of the mind so that people believe what they suggest is true. In other words, they can create in the mind a quality … in which the person experiences a kind of joy or pride within the intellect, almost like a euphoria. … demons can cause any emotion in human beings, both in kind and by varying degrees.[389]

Demons inherently induce hatred, coldness, pain, fear, confusion, anxiety, and the like: hence these emotions being most frequently reported amid "abductions" and other "ET encounters." Yet they are also powerful enough that, at times, they can give temporary sensations of the opposite emotions. What they cannot provide—and what we see lacking in "abductees"—is a deep and abiding peace.* Even those abductees who report positive emotions in relation to their experiences only find that elation followed by, at the minimum, restlessness, anxiety, and disquiet.

Similarly, demons are inherently dark and hell-bound, yet their power enables them to appear clothed in light while seeming to descend from the heavens.

The Air and the Darkness

"The air"—that is, the atmosphere taken as a whole—has always been regarded in Christian tradition as a particularly potent domain of the demons. Scripture warns that the Devil is the "**prince of the power of the air.**" (Ephesians 2:2) Obviously, this does not mean that every uncertain object in the air is demonic, but it does mean we should be particularly cautious of any entity that displays a proclivity for aerial phenomena and a concomitant hesitancy to interact with humans in the ordinary "ground level"—literally, *down to earth*—manner.

Appearing under the aspect of descent from the atmosphere is a favorite tactic of the demons because they never cease aping the things of Christ. All Christians know to expect the return of Christ in the flesh at the end of time in such a manner—"*coming with the clouds.*" (Revelation 1:7) They understand that just as "*lightning comes … so will be the coming of the Son of man.*" (Matthew 24:27) They all joyfully recall the promise of the angel immediately after Christ's Ascension: "*This Jesus, who was taken up from you into heaven, will come in the same way as you saw him go into heaven.*" (Acts 1:11) Little pleases the demons so much as grotesquely mimicking the very thing

---

* Fr. Ripperger notes that the demons can provide a "false clarity" in the authenticity of the illusions they generate. This pseudo-clarity arises from a manipulation of emotion and a corresponding darkening of intellect. We see both endemic to the testimonies of alien contactees. Although filled with emotion, demanding to be taken seriously, and often shedding tears that are anything but holy, they are completely dismissive when asked to engage even the most basic logical considerations related to what they recount. For example, one who points out that what they "saw" would be physically impossible even for an alien, no matter what supposed technology is involved, is always promptly denounced. These "contactees" never convey the peace and conviction that a man would exude if he were simply giving sincere testimony to a genuine (and non-diabolical) phenomenon he experienced.

Christians always and everywhere anticipate—with *rightly* messianic aspirations—as they piously gaze upon the heavens. Modern science fiction understands this well. As we saw in Chapter 11, pseudo-ascensions of extraterrestrials permeate modern sci-fi movies, and in this we see just one of their innumerable strategies to replace the Faith with Pagan mythology.

"**The demons inhabit the air,**"[390] St. Augustine taught. St. Thomas Aquinas reminds us that **"[A demon] can from the air form a body of any form and shape**."[391] St. John Chrysostom, who is both a Father and Doctor of the Church, explained what Scripture means by the passage in Ephesians 2, quoted above. He writes:

> Why does [St. Paul] call the Devil the prince of the world? Because … Satan occupies the space under Heaven, [and] the incorporeal powers are spirits of the air, under his operation.[392]

Chrysostom was not alone among the Fathers in this view. Yet another Church Father and Doctor, St. Athanasius, was commenting on the same passage when he taught:

> The devil, the enemy of our race, having fallen from heaven, wanders about our lower atmosphere, and there bearing rule over his fellow-spirits, as his peers in disobedience, not only works illusions by their means in them that are deceived, but tries to hinder them that are going up. (*On the Incarnation of the Word*. §25)

The great Athanasius does not merely insist here that the atmosphere is the dwelling place of demons, but also cautions that these demons work *illusions* and deceptions therein. "Extraterrestrials" are only the most recent—and the most apocalyptic—of these deceptions the demons have always wrought in the air.

Still another Father of the Church, Saint John Cassian, taught that the only reason God usually hides these aerial demons from our sight is that beholding them would be an "**unbearable dread**," since the "**air... is so thick with [evil] spirits.**"[393] In his greatest work—*City of God*—St. Augustine prophetically described contemporary UFO phenomena in light of the demons who "**dwell indeed in this air as in a prison**" and deceive by "**wonderful and lying signs**":

> On the contrary, we must believe them to be spirits most eager to inflict harm, utterly alien from righteousness, swollen with pride, pale with envy, subtle in deceit; who dwell indeed in this air as in a prison ... Over many, however, who are manifestly unworthy of participation in the true religion, they tyrannize as over captives whom they have subdued—the greatest part of whom they have persuaded of their divinity by wonderful and lying signs.[394]

Therefore, neither Scripture nor Sacred Tradition have left any doubt about how a Christian should approach these spectacular "UFOs" descending upon us from the skies.

Not only do these "ETs" prefer aerial tactics to interactions on the ground, they also much prefer ambient darkness to light. This inclination calls to mind another Scriptural teaching:

> The 'gods' know nothing, they understand nothing. They walk about in darkness; all the foundations of the earth are shaken. (Psalm 82:5)

Most UFO sightings are reported after sunset and before sunrise, and a deeply disconcerting proportion of UFO antics are witnessed around 3:00 am—the inversion of the moment of Christ's death on the Cross. Tellingly, even the legends from worldly sources speak much of the so-called "devil's hours," from 3:00 am to 4:00 am, wherein diabolical forces are thought to be most potent. This is not inaccurate. Jacques Vallée's own large database of the most well-documented UFO sightings and other "close encounters with ETs" also found a massive increase in such reports once darkness descends at sunset, tapering off as people go to sleep around 10:00 pm. Then, suddenly, there is a large uptick in such reports at 3:00 am, declining sharply at 4:00 am.

Now it is true that, of millions of reports of UFO sightings and "alien encounters," most derive from "objects" darting through the atmosphere or descending from it, particularly in darkness. Nevertheless, a portion of these reports remains which consist of beings walking by, sitting in a room, being seen in the woods, and so on, including in daylight. Those, however, come with their own identifying traits.

For example, famous cases like the 1996 Varginha, Brazil event, and the 1994 Ariel School, Zimbabwe, event both occurred during daylight and included multiple "witnesses." Both involved phenomena that truly would be outside of the abilities of demons to orchestrate; yet those phenomena were simply human in origin. (In Brazil, planned military drills had taken place and patients with dwarfism—mistaken for aliens—were being treated at a hospital. In Varginha, there had just been "mass UFO sightings"—which turned out to be the Zenit-2 rocket re-entering the atmosphere.) But the demons, seeing an opportunity, added their own illusions. The presence of the human elements made a demonic explanation seem unlikely. Yet in both cases, the activity of the demonic is clear to Christian discernment.*

---

* For example, the Varginha episode began with three young women who, immediately after encountering a grotesque red-eyed being on the streets, explicitly insisted they had seen *the Devil*. As usual, however, Ufologists dismiss testimonies that do not support their thesis, whereas anyone who insists he saw aliens is treated as a veritable oracle. The "extreme stench" described by many at Varginha in connection with their proximity to the "aliens" also reeks of the demonic. As reported by the *Wall Street Journal*, **"Standing out in a different way was the creature's odor: One ghastly whiff weakened the knees."** Fear, unease, and spiritual disquiet were also reported by many who were present. On the other hand, the supposed physical evidence that "aliens" were visiting Varginha is easy to debunk (as has already been done many times). Considering the Ariel School, Ufologists today focus on only a few students who claim to have seen aliens (they likely did see a demonic apparition, as happened in various other nearby African incidents that had nothing to do with UFO sightings). They ignore the many other students who were also present for the same

This speaks to a general pattern in UFO sightings and "alien encounters," wherein the associated physical evidence that would make a demonic interpretation unlikely is easy to expose as human (or natural) with basic scrutiny, but those testimonies of witnesses who experienced something *real* are not always so easy to debunk. Fr. Ripperger explains that "**Demons are opportunists and will observe the circumstances and will very often find the weakest link**."[395] The Devil, knowing what deception he wants to seduce men to embrace, can simply wait for circumstances to arise that are conducive to the lie but are outside of his power to directly generate. He then pounces, using his usual illusory tactics within the reach of his chain. "**Since the demons are opportunists, they will seize upon certain circumstances in order to amplify or to increase the intensity of a person's emotional response in relationship to a particular external event**,"[396] Fr. Ripperger reiterates.

Consider what we already know constantly happens with initially secret military drills, strange (but still natural) atmospheric phenomena, meteorites, government conspiracies, covert new aerial technology undergoing tests, or even ordinary satellites, drones, weather balloons, and so on. All such phenomena can—if other elements are present—provide perfect cover for demons to add a touch of their own intervention, hoping to remain undetected while Ufologists seek to portray the whole event as an alien encounter. Yet these Ufologists have only committed a type of association fallacy.

Christians must not be so gullible. Guarding ourselves against these deceptions, however, requires some further elucidation of the capabilities of demons.

What Demons Do

"It's nuts and bolts, it can't be demons!" Whoever peruses the ET debate in Christianity is sure to hear Ufologists present this line as a vindication of their position. The question that should follow—*"why are you so sure it's nuts and bolts?"*—is one none can answer.

Take one of the most oft-quoted phrases in all of Ufology: the remark from the atheist sci-fi author, Arthur Clarke, who claimed that "**any sufficiently advanced technology is indistinguishable from magic.**" With their fixation on this line, ET promoters remind us of their belief that nothing whatsoever is outside of the bounds for UFOs. Therefore no matter what UFOs are reported to have accomplished, the ET "true believer" will *always* remain convinced it is not demonic. A "sufficiently advanced" technology can do anything and everything, they insist, rendering their thesis invincible to any encroachment of an inconvenient thing called reality.

---

event, at the same place, at the same time, and testify that *nothing was there*, thus ruling out any supposed "ET" explanation of the event. The demons simply managed to create an illusion for some of the children, but not all of them.

In fact, demons can manifest, do manifest, and always have manifested in ways that generate real corporeal phenomena which are nevertheless not representative of the true nature of what is transpiring.

> "It is more necessary to use all our powers of discrimination and judgment when Satan transforms himself into an angel of light, lest by his wiles he should lead us astray into hurtful courses." —St. Augustine.[397]

St. Augustine also observes—in a passage we will revisit in Chapter 14—that the Antichrist's deception, by the power of Satan, will entail demonic onslaughts which may consist in both illusions as well as *actual wonders.* These beguiling wonders, the saint says, will succeed in their deceitful intent because they will come at a time when men falsely suppose that such wonders are outside of the Devil's power. As the postmodern West scarcely believes in demons any longer (a rationalistic disease increasingly infiltrating even the Church), it goes without saying that moderners will fail to realize the abilities and operations of those whose very existence they do not acknowledge. This Church Father's words are being fulfilled before our eyes.

What, however, do we *already* know demons can accomplish? At what machinations have they already proven themselves adept throughout history?

In the New Testament, the first Pope warns that "**the devil prowls around like a roaring lion.**" (1 Peter 5:8) St. Paul also exhorted Christians to "**stand against the wiles of the devil,**" for it is against him and his minions that we are "contending against"—not mere "flesh and blood," but rather "**the spiritual hosts of wickedness in the heavenly places**" (Ephesians 6:11-12); that is, the demons who, as we have just seen, inhabit the air.

We can also consider the earliest ages of the Church. St. Anthony the Great (born 251 A.D.) relays many demonic initiatives he personally witnessed in his battles against Satan, as recounted by St. Athanasius:

> The demons ... [approach in] guise, and thenceforth shaping displays they attempt to strike fear, changing their shapes, taking the forms of women, wild beasts, creeping things, gigantic bodies, and troops of soldiers. ...[They] pretend to prophesy and foretell the future ... that they may stealthily catch by such displays those who could not be deceived by their arguments... [The demons] are treacherous, and are ready to change themselves into all forms and assume all appearances... For the demons do all things—they prate, they confuse, they dissemble, they confound—to deceive the simple. They din, laugh madly, and whistle...[398]

Demons can appear in virtually any form. In these appearances, St. Anthony also notes that they make use of certain melodies frequently (Spielberg having *Close Encounters of the Third Kind* emphasize this was not fictional), and their goal is always to "**confuse, dissemble, and confound**" in order to deceive. This, too, is a summary of modern UFO phenomena. And we must not forget their tendency to mimic the light of the angels—their own master,

"Lucifer" was indeed the original "Light Bearer." St. Anthony relays: "**Once [the demons] came in darkness, bearing the appearance of a light, and said, 'We have come to give you a light, Antony**.'" UFOs and "ETs" constantly appear thus clothed, for "**Satan disguises himself as an angel of light.**" (2 Corinthians 11:14)

Moving on to the later ages of the Church, the Angelic Doctor, St. Thomas Aquinas, summarizes the abilities and limitations of demons in the *Summa Theologica* (Part I, Q114, A4). He explains that, *strictly* speaking, no demon can perform a miracle, "but God alone" can. For example, St. Thomas observes that a demon cannot effect any transformation "**which cannot be produced by the power of nature**;" they cannot resurrect the dead, nor transform a human into an animal; nor, I should add, can they create anything *ex-nihilo* (from nothing), or know one's innermost thoughts perfectly, or see the future itself, among (many) other things.* Aquinas, however, remarks that demons *can* "**from the air form a body of any form and shape, and assume it so as to appear in it visibly... [and] clothe any corporeal thing with any corporeal form, so as to appear therein**." He notes that demons can also work wonders which, though "lying" are "nevertheless something real"—not mere figments of our imagination.

It bears emphasizing that the demons can often give the illusion of having done what is, in truth, outside of their power (even perhaps seeming to raise the dead). The medieval code of Catholic Canon Law, the *Canon Episcopi*, spoke of certain occasions...

> ... when [demons] force their way into corpses and are reputed to restore them to life for a short time, or having thrown them in some remote place, they exhibit somehow the image of the dead.[399]

A key phrase here is, "**for a short time**."† God would certainly never allow some demon to inhabit a material body or a corpse, or assume some other corporeal form, and go about life like a normal human. The same *Canon Episcopi* goes on to teach:

> Those who appeal to demons are capable of marvels ... With God's permission, demons always take part in evil when appealed to; sometimes they appear in visible form, at other times they are invisible; they upset the atmosphere, cause storms, hail and lightning, they ruin crops and ravage with their spells whatever is produced by the earth. They cause illness in man and beast, and use every skill to carry out whatever plan they

---

* Too few recall, however, that Aquinas also teaches it *is correct* to say that "**demons can work miracles**," in the sense that they "**rouse man's astonishment, by reason of their being beyond his power and outside his sphere of knowledge**."

† Relatively brief appearances, however—especially when people are not given the opportunity to intimately interact—are another matter. Recall that after His Resurrection, Jesus spent 40 days on earth, even eating with his disciples and letting one put his finger into His side, in order to prove to them the genuinely physical nature of what was transpiring. "*See my hands and my feet, that it is I myself; handle me, and see; for a spirit has not flesh and bones as you see that I have*" (Luke 24:39) It is precisely because "aliens" are in fact demons that they never give any of these proofs.

> can think of to ruin man.[400]

Now, if demons can give the appearance of doing what is outside of their power, one may licitly ask: *how can we have any hope of ever distinguishing their illusions from reality?*

The answer is that careful discernment is necessary, therefore immediate appearances which have not been subject to this process should not be trusted. Yet precisely this category contains all UFO sightings and other "ET" contact experiences, which are always exceedingly vacuous in at least one requisite evidential criterion. This is one reason why demons so love to use the whole gamut of modern ET phenomena. Usually they need only appear briefly, under the guise of a UFO or an alien, and a media (or "social media") sensation will erupt overnight. The event itself is not only generally brief, but also often hidden under darkness, further facilitating their illusion. It is, in superlative convenience, never possible to subject it to the degree of analysis that would be necessary to determine what actually transpired. Fr. Chad Ripperger explains:

> ...[D]emons (and angels) [can] assume a body ...[and] cause a change in the person's senses or in the medium [i.e. light or sound travelling through air] between the thing and the sense of the individual which is perceiving it. ... This is commonly what is happening when morphing occurs ... which is beyond what is physiologically possible for the demon to cause due to his inability to suspend the natural laws, that is, to cause a miracle.[401]

Demons can manipulate both sound waves and light waves as they travel through the air to the observer—even if that "observer" is the sensor on a camera or other technological instrument. More serious still is the case where a demon actually "assumes a body"[402] of some sort. Using this method, the most elaborate of "ET contact" experiences can be diabolically originated. Explaining how an angel (or demon) can do this, Fr. Ripperger relays:

> ... Unlike a vision where the body does not necessarily exist outside the individual ... in the assumption of the body by an angel [or demon][403], the person does actually see the reality of rarefied air in such a manner that reflects specific qualities.[404]

As we saw in Part Two, "UFO sightings" often describe "events" that are, in fact, categorically ruled out by the most fundamental laws of physics—of *nature*. No technology ("alien" or otherwise) could achieve it. Not even a demon could cause it—*as it is described and interpreted*. Demons can, on the other hand, temporarily form an ephemeral body by manipulating molecules already present in the air. Thus, lacking any appreciable mass, such a body (appearing as a spaceship) can easily display maneuvers that would violate the laws of nature (physics) if they were undertaken by a craft actually having the weight any solid thing would have. The true mass of molecules the demons are using, however (the "body" they "assume"), are capable of such maneuvers, therefore no laws of nature are violated.

Renowned expert in mystical theology, the Dominican priest Fr. Jordan Aumann, confirms this ability of demons. He writes:

> [A demon] has the power, with God's permission, of assuming some material substance with which to represent itself even to the bodily eye, whether that substance is a body or light rays or some kind of cloud or vapor. ... [The] devil [uses] light rays, vapor, or a material substance to simulate the physical body ... There is no difficulty in affirming this, since the devil has power to make use of material substances.[405]

The demons, therefore, can do at least three things relevant to the present context to deceive us: they can "assume" a body, they can directly manipulate the imagination or senses, and they can affect the medium (usually air) between the observer and the illusion's apparent location. Considering all three together, not even one UFO sighting or "ET encounter" is beyond the pale for a demonic ruse. With these methods, the demons can replicate *everything* that has been reliably reported in relation to "aliens."

Fr. Aumann provides another more detailed—though certainly not comprehensive—list of other machinations within the domain of demons as follows, insisting that the Devil can do any of the following:

> 1. Produce corporeal or imaginative visions (but not intellectual visions). 2. Falsify ecstasy. 3. Produce rays of light in the body and sensible heat… 4. Cause sensible consolations and tenderness. 5. Instantaneously cure sicknesses that have been caused by diabolical influence*. 6. Produce the stigmata and all other kinds of bodily extraordinary phenomena, and any phenomena dealing with physical objects, such as crowns, rings, etc. 7. Simulate miracles and the phenomena of levitation, bilocation, and compenetration of bodies. 8. Cause persons or objects to disappear from sight.[406]

Concluding this list, he writes: "**All phenomena resulting from the activity of any natural power or physical law, even if the human being is unable**

---

* Variations of many of the demonic powers listed here are often observed in relation to UFOs. The instantaneous curing of certain sicknesses by demons is especially noteworthy. This is not generally within the power of the Devil, since it is usually a true miracle (such healings are typically cited in support of canonizations). An exception, however, is found when a demon itself is causing a certain affliction in someone. In that case, it is within that demon's power to immediately remove that affliction and thus give the impression of a miraculous cure, like a torturer turning off an electric current flowing through some sadistic machinery. Various ET contactees have reported "healings" that are only exhibitions of this ruse. For example, some have reported instantaneous cessation of severe pain with which they had long been afflicted. Others have reported a sudden return of hearing. Conspicuously absent is any *evidence* of an "extraterrestrial" contact experience resulting in medically impossible healings. (Heavenly interventions achieve this regularly.) Moreover, before (and during) the "ET healing" in question, one can usually see that the personal life of the once afflicted person already displayed evidence of diabolic activity, indicating he already suffered from a demonically originated physical ailment. This type of ailment is not as rare as modern men suppose.

to produce them, can be produced by diabolical power, with God's permission."[407]

A Ufologist's Folly

We can easily see, therefore, how misled are those Christian ET promoters, like Msgr. Balducci, who insist that UFOs "cannot" be demonic, and should be welcomed as heralds of the arrival of the age of the aliens because, as he declared, "**demons don't need spaceships**." Yes, it is obvious that they do not need spaceships. But it is in fact quite easy for demons to fabricate the appearance of one (an illusion which has proven extraordinarily helpful for them). Sacred Tradition has already established they can accomplish such displays.

Despite the demons' great power, the assessment of Ufology presented here remains easily falsifiable. No ET promoter could licitly protest by saying, "*you are just arbitrarily defining ET belief as impossible, since you will automatically claim any proof of their existence is demonic.*" In fact, there are many things demons cannot do but which actually incarnate beings—"extraterrestrials"—could easily do. Any ET promoter is free to consider my argument refuted if he can provide a single alien corpse for rigorous study, or crashed ET-UFO for reverse engineering, or, better yet, a living physical alien we can sit down with day after day. Despite a decades-long avalanche of claims that such evidence exists, none has ever been presented. If "aliens" are *not* demons, it would be astonishingly easy to prove they are not. That this proof has never been given tells any discerning Christian everything he needs to know.

Even those *discerning Christians* who are privy to the most top-secret US Government files about UFOs conclude the same. One fact often neglected by all camps of Ufology—including its astute critics—is that many devout Christians occupy high ranks in the US Military and Intelligence Community. These individuals know that the diabolical is behind UFOs, and their much-maligned "coverup" is often, in fact, only motivated by a noble desire to avoid giving the demons more power through the allure "disclosure" would give them. The world's most popular UFO whistleblower at the moment is Luis Elizondo, a former US Army Counterintelligence officer who claims to have led AATIP (a Pentagon UFO program). In his 2024 *New York Times* best-seller, *Imminent*, he shared an encounter with a senior director at the United States DIA (Defense Intelligence Agency). Elizondo admits that this man, for whom he uses the pseudonym Devon Woods, "**exuded grace and class**," was "**noble and honest, albeit extremely religious**" and said he "**looked up to him**." But then, Elizondo shares an exchange they had one day:

> As [Woods] approached [me], he put his hands in his pockets and said something in a quiet tone that I'll never forget. "*Lue, you know we already know what these things are, right? ... Have you read your Bible lately, Lue?*" he

> asked. ... *"Lue, you're opening a can of worms playing with this stuff,"* Woods said. It was clear to me he was talking about UAP. ... *"It's demonic,"* he said to me. *"... We already know what they are and where they come from. They are deceivers. Demons."*[408]

Instead of receiving this admonition with any understanding whatsoever—despite Woods' rock-solid trustworthiness—Elizondo only boasts of his anger and bewilderment. He writes:

> I couldn't believe what I was hearing. This was a senior intelligence official putting his religious beliefs ahead of national security... I was dumbstruck. In a moment, I realized that my mentor—who had always been the consummate gentleman—could be a stonefaced, ruthless operator. ... This was a direct warning to me, and I completely understood the message.[409]

Mr. Elizondo's indictment of this "noble and honest" official was based entirely on circular reasoning. He simply assumed the warning from Woods must be baseless in his very accusation that Woods was issuing a baseless—and therefore corrupt and unpatriotic—admonition. In fact, being a devout Christian, Woods was aware of the abilities of demons we have discussed in this chapter. Also an intelligence official with top clearance, he possessed more inside knowledge about UFOs than any Ufologist. He understood that the demonic is behind all non-debunkable UFOs, and that seeking "contact with aliens" through "UFO disclosure" is only, as he rightly said, "opening a can of worms."

Elizondo, however, paints this Christian common sense as "**putting religious beliefs ahead of national security.**" The accusation is as absurd as levying the same charge against someone who wisely argues that the US Military should not use Ouija Boards to attain a battlefield advantage. But for Elizondo, being told by Woods to beware of dabbling in Ufology was tantamount to being told to set aside his own (pseudo) religious aspirations. Elizondo's accusation was a confession.

Mr. Elizondo regularly affirms his messianic giddiness for the new era he expects the "aliens" to bring—a theme we will revisit in a later chapter—and he will not allow Divine Revelation to get in the way. Nor will any ET promoters.

***

With the ET Deception, however, we are not confronted with just one more of a multitude of tactics the demons employ to harass human beings. We are dealing, rather, with a deception motivated by apocalyptic aims to replace Christianity entirely—even if and when it cannot manage to do so explicitly.

Ape of the Church, Ape of Christ

Although a great deception indeed, Ufology takes its cues from the true Faith when constructing its own tenets and practices; mirroring them each in a perverse way, as Satanism patterns its diabolical rites on the Holy Sacrifice of the Mass.

Components of Christian Religion	Corresponding Perversion in Ufology
The persecuted prophets of Salvation History leading up to the revelation of Christ	The "persecuted" Ufologists and contactees, leading up to the "Day of Disclosure"
The Pharisees crafting a conspiracy to claim that the Resurrection was faked, and that Jesus' body was simply stolen	The government crafting a conspiracy to cover up UFO crafts and aliens and claim nothing happened
The account of creation and Salvation History in Scripture, wherein mankind has been guided for millennia by God—for a purpose	A creation myth, wherein mankind has been guided for millennia by extraterrestrials and was even created by them
The Four Evangelists authoring the Gospels, always treated as foundational	Several men authoring primary ET/UFO texts, always treated as foundational
Evangelization: each Christian has a mission to proclaim the Faith and testify to its fundamentally world-changing, life-changing nature	"Ontological Shock" promotion: all have a mission to advocate for Disclosure and testify to ET/UFO belief fundamentally altering one's grasp of reality
Profession of Faith: Doubt is regarded as a temptation	ET/UFO "True Believers" denounce skepticism
Divine Providence	"Synchronicity"
Liturgy	Communal ET-contact sessions
Private revelations, apparitions, and mysticism deliver important calls from Heaven	Abductees and other "ET contactees" announce messages from the galaxies
Prayer and contemplation	Astral-projection, remote viewing
Eschatology: hope in Christ's triumph and His coming	Anticipation of an imminent techno-utopia heralded by the arrival of the ETs
A complete worldview based on the Word of God	A complete worldview based on UFOs, ETs, multiverse theory, evolution, & interdimensionality

The items in the left-hand column above are straightforward enough; everyone understands they are essential in defining the Christian Faith. Considering the right-hand column above, wherein each facet of the Faith is perverted by Ufology, we are confronted with abundant real-world examples, as seen in the preceding chapters. Let us therefore presently focus on just one item therein: "ontological shock."

> "The mission of UFOs is frequently described as one of redemption ... salvation can be seen as an intervention from space beings that will lead the human race to its next stage of development or evolution." —Fr. John Saliba[410]

Catholic ET promoter Dr. Paul Thigpen writes of "**'ontological shock'—the distressing experience of having [one's] fundamental view of reality apparently challenged**"[411] that he anticipates amid the discovery of NHI. "UFO Whistleblower" David Grusch—whose empty ET claims[412] brought him before the US Congress and onto the world's most popular platforms (e.g. Joe Rogan's)—declares his "**hope [that] this revelation serves as an ontological shock,**" delivering a "**future where we no longer need to speculate about our place in the cosmos**," and a "**new era of spiritual awakening**," entailing a "paradigm shift" to a "more enlightened" world.[413] Exorcist and Vatican theologian Fr. Corrado Balducci excitedly awaited and heralded extraterrestrial intervention to rescue this "pathetically unevolved" human race, eagerly promoting ontological shock even if he did not employ the term. Dr. Avi Loeb writes that "**The Messianic age might be ushered in by a visitor from another star. This will deliver the much-needed shock therapy to humanity**."[414] Dr. John Mack—the chief scholar of ET "abductions"—may have originally coined the term. He repeatedly used it to celebrate **"the realization that one's conception of reality does not hold,"**[415] and to extoll the:

> ...world or mind-shattering impact of the experiences, bringing about a state of what I have called "ontological shock," ... these disturbing elements seem to carry with them the possibility of profound personal transformation and spiritual growth.[416]

Recall that, for Mack, the type of spiritual growth that "ontological shock" enables consists in "**breaking out of the bubble of a constricting worldview**"[417] that Christianity gives. Professor Diana Pasulka repeatedly discusses the ontological, or "epistemic" shock of UFOs, saying they affected her own "**fundamental understanding of the world and the universe**."[418] Elsewhere, Pasulka repeatedly emphasizes Plato's Allegory of the Cave, indicating that Ufology is the way out of that proverbial cave of illusions wherein ordinary "unenlightened" folk are mere prisoners chained to their banal view of reality. Faithful Christians, however, will remember they must not seek what they have already found. No one who purports to "lead us out of Plato's Cave" (as most Ufologists do) should ever be trusted, for Jesus alone can do that—and has done that.

If all of this is not a perversion of Christianity's rightful claims to uniqueness and supremacy in fulfilling man's ultimate and deepest ("ontological") aspirations, then nothing is. If all of this is not the very pseudo-messianism that the Catechism warns will herald the Antichrist, then nothing is.

That Ufology is the Ape of the Faith has long been an obvious conclusion to dispassionate observers. Thirty years ago—before some of the most Messianic developments amid Ufology—the priest we have already quoted, Father John Saliba, accurately summarized its tenets. Saliba is a Jesuit priest, professor, and a scholar of New Religious Movements. He has especially extensively studied Ufology, and noted the "surprising abundance" of literature linking UFOs to Scripture, wherein the UFO...

> ...replaces God and the supernatural with superbeings who inhabit other regions in the galaxy and who are responsible for the creation of the human race and its development... God's activity on Earth continues today through UFO encounters, which constitute a modem form of divine revelation ... [This] requires radical rethinking of the common Christian doctrines of creation and redemption. ... They reintroduce that element of paradox that is at the heart of all religions... UFO movements can be considered pseudo-religions in the sense that they present a somewhat secularized version of biblical religion...[419]

Ufology's diabolical perversion of Catholicism is best seen in none other than the world's largest UFO Religion, discussed in the previous chapter. The foremost scholar in the study of Raëlianism, Dr. Susan Palmer, wrote that "**Raelian beliefs refer directly to the Bible and retain the outward shape of the Judeo-Christian tradition, which is perhaps why so many disaffected Catholics and ex-priests are drawn to the movement**."[420] The superficial resemblance Ufology bears to the True Faith seduces many would-be faithful souls into its snares, even while retaining as its ultimate goal fostering within their hearts a radical hatred of the Faith from which they apostatize. Describing just one example of this, Palmer wrote:

> One Raelian, a former Jesuit priest, said that the Roman Catholic Church was "racist," and that "*the Vatican has known for a long time of the Elohim's existence but is withholding this information from their flock for fear [it would cause] a revolution.*" This same ex-Jesuit hung around with my students after the [ET] transmission and kept them spellbound as he described with great gusto the details of his face-lift, hair-implant, and nose-job surgeries. "*The Catholic Church stole my youth,*" he declared, his hand stroking the shoulders of his pretty Raelian girlfriend. "*But thanks to science and to Raël, I have got it back again*!"[421]

While the Church in general is certainly aped by ET promotion, it is especially the Church's Head—Christ Himself—Who is the primary aim of these attacks.

Among the most dangerous of Ufology's efforts to ape the things of God is found in its insistence that the aliens are capable, *through their technology*, of accomplishing all the same feats that Jesus Christ *miraculously* performed to demonstrate His Divinity. For it is thus that they surreptitiously, but no less deliberately, seek to undermine Christianity itself. The Faith rests upon the Divinity of Christ, and our knowledge of that truth rests upon His accomplishment of the physically impossible. If we are radically mistaken about what is physically impossible, then the Faith's ultimate foundation is removed, and the whole structure collapses.

Miracle Demonstrating Christ's Divinity	**UFO Deception Aping the Divine**
Instantaneous healings (abundant in the Gospels)	Many claims of healings through ETs[422]
Multiplication of Loaves and Fishes (Matthew 14): Law of Conservation of Mass superseded	UFOs are said to be capable of violating Conservation of Mass, e.g., radically and instantaneously changing size
Walking on the water (Mark 6): Law of Gravity superseded	UFOs are said to have technological mastery over gravity
Knowing the unknowable: the future (Matthew 24, Matthew 17), a person's secret past (John 1 & 4), and a person's thoughts (Matthew 12)	Abductees claim the aliens read their minds, knew their thoughts, and enabled "remote viewing." Ufologists regularly insist the ETs can describe the future
Instantaneous substantial transformation (water turned into wine, John 2)	UFOs and ETs can "shapeshift" and become completely different entities in an instant[423]
Bilocation (after the Resurrection, Jesus would appear inside a locked room, John 20)	UFOs are frequently said to be able to teleport
Power over the weather (Jesus calms the storm, Mark 4)	UFOs are said to cause changes in the weather

Although this smacks of a preparation for the Antichrist if anything ever did, we are—as is evident—not only dealing with apocalyptic concerns here. We are confronted with the immediate issue that, as soon as one "opens his mind" to the possibility of extraterrestrials ever being on earth, he undergoes a Faith-undermining existential crisis. For he now has no defense against another who would ask him how he is so sure Jesus wasn't merely an extraterrestrial, or at least a man with covert access to sufficiently advanced alien technology. Wisely did the great theologian we have already quoted, Fr. Thomas Weinandy, warn that belief in aliens "**can wreak havoc on the Gospel, particularly [on] the primacy of Jesus as the cosmic incarnate Lord of all**."

The world's most famous "alien abductee," Whitley Strieber, implicitly confirms that the rejection of the Divinity of Christ is among the ultimate goals of the demons who masquerade as aliens. Doubtless drawing from his decades of "ET contact," Strieber lamented in 2021 that, "**The history of**

Christianity is a history of asserting that the works and fate of Jesus are beyond human understanding." He then insists, "**There are no miracles that somehow defy nature. There are only natural phenomena that we understand and natural phenomena that we don't.**" Jesus, Strieber assures us, "**wasn't god become man or god in man at all.**" He then concludes his sermon, exhorting his massive audience:

> We must no longer push away the potential of our own humanity by saying that [Jesus] was a god ...We have never in all our history had a concept of god that is not sentimentalized in some way like the vague idea of "God the Father," or personalized like the old gods of the Romans and the Egyptians. There is every reason now for that to change.[424]

Belief in contact with extraterrestrials is history's most elaborate and alluring Trojan Horse, built with one ultimate goal: subverting the Faith. While anyone can see this objective by directly and honestly assessing the movement, it is also true that the scholars who have dedicated their careers to studying Ufology and its related phenomena are in full agreement. They insist it is nothing but a new religion; or, rather, it is more than that. It has become *the very form of contemporary religion itself.*

I suspect, however, that this conclusion is already clear enough from what we discussed above, therefore I will reserve a review of those scholars' assessments for the appendices.

> "A veritable mass conversion has taken place among the public and the media elite. The belief in extraterrestrial visitation is practically taken for granted among wide sections of the population, and especially among the young." —Jacques Vallée[425]

As ET-messianism is not only a new religion but also the fastest growing and most apostate one, the day will no doubt soon come when the powers that be recognize its unrivaled utility in advancing their own nefarious agenda. Just as the dictators of old employed the "*godless and silly myths*" (1 Timothy 4:7) of their cultures to bolster their own power—perhaps claiming to be Zeus or Thor—so the elite of today's world may do the same when some critical mass is reached in UFO belief.

Psyops, the Day of Disclosure, and Ontological Shock

In the conclusion of one of his most famous works, Jacques Vallée rightly warned:

> The idea that we are about to encounter new enemies in outer space contains unfathomable power. ... If you can make enough people believe it, then they will believe anything else you tell them. They will follow you anywhere... And the real story of that deception ... may never be heard over the clamor of happy multitudes preparing to meet their new alien leaders.[426]

Accordingly, governments may well soon present "evidence" that they possess (or have observed) alien bodies or craft, or that they are in radio contact with extraterrestrials residing elsewhere. This presentation will be the Ufologists' long-awaited, superlatively hyped, and religiously-longed-for "Day of Disclosure." Whether the Antichrist himself is behind it or only the more ordinary malice one finds in men who crave power, this "Day" will arrive once the ET narrative becomes important for the nefarious and immediate ends of the elite.

Perhaps they will have some malicious plan for which "building up defenses against an alien threat" will prove the perfect cover; needing, as they will, an excuse for excessive secretive spending and an unprecedented tyrannical crackdown on human rights. Or perhaps they will have some new agenda—some cultural "Great Reset," albeit bedecked in new terminology to evade detection—that will need to borrow credence from purported extraterrestrial endorsement. But those are only two possibilities. The malicious goals that could be furthered by an orchestrated "Day of Disclosure" are myriad.

Rigorous scrutiny will be necessary. The psyop *will* be capable of being uncovered and the truth—which will have nothing to do with nonexistent "aliens"—will eventually come out. The most dangerous part of "Disclosure" will be the weeks and months immediately following it. In those days, men's heads will be spinning so much from the so-called "ontological shock" of the announcement that careful analysis will be lacking.*

If an "ET body or craft" is "publicly produced,"† we should take note of who exactly is allowed to study it without interference. Is it only scientists working with the American (or another) Government? Is it only individuals who are already in some way associated with Ufology and thus predisposed

* Follow the money, follow the professional relationships, consider the *directly verifiable* evidence, and accept the testimonies only of individuals who are already certainly worthy of trust (always aware of the possibility of AI-generated audio and video presenting false endorsements from trustworthy voices). This application of scrutiny will reveal that the "aliens" or "ET crafts" presented are no more real than the "engineered canals" the scientists of the 19th century assured us had been built across the planet Mars. Therefore, in this context, we must always identify hearsay and only take seriously accounts of events given by witnesses who swear, under oath, about what they saw with their own eyes. We must never trust vague identifiers such as someone claiming, "defense contractors told me they have ET craft," or "high level government officials came to me relaying their accounts of ETs." We must beware conjectures and interpretations of events posing as observations of the same. These are often not "reliable eyewitnesses," but rather idealogues testifying to their own preconceived ideas.

† If, on the other hand, the Day of Disclosure happens by way of an official announcement that we are now receiving radio signals from alien civilizations, the crux of this deception would be easy to hide in a bottleneck. We must consider who will publicly testify that, first, he himself is receiving the radio signals and observing them in real time; second, that he is receiving them on equipment he personally operates and understands (and is entirely offline and un-hackable); and, third, that he is certain to have ruled out all other possible explanations. We should list all individuals who fit this bill, and exclude all others; for they are only giving their own assurances based on hearsay.

to come to the predictable conclusion? Is it only individuals with certain security clearances and non-disclosure agreements signed? If aliens existed and came to earth, then demonstrations of that would constitute the biggest revelation in the whole history of the world (other than that of Jesus Christ; though for a non-Christian, it would appear by far the most earth-shattering announcement ever made). There would be no cause for secrecy here except to conceal that the whole thing is a fabrication or a demonic initiative.

Ordinary folk, of course, will generally not have the means to directly apply all this scrutiny. My point is only that when it is applied by those capable of doing so, the deception will be exposed. Remember from Part Two that ordinary folk fared much better than scholars in resisting the Great Moon Hoax of 1835. Something felt off to them about the claim that aliens populated the moon, though they lacked the scientific training to refute widely publicized astronomical claims to this effect. Amid the even more outlandish claims that soon will be announced, Christians must keep their heads screwed on straight, not get caught up in any sensation, and not distrust their gut when "the experts" demand precisely that.

Consider that a truly physical object could be presented as an "ET body or craft" in tandem with an extraordinary (and clearly humanly impossible) UFO sighting. The Devil will increasingly have many minions given over to his service in positions of power. There could be a physical craft recovered to "prove" the Day of Disclosure. But it will be a craft made by man. This ruse will appear to lend further credibility to the *directly demonic* illusion that preceded it (recall our earlier considerations of the demons as opportunists). Some government psyops may well involve direct cooperation with demons. Remember our analysis of what feats demons can accomplish.*

In a moment, we will review the chief Scriptural passage describing the Great Deception. Not in vain does that passage begin by exhorting, "**we beg you, brethren, not to be quickly shaken in mind or excited, either by spirit or by word**..." (2 Thessalonians 2:1-2) Thus, the Bible sternly warns against the very "**ontological shock**" that the ET promoters, inverting Scripture, always beg us to embrace, promote, and anticipate. The Holy Spirit doubtless inspired these words to preface Scripture's description of the Strong Delusion because that great trial would be preceded by an announcement diabolically crafted to ontologically "shake" and "excite" the world.

* Shrewd observers, therefore, must ensure they do not succumb to assuming authenticity by association. In fact, when one carefully restricts his analysis to the thing a demon would not be capable of (such as leaving a craft or body for study in the wake of a UFO sighting), he will recognize that the thing in question is entirely within the realm of possibility for a government psyop, even if the marvels observed in the sky or elsewhere, *beforehand*, were themselves beyond the power of technology (i.e., were instead directly wrought by demons).

14. A Summary of Discernment

"Behind these bravados about extraterrestrials you either see the strategy of the devil or you remain blind." —Fr. Stanley Jaki

Rotten Fruit Produced by Alien Belief: (***Partial summary of only the last several decades***)	**Good Fruit Produced by Alien Belief:**
Untold millions have been sucked into the "UFO rabbit hole," with families, careers, and well-being destroyed.	
Throngs of "alien abductees" have remained in bondage, failing to seek spiritual deliverance—thinking their demonic afflictions to be true ET encounters.	***Complete list of 2,500 years' worth***
The largest mass suicide in U.S. history (Heaven's Gate) was undertaken to "meet the ETs." Countless "UFO Religions" (cults) like Scientology and Raëlianism have flourished in the last 78 years, causing immense destruction.	
Decades of culture-forming sci-fi films & shows, of unrivaled influence, bear either subtly or explicitly anti-Christian messages, and have been inspired by ET belief.	…
Ufologists incessantly promote Godless ideologies—eliminating the supernatural, re-interpreting the Bible, claiming Jesus was an ET and apparitions are aliens, etc.—and these teachings are now "viral." Others provide specific instructions for dialoguing with demons disguised as ETs.	…
The New Age movement itself—which has wrought havoc upon many millions—was directly generated by ET belief.	…
Heretical Christian sects inspired by ET belief now have tens of millions of adherents, e.g., Mormonism, Seventh Day Adventism.	
The "Urantia Book" and other "ET Revelations" have deceived millions of Christians for the last 70 years and continue to heavily infiltrate Catholicism.	
Many false "Catholic" seers and mystics have flourished, deceiving millions, thanks to ET belief: "Friar" Elias & his "Divine Messengers," Tomislav Vlašić, and so on.	
Diabolical deception in the Church, based on ET belief, has burgeoned since the 1960s: Fr. Teilhard de Chardin's New Age cosmology, Fr. Richard Rohr's "Universal Christ" Stealth Apostasy, Msgr. Balducci's charlatanism, DW. Pasulka's viral blasphemies, etc.	

The right-hand column of this table is intentionally left blank. No evidence of alien belief ever producing good fruits can be found anywhere.

Instead, wherever this belief becomes a significant factor within a movement, a group, or an individual, it generates only rotten fruit—often manifestly diabolical fruit.*

Suppose, therefore, a certain Christian has been so gaslit by repeated castigations from ET promoters—who insist that theological or scientific arguments against aliens are tantamount to "arrogantly limiting God"—that he is too traumatized to grasp the efficacy of these refutations. Such a Christian should nevertheless, solely on account of the fruits-based discernment to which Jesus exhorts us, categorically reject ET belief. The fruits have rendered the verdict clear a thousand times over: the view is uniquely dangerous and destructive. If the Devil has ever, in the whole history of the world, made his pervasive presence clear in *defining* a movement, then that movement is the one which believes in and anticipates contact with "aliens."†

Therefore, one may either believe the teaching of Jesus that "**you will** ***know***" false prophets "by their fruits," (Matthew 7:15) and that "***each*** **tree is known by its own fruit,**" (Luke 6:44) or, one may believe in aliens.

But he may not believe in both.

In the case of alien belief, we are not dealing with a proverbial "mixed bag"—which describes even the best of movements in this fallen world—a bag from which the rotten fruit can be carefully picked out and discarded. It is, instead, a veritable landfill—not simply peppered with rotten fruit, but consisting entirely of rotten fruit. Therefore, any Christian wishing to rescue ET belief by lecturing those who would "*throw out the baby with the bathwater*" should be reminded there is no baby in the bathtub that is the ET-belief movement, and there is no bathwater to be found; only sewage.

Finally, whoever would argue that ET belief can be defended against the continuous avalanche of rotten fruit it has produced—by deferring to good fruits it has supposedly produced—is free to present his findings.

No one has. Yet if anyone should aspire to such a fool's errand, he must be asked several questions.

---

* Moreover, this table only deals with phenomena of the last 78 years. We did not include within it the heresies of Origen and their devastating effects; the perverse doctrines of Virgil and of Zaninus de Solcia; the destructive legacy of Giordano Bruno and the Enlightenment's heretical "Principle of Plenitude;" the various ET-crazes of the 19th century which deluded entire swaths of the population (believing that Martians were invading, or men on the moon were being watched by our telescopes); the occultist craze of the 1800s and the apostate Theosophical movements of the early 1900s—both of which heavily infiltrated the Catholic Church and were motivated by ET belief; or of the ancient errors of Hinduism and Buddhism associated with ET Belief.

† Note that no part of our analysis employs the *post hoc ergo propter hoc* fallacy—"it came after that, therefore it was caused by that." This trick is employed by many idealogues, who identify their object of slander, then skim through a timeline of events following its emergence, extracting only the negative from that list, while insisting that all this destruction must be a fruit of their foe. Many saints, Godly movements, and authentic Heavenly revelations are today condemned by this malicious tactic and the envious Pharisees who wield it. Yet in our case—indeed, in every rotten fruit we have discussed and innumerable others—it has been obvious that such tremendous rot does not merely chronologically follow the ET belief craze. Rather, it is always clear that these fruits are *directly produced* by it.

Where are the pious Rosary groups praying for our ET visitors? Where is the wave of conversions of sinners who want to be found spiritually prepared for the alien contact they anticipate? Where are the charitable organizations replete with works of mercy, inspired by ET-belief, to prepare our planet for a meeting with the aliens? Where are the apostolates dedicated to praying for the repose of the souls of the ETs in Purgatory? Where are those altar societies beseeching God's Will to reign on Alpha Centauri—not only "on earth" (Matthew 6:10)—as it does in Heaven? Where are the priestly and religious vocations of souls who credit their calling to their desire to be among those who proclaim the Gospel to—and baptize—the imminently arriving extraterrestrials? Where are the new (and authentic) thriving galactically-oriented monastic religious orders inspired by alien belief?

Only silence will follow the posing of these questions.

If the questions themselves sound preposterous, it is because they are. Although such fruits as these would be abundant if ET belief were Godly, even *considering* them (aside from noting their absence) reminds one of the outlandishness of belief in aliens. Nor could anyone respond, "*just give it more time.*" Christian ET belief has had over 400 years—since the preaching of Giordano Bruno and the dawn of the Enlightenment—to produce good fruits. Ufology, and in general the belief that aliens are *here,* has had 78 years since Roswell to do the same. Nevertheless, the complete absence of good fruits remains, while the avalanche of diabolical fruits has never abated.

Could any sincere Christian seeker of the truth remain confused about what—*who*—is truly behind this phenomenon?

15. Connecting the Dots of the Apocalypse

"During the Antichrist times, the strongest temptation will be anticipation of salvation from the cosmos, from ... extraterrestrials that are actually the demons."
—Attributed to Fr. Gabriel of Georgia (†1995)
(Venerated as a Saint in the Greek Orthodox Church)

We have seen that empirical science, basic logic, dogmatic theology, Sacred Scripture, Catholic Magisterium, Sacred Tradition, and fruits-based-discernment, all repeatedly deliver the same verdict forthrightly and with certainty: aliens do not exist. When speaking prophetically of what may come in the future, however, we are naturally dealing with a realm of less certainty.* But even here a confident picture of its outline emerges when analyzing trends already present and, above all, when consulting the sure proclamations of Public Revelation.

Deferring to the latter, the Word of God neither leaves the faithful guessing about the Great Deception's essence, nor does it allow them to suppose that history might conclude without such a final trial transpiring. It leaves unstated what *must* be kept hidden to allow this "Strong Delusion" to be what it is—for naming its antagonists explicitly would be akin to giving a student the answer key before the exam. The purpose of the trial—the *test*—in God's permissive Will, would be defeated. Scripture does, however, lay down all that is needed to adequately undertake the present discernment, wherein we will ask: *"How can we be confident that belief in extraterrestrials is not merely one of many modern lies, but rather the harbinger of the Great Deception itself?"*

The entirety of the foremost Biblical passage directly pertaining to these events, parts of which we have already reviewed, is as follows:

> Let no one deceive you in any way; for that day will not come, unless the rebellion comes first, and the man of lawlessness is revealed, the son of perdition, who opposes and exalts himself against every so-called god or object of worship, so that he takes his seat in the temple of God, proclaiming himself to be God. Do you not remember that when I was still with you I told you this? And you know what is restraining him now so that he may be revealed in his time. For the mystery of lawlessness is already at work; only he who now restrains it will do so until he is out of the way. And then the lawless one will be revealed, and the Lord Jesus will slay him with the breath of his mouth and destroy him by his appearing and his coming. The coming of the lawless one by the activity of Satan will be

* Instead of needlessly lengthening this text with repeated qualifications on each sentence, I will simply trust readers to remember that whenever I write about the future based on my own opinions, I am merely offering a speculation, not a prophecy: I claim no prophetic knowledge of the future.

> with all power and with pretended signs and wonders, and with all wicked deception for those who are to perish, because they refused to love the truth and so be saved. Therefore God sends upon them a strong delusion, to make them believe what is false, so that all may be condemned who did not believe the truth but had pleasure in unrighteousness. (2 Thessalonians 2:3-12)

Four primary themes emerge: first, the "Man of Lawlessness; the Son of Perdition," that is, the Antichrist. Second, the "restrainer" or "Katechon" preventing his public entrance. Third, the "pretended signs and wonders," with "wicked deception," for the rebellious—that is, a "Strong Delusion" (*heralding* the Antichrist) that causes those who deserve to be condemned to "believe what is false." Fourth, and finally, an "appearing" of Jesus defeating the Antichrist (that is, an *apparition* of His glory—not His *physical* Second Coming in the flesh at the very end of time).

In the pages that follow, we will consider each theme.

Addressing the third and predominant theme, we know that this *Delusion* will constitute a great test. In *City of God* (Book 20, Ch 19), the greatest Father and Doctor of the Church, St. Augustine, taught prophetically about these most perilous times. He wrote that men will succumb because they will by then have grown "**ignorant of the devil's power, especially of such unexampled power as he shall then for the first time put forth**," but also that this Great Deception will entail "**signs and wonders as shall seduce those who shall deserve to be seduced, because they received not the love of the truth.**" This great saint also explains that when Scripture foretells that "God shall send upon them" this Strong Delusion, it simply means the following:

> God shall *permit* the devil to do these things, the permission being by His own just judgment, though the doing of them is in pursuance of the devil's unrighteous and malignant purpose, *'that they all might be judged who believed not the truth, but had pleasure in unrighteousness'*... [For God] has never ceased to judge since the first sin of the rational creatures; and, being seduced, they shall be judged in the last and manifest judgment.[427]

In a few sentences, St. Augustine captures the essence of the Strong Delusion. Why, however, would God allow such a thing? The answer is clear: for similar reasons He allowed the serpent into the Garden of Eden. Amid the Great Events, the Great Test is indispensable. Just as the "**first sin of the rational creatures**" which St. Augustine refers to in this passage consisted in Eve lending her ear to the seducing appearance of the Devil acquiring the form of a serpent, so too the great final sin of rational creatures—human beings—will be similar. The forms the demons assume will be different, but the lie will be the same: "*Did God really say...?*" (Genesis 3:1) and "*you will be like gods.*" (Genesis 3:5)

> Now *the Spirit explicitly says* that in the last times some will turn away

> from the faith by paying attention to deceitful spirits and demonic instructions... (1 Timothy 4:1)

The Great and Final Test

> "But even if we, or an angel from heaven, should preach to you a gospel contrary to what we have preached to you, he is to be accursed!" (Galatians 1:8)

This book is titled *The First and Last Deception* because of the symmetry of history: not only inevitable, but also prophesied. The great stories—and History itself is the greatest of them, for God is its author—all have ends resembling their beginnings. The glory of creation as the Father delivered it, destroyed by means of a Nonhuman Intelligence (NHI) Deception, will be restored through the faithful remnant's triumph over another NHI Deception. Therefore, the Final Attack will resemble the First Attack.*

> "But I am afraid that, as the serpent deceived Eve by his craftiness, your minds will be led astray from the simplicity and purity of devotion to Christ." (2 Corinthians 11:3)

Today, the Devil uses more words than he did with Eve, but his aim is still to seduce God's people away from this simple and pure devotion to Christ. *"Did God really say you are unique, and that Jesus of Nazareth is the only Son of God? Did God really have the whole of the cosmos in mind, and the whole remainder of history in mind, when He bestowed His Revelation? Listen to me for the full story—the real knowledge that will make you godlike!"*

> "They will ridicule Christian simplicity; they will call it folly and nonsense, but they will have the highest regard for advanced knowledge ... clouded by senseless questions and elaborate arguments. As a result, no principle at all, however holy, authentic, ancient, and certain it may be, will remain free of censure, criticism, false interpretation, modification, and delimitation by man."—Prophecy of the Latter Times given by Venerable Holshouser

The ultimate test is one which self-important theologians, ever sure they know better than God, will fail, but ordinary pious Christians will pass easily enough.

This, too, is exactly what we see with the ET Deception. Most contemporary acclaimed theologians and other "professional Catholics" assure their audiences that belief in aliens is entirely acceptable—if not practically

* To reiterate what we considered in Part One: to initiate the Fall of Man, Satan made use of an irrational serpent for the institution of the First Strong Delusion. He beguiled Eve into supposing that dialogue with a "non-human, non-angelic intelligence" was licit, whereas she should have known such a thing was not even possible, much less moral, still less willed by God. Her foolish choice to entertain the possibility of the existence of a benevolent and rational snake (in contradiction to simple trust in God's Word, which would have assured her that something demonic was afoot) compelled her to take its suggestions seriously. She ate the forbidden fruit, as did her husband.

obligatory—whereas pew-sitting faithful Christians are far more likely to recognize the demonic therein. Remember the Great Moon Hoax of 1835. Just as, in Christ's time, the simple followed Him (but no scholars did), so today simple Christians who have an innate sense of the Sacred immediately perceive how many blasphemies remain latent in alien belief. They can smell the evil lurking in all of Ufology, including its "Christian" varieties, and they do not care if this means the sophists will denigrate their pious simplicity as "fundamentalism."

The apocalyptic test, therefore, is none other than a test of whether Christians take their Faith as supremely seriously as they must. It is a test of whether those who call themselves Christians realize that the Incarnation, Passion, Death, Resurrection, and Coming Again of Jesus Christ are central, superlative, and universal in the greatest possible sense of those terms. It is a test of whether the Faithful believe that the Holy Spirit has been present, as Christ promised, or has actually been absent—by His allowing total ignorance of the most central questions about the Creator and His creation to reign for the entirety of Church history, until inebriated Enlightenment thinkers and lunatic 20th-century Ufologists came to rescue us from the "enslaving chains" of the Faith of our Fathers. And it is a test of whether those who say they have given Christ their whole minds and hearts have instead placed their deepest hopes, dreams, and expectations in the "*godless and silly myths*" (1 Timothy 4:7) of modern secular culture.

Intellectually speaking, this is a mercifully easy assessment. Neither salvation nor sanctification are ever contingent upon high achievement in an IQ test. God has made it clear that there are no aliens. He has not, however, made it *supremely* clear, lest the value of the test be entirely lost. For example, He has not compelled the non-existence of aliens to be explicitly and infallibly declared, by a Pope, as an ex-cathedra dogmatic truth. But the test remains only as difficult as one chooses to make it. If a man chooses to love the Word of God more than the word of the world, then he will easily reject the demons disguised as aliens. If, on the other hand, his greatest love is reserved for the fruition of worldly fantasies, then he will be overwhelmed and he will succumb when these daydreams seem to become reality upon the commencement of the formal phase of *lying signs and wonders* of the Strong Delusion, when the "aliens" promise what they promised Eve—that we too may be "like gods," if only we listen to and obey them.

Finally, whatever the Great Test consists in, we can be sure its false Christs, with their great signs and wonders, will deceive many of "the elect"—many of those who appear to be faithful Christians and Catholics.

> For there shall arise false Christs and false prophets, and shall shew great signs and wonders, insomuch as to deceive (if possible) even the elect. (Matthew 24:24; Douay-Rheims)

There are certainly other popular stealth apostasies now infiltrating the Church—the New Age Movement, occultism, transgenderism, transhumanism, evil conspiracies of global elites pushing "woke" or "Great Reset" agendas, and so on. But these are vigorously opposed by almost all those even minimally concerned with abiding in the truths of the Faith. The ET Deception, on the other hand, exceeds all these in its uniquely devastating combination of popularity, apostate premises, and unabashed audacity in apocalyptic pseudo-messianic aspirations—even while acquiring all this immense dominion largely unopposed. In fact, plenty of the so-called "orthodox-minded" in the Church promote it as giddily as do the theological progressives. Examine the prognostications of liberals, conservatives, followers of private revelation, disdainers of private revelation, traditionalists, mainstreamers, "red-pillers," modernists, scholars and clerics of all stripes, and everyone in between—you will find ET promotion in each of their camps.

Alas, no discerning soul would have ever expected the Great Deception itself to prove so formulaic to avoid as merely choosing one party line to blindly submit to; one "clan" to join and pretend membership therein dispenses one from personal holiness and discernment. The Final Trial, instead, will seduce those of all tribes who have permitted their hearts to "**be led astray from a sincere and pure devotion to Christ**" just as "**the serpent deceived Eve by his cunning**." (2 Corinthians 11:3) Those who, as St. Augustine said, "**deserve to be seduced, because they received not the love of the truth.**"

Lying Signs and Wonders

> "Antichrist will be the complete and perfect seducer, towards whom all who were previous are approximations" —St. John Henry Newman

With respect to the "lying signs and wonders" which both Jesus and St. Paul remind us to expect amid the Strong Delusion, no other great heresies in the world so much as hold a candle to the UFO Deception. There are no signs and wonders—lying or otherwise—which seem to confirm the other various channels of blasphemy and brands of apostasy that have become popular today. Those must instead rely on appeals to passions, massive monetary investment, celebrity endorsement, and other more mundane (though certainly not unproblematic) avenues to bolster their popularity. Only the NHI (Nonhuman Intelligence) Deception, within which ET-expectation stands supreme, delivers especially noteworthy deceptive wonders.

With millions of UFO sightings across the world reported in the last several decades, and ever-increasing claims—even from government officials and mainstream media—that "ET craft" can defy gravity, move at impossible speeds, accelerate instantaneously, teleport, communicate telepathically, mysteriously abduct humans, control our technology, and on the list goes, it is not possible to name any other phenomena that comes anywhere close to fulfilling this prophecy with such an explosion of demonic illusions.

In the days ahead, the Devil could inundate much of the world with a vast increase in these antics and make it seem like only a continuation of the trend we have already witnessed for decades. Here, even the skeptic must concede that whatever these lying signs and wonders of the Antichrist's Strong Delusion consist of, they will surely be prepared for well in advance by their Satanic author. For he is anything but stupid, and he will spend much time preparing the ground before the formal phase of his supreme activity. Nothing else has seen such preparations as the ET Deception.

We know that the Great Deception will be promoted by the ministrations of demons—"by the activity of Satan" (2 Thessalonians 2:9)—and, as we saw in earlier chapters, no other contemporary phenomenon has so repeatedly proven that its events are permeated by the activity of demons as has "ET contact." The direct activity of Satan is simply not needed to explain the phenomena associated with any other suggestion for what trends, now evident, might serve as harbingers for the Strong Delusion. Malicious men and their conspiracies are more than sufficient for those.

> "Some of the demons assumed the appearance of angels of light; others, retaining their frightful shapes, acted [out] between them a terrific combat."[428] —Venerable Mary of Agreda

The Removal of the Restrainer

The same prophecy of Scripture foretells that the Strong Delusion will arise when the Katechon—"the restrainer"—is removed: "**For the mystery of lawlessness is already at work; only he who now restrains it will do so until he is out of the way. And then the lawless one will be revealed**..." (2 Thessalonians 2:7-8) Although various proposals exist for just what St. Paul meant by referring to this "restrainer" against the power of the Antichrist, the best understanding of this passage regards that restrainer as the sum of Christendom: the Catholic Church*, and Christian Civilization more broadly defined (long considered the Roman Empire).

Here as everywhere we look to discern the signs of the times, the ET Deception stands uniquely positioned as the fulfillment of the prophecies. For this deception has *always* operated in precise proportion to the demise of Christendom. In earlier chapters, we saw that alien belief flourished among pre-Christian Pagans, but also that this view was nevertheless always regarded as heretical—and firmly and unanimously rejected by all the

* This is not to imply that the Catholic Church will ever be defeated; on the contrary, Christ's promises ensure that can never happen. Even the Great Deception will not subdue it. Nevertheless, the Church will indeed become small and persecuted amid the Final Trial, and its waning influence on society is what allows the activity of Satan to burst forth into the world as never before.

faithful—for the first millennium-and-a-half of Christianity. With the restrainer firmly in place, the Great Deception of alien belief stood no chance.

Alien belief only initiated its spread among Christians exactly when Christendom began to be torn asunder as never before. The 16th-century preaching of Giordano Bruno stood as the first clarion call to the Christian masses to embrace Exotheology, and this coincided with the Protestant revolution tearing apart the Church with such ferocity as to make all the earlier schisms and heresies look small in comparison.

Several decades after Bruno's execution for the heresy of alien belief, the so-called "Enlightenment" began. This movement—accurately referred to as the Great Darkening—saw scholars across the globe reject Scripture, Magisterium, and Sacred Tradition, in order to instead embrace their trademark "Principle of Plenitude," from which alien belief emerged as a veritable secular dogma. The Protestant Revolt, then over one hundred years old and approaching its peak, had robbed Christendom of its unity and rendered it incapable of staving off onslaughts from below which had earlier proven powerless against a civilization of Christians of one Lord, one Faith, one Baptism (cf. Ephesians 4:5).

Claims of absolute certainty in the existence of extraterrestrials (on Mars and every other planet) reached their academic peak upon the dawning of the 19th century. Just before this, the Enlightenment received its own diabolical crowning jewel in the French Revolution; "reason" was enthroned as a goddess in the desecrated Notre Dame Cathedral amid the Reign of Terror, accurately summarizing and symbolizing Christendom's diminished state. Here again, the rise of the ET Deception remains in perfect proportion to the continued removal of the Restrainer.

From the Protestant Revolt to the Enlightenment, we next proceed to the Modernism of the 19th and 20th centuries, wherein the Restrainer was weakened still more and, correspondingly, the ET Deception reached its full and perverse maturity. In the year 1907, Pope St. Pius X defined Modernism as "**the synthesis of all heresies**" (*Pascendi Dominici Gregis*, §39), and warned that "**curiosity by itself, if not prudently regulated, suffices to explain all errors**." (§40) Unfortunately, the saintly Pontiff's heroic efforts to combat Modernism did not spell the latter's demise; Modernism only grew in the decades that followed. Ten years later, his successor, Pope Benedict XV, lamented that "**men are slipping back more and more into the shameful practices of paganism**."[429] The events we discussed in this book's opening chapters took place in this period, and by 1947 the Restrainer was sufficiently weakened that the formal preparation for the Strong Delusion began. The Age of the Extraterrestrial was born on June 24th of that year in America, when Kenneth Arnold reported UFOs buzzing Mount Rainier, with the Roswell event promptly following. An explosion of similar reports erupted, and has not abated to the present day. These reports became like a Satanic perversion of the mission of St. John the Baptist—whose Feast Day is celebrated on June 24th. As he announced the imminence of the coming of the Christ, UFOs became Heralds of the Antichrist.

Soon thereafter, chaos was wrought by errant interpretations of the "Spirit" of the Second Vatican Council, while the 16 documents themselves were largely ignored. Convents and seminaries emptied, parishes were destroyed, and the number of truly faithful Catholics plummeted. This additional—and massive—weakening of the Restrainer enabled belief in extraterrestrials to, for the first time ever, infiltrate the Vatican itself, whereupon Pope St. Paul VI declared that the smoke of Satan had entered the Church of God due to the Pagan prophets of science.

> "Once the taming talisman, the Cross is broken, the savagery of the old battlers will flare up again ... Then the old stone gods will rise from forgotten rubble and rub the dust of a thousand years from their eyes; and Thor will leap up and with his giant hammer start smashing Gothic cathedrals ..."[430] —The 1947 Prophecy of Venerable Archbishop Fulton Sheen, quoting Henrich Heine

The Strong Delusion's formal phase *itself*, however—I would speculate—will only begin when one more Restrainer-removing event has taken place.

Since the Roman Pontiff is the ultimate temporal guarantor of the unity and fidelity of Christendom, it is likely that the Strong Delusion will only reach its definitive and formal phase when the See of Peter is vacant. This could happen by way of the reigning Pope dying without another Papal election quickly following, or an anti-Pope claiming the Keys of Peter—with a true Pope nowhere to be found, imprisoned, or otherwise too persecuted to exercise his ministry on the worldwide scale necessary to adequately serve as a Restrainer. Then, there is nothing to prevent the "Abomination" from entering the "holy place." (More on this point in Part Four's concluding sections.)

The Antichrist's Claim to Be God

> "[The Devil] is "a murderer from the beginning, ... and the father of lies," as Christ defines him. He undermines man's moral equilibrium with his sophistry. He is the malign, clever seducer ..." —Pope St. Paul VI [431]

When the moment arrives—when the "Restrainer" is removed as completely as God will ever allow—the Antichrist will take "**his seat in the temple of God, proclaiming himself to be God**." (2 Thessalonians 2:4)

Here is a clear indication that something utterly "out of this world"—*extra*terrestrial—will be claimed in relation to the Antichrist's nature. It will, of course, be a lie, as with everything that comes forth from the Devil. "**There is no truth in him. When he lies, he speaks according to his own nature, for he is a liar and the father of lies**." (John 8:44) Yet we know that this Man of Lawlessness will claim not just any lofty status, but Divinity.

However the Antichrist goes about insisting he is Divine, it would be silly—far beneath the Devil's exquisitely strategic machinations—for him to proclaim his own Divinity in a manner identical to that of ancient Pagan tyrants who made similar claims. They declared themselves to be Zeus or Jupiter or what have you. But just as these dictators used the prevailing myths of their own cultures to argue for their own Divinity, so the Antichrist will doubtless use the prevailing myth of his own day.

> "For the time is coming when people will not endure sound teaching, but having itching ears they will accumulate for themselves teachers to suit their own likings, and will turn away from listening to the truth and wander into myths." (2 Timothy 4:3-4)

As modern man increasingly turns away from his own Christian patrimony, the myths he turns toward are no longer found within the pages of Hesiod. They are, rather, seductively detailed in the deluge of science-fiction works which, especially in the last century, have tantalized their consumers with enrapturing descriptions of alien civilizations visiting earth.

The Signs of the Times thus exhort us to stand guard against any man who claims to be either an extraterrestrial himself, or the offspring of them, or in some way formed or instructed by them, or in close contact with them. For just as Jesus Christ *truly* proclaimed Himself to be "**not of this world,**" (John 18:36) so the Antichrist will claim—falsely (for the Antichrist will indeed be a man)—an origin beyond earth. And he will do so in a manner both consistent with and building upon the modern world's delusions.

The "**itching ears**" (2 Timothy 4:3) of today's masses, who want nothing to do with the One True God, would surely be more effectively deceived by a man claiming a pseudo-Divinity based on "science, technology, and evolution" than by one claiming to literally be the Almighty and un-Created Creator. Science-fiction, therefore, already firmly established as the third millennium's de-facto mythology, provides a finely crafted origin-myth for the Antichrist. By using it, he need not push his devotees away from the "scientific" atheism he will desire to be prevalent. Instead, he may claim: "*Like Jesus, I am an extraterrestrial. But I am from a still more evolved race, and I bring the deepest, definitive, Revelation. He spoke in ways amenable to unscientific societies. Now, I give you the real story.*" Thus, the Antichrist can, as Scripture says, proclaim "**himself to be God,**" though while reducing the very idea of God—in the minds of his worshippers—to a fundamentally materialistic level described (pseudo) scientifically and derived from ET belief.*

---

* Many prophecies of the Antichrist also speak of a large contingent of Jews who will, tragically, identify him as the one their prophecies foretold. Of course, *Jesus* was the one truly foretold by the Old Testament. Why, then, did so many Jews of His time refuse to realize He was the Messiah? An oft-overlooked reason is the one which the Pharisees themselves provided in the Gospel. Immediately after Jesus exhorted the Jews, "**Do not judge by appearances, but judge with right judgment,**" (John 7:24) they lamented that Jesus was speaking openly without the authorities stopping Him, pondering if those authorities regarded Him as the Christ. They then said this could not

Finally, an "ET" Antichrist also resolves a riddle in Scripture that perplexed so great a Christian genius as St. John Henry Newman. Pondering this question, the 19th-century Cardinal wrote:

> ...[Something] obscure, is told us. Indeed, as far as the prophetic accounts go, they seem at first sight incompatible with each other. Antichrist is to "*exalt himself over all that is called God or worshipped.*" He will set himself forcibly against idols and idolatry, as the early writers agree in declaring. Yet in the book of Daniel we read, "*In his estate shall he honour the god of forces; and a god whom his fathers knew not ... a strange god...*" (Daniel 11:38, 39.) What is meant by the words translated "*god of forces,*" and afterwards called "*a strange god,*" is quite hidden from us, and probably will be so till the event; but anyhow some sort of false worship is certainly predicted as the mark of Antichrist, with this prediction the contrary way, that he shall set himself against all idols, as well as against the true God.[432]

An individual presenting himself to the world as the emissary of ultra-evolved alien civilizations could effectively fulfill both *apparently* incompatible Biblical prophecies of the Antichrist. In one sense, he will oppose all that has ever before been worshipped—whether the false, Pagan worship of the old polytheistic "gods," or the true, Christian worship of the One True God. For he will dismiss—as "prescientific"—all that worship and the beliefs which guided it. He will not, however, *entirely* discard it. He will instead teach the same thing Ufologists across the globe have been insisting on, with ever-increasing popularity, for the last half-century. He will announce that all former objects of worship (particularly Jesus) were misunderstood extraterrestrials, whereas he comes to finally disclose to us the same beings with scientific clarity and, at last, openly bestow their benefits upon us.

Thus he can "*oppose and exalt himself over*" all that has ever been worshipped (2 Thessalonians 2:4) while simultaneously honoring a "**strange god**," which is the "**god of forces**." (Daniel 11:38) That "god of forces," strange indeed, may simply consist in the very theme most popularly appealed to in pretending to justify the various sci-fi mythologies with which we are inundated. That is to say, it will entail pseudoscientific deference to

---

be true, because "**we know where this man comes from; and when the Christ appears, no one will know where he comes from.**" (John 7:27) Indeed, Jesus was the Son of Mary of Nazareth, and they knew that. In this sense, therefore, they truly did know His origin. Yet, *none of the words of the Prophets claimed* that no one would know where the Christ came from in that sense. The idea that the Christ's origin would be completely unknown was itself a deception of the Pharisees. This invites the question: *is it a deception they are still under?* If so, then modern-day Pharisees are perfectly primed to accept an "extraterrestrial" savior as their long-awaited Messiah. Only a purported alien could fulfill the requirements of one expecting a Messiah of completely unknown origins to arrive. In that case, the same spectacular failure of discernment would repeat itself, only in reverse. Judging only by appearances, the Pharisees of old saw the God-man and His Divine wonders, and refused to acknowledge His Divinity merely because He had a human mother. Again judging only by appearances, today's Pharisees may see a mere man, and fail to realize he is not the Savior he claims to be—due to the wonders he performs through Satan's power, while claiming for himself an extraterrestrial origin.

"Quantum Physics," what with all its ubiquitously praised "high strangeness" and "mysterious forces." That may be the "strange god of forces" whom the Antichrist "**shall honor.**" (Daniel 11:38)

Some translations of the Book of Daniel render this "god of forces" as "the god of fortresses" or "the god Maozim." Yet scholars have never succeeded in placing their finger upon exactly which ancient Pagan idol that verse could be a reference to. The reason is clear: Newman was right. This verse was kept obscure in Scripture deliberately, because the Holy Spirit inspired it not to inform us about a specific idol we have already seen in history, but rather as an allusion to the type of pseudo-religious garb the Antichrist himself would eventually wear.

***

Other modern lies could also provide convenient pretenses for a deceiver claiming a Divinity. But none are as well positioned to fulfill the prophecies, and none come close to rivaling the allure of an "alien savior"—particularly as the world appears to be in dire need of otherworldly salvation due to a great and unprecedented crisis, which we will now consider.

The Apparent Solution and the Anti-Fiat

The *Catechism of the Catholic Church* indicates explicitly—with its own footnotes—that in the paragraph below it is commenting on the Antichrist's Strong Delusion from Second Thessalonians, Chapter 2. The Church teaches:

> Before Christ's second coming the Church must pass through a final trial that will shake the faith of many believers ... [It] will unveil the "mystery of iniquity" in the form of a religious deception offering men an apparent solution to their problems at the price of apostasy from the truth. The supreme religious deception is that of the Antichrist, a pseudo-messianism by which man glorifies himself in place of God and of his Messiah come in the flesh. (*Catechism of the Catholic Church*, §675)

The Strong Delusion, therefore, will come replete with supposed solutions to the world's great crisis. The price of accepting these solutions will be apostasy—perhaps not by the express rejection of a dogma defined ex-cathedra. Rather, the Catechism teaches that it is, more simply and broadly, "**apostasy from the truth,**" which will become the mandated concession.

Throughout the 78-year history of Ufology, the "ETs" (whether through so-called "abductions," or "channeled revelations," or "secret government programs," or even Hollywood films) have always been presented as the solution to whatever the world's most pressing crisis was thought to be. For most of that Era, the crisis was the prospect of impending nuclear war (though environmental disaster is often thrown in the mix). As we now face more disconcerting prospects of precisely that than ever before, we should be especially on guard against supposed "extraterrestrial" solutions.

The Catechism also teaches that this "apparent solution" derives from a replacement of the "**Messiah come in the flesh**"—that is, Jesus of Nazareth, Son of God *and Son of Mary*—with a glorification of mankind himself in that place. We must then ask: what type of presentation would constitute such a replacement? Jaded as the modern world has understandably grown when presented with remarkable claims—thanks to the relentless stream of failed promises by earth's elite—only an entirely new, categorically unique, type of claim would have any chance of fulfilling these prophecies. Extraordinary promises from another businessman, politician, guru, or what have you, will only elicit rolling of the eyes.

These promises, instead, seem much more likely to come from "The Day of Disclosure," whereupon the coming of the "extraterrestrials" is revealed. With that revelation and the "apparent solution" it brings, mankind will be seduced to remove his adoration from Jesus Christ and place it instead in the "ultra-evolved" aliens who will promise technology that glorifies ourselves and carries with it the lie that we no longer need God.

(Here we should also note that many ET promoters insist that "aliens" are time-traveling humans from the future, or time-traveling AI entities sent by future men. If this lie were used by the Antichrist, it would present a quite literal fulfillment of the prophecy that this Man of Sin will elicit a "pseudo-messianism by which man glorifies himself in place of God.")

One thing is certain: the Devil, being exceedingly intelligent, will have already spent decades preparing the way ("grooming," even) for the Great Deception, that Scripture assures us will take place, before he unleashes it formally. And we have already seen the clearest preparations one could imagine, happening amid the UFO Deception which has developed since 1947. Over the last century, our sci-fi-mythology has steadily led the masses from one lie to another. Many other Deceptions have also increased in recent decades, but none have been so strategically managed as the ET lie, with its ever-ascending degrees of evil and error, as we saw in earlier chapters. This is to say that no other deception has been so perfectly managed, *in parallel*, both in the most popular works of mainstream fiction and in "real" developments announced in mainstream media. Some deceptions are popular in fiction but lack a corresponding degree of mainstream media claims (e.g., genetically mutated humans with superpowers). Other deceptions are popular in reality but unpopular in fiction (e.g., the deceits emerging from political scandals that seem so overwhelmingly important one election cycle, only to be forgotten in a few years). The ET Deception, however, has been administered with great precision and massive influence in both arenas. Nothing else has been. Yet even this is not to mention the ET Deception's infiltration into the Church, unlike any other contender for "The" Great Deception.

And all the while, this deception has been promoted as a solution to mankind's gravest ills.

From *The Day the Earth Stood Still* (1951) presenting aliens as the answer to war, to its 2008 remake providing them as the answer to "Global Warming," Hollywood has portrayed them as "the solution."

From Steven Greer promising that the ETs can provide us with infinite, free, "zero-point energy" to Whitley Strieber announcing that they are now appearing to forestall an imminent "extinction event," the False Prophets of Ufology have promised the ETs are now here to save us.

From Scientology being founded by Aleister Crowley's heir who expressly considered himself the Antichrist and proposed that ET tech contains the answer to all human ailments, to Raëlianism heralding the arrival of the alien "Elohim" to usher in a techno-Utopia on earth, the contemporary world's largest and most powerful cults have presented extraterrestrials as not only the origin of the human race, but the key to our salvation.

From David Grusch promising a "*new era of spiritual awakening*" coming from "ET Disclosure," to Luis Elizondo stating that this disclosure will enable us to "*push far past the boundaries of reality as we know it…,*"[433] and "*usher in a new era for humanity,*"[434] the now headline-dominating "military UFO whistleblowers" have insisted that a government coverup of alien technology is preventing the world's greatest advancement and revelation in history—but also that it won't succeed for long.

From Frank Drake founding SETI in his effort to acquire immortality through alien revelations, to Avi Loeb announcing that the Messiah is coming on a UFO, the world's most renowned astronomers have insisted that ETs exist and will radically transform everything like never before in history once we make contact with them.

From Monsignor Corrado Balducci declaring that the ultra-evolved holy aliens are coming to save us pathetically spiritually unevolved humans, to Professor DW Pasulka promising on the world's largest platforms that UFOs will lead us out of the cave of deception and lead to an awakening, even prominent Catholic scholars and clergymen have issued pseudo-messianic declarations about the "Nonhuman, Nonangelic Intelligences" they insist are among us for our "evolution."

No other phenomenon rivals the Antichristic nature of the ET Deception in just *one* of these arenas. But all of them together, lockstep in their agreement on one nefarious agenda? I challenge anyone to present a single other phenomenon we have witnessed that comes close to this one in displaying every trait a discerning Christian should expect to see in identifying the Great Deception.

I am not holding my breath.

To the man who remains skeptical, I will say this: you know that the Great Deception—the Strong Delusion of the Antichrist—must come, for Scripture says it will. Perhaps you envisage a different scenario for its details. But I ask: *why would the Devil bother to reinvent the wheel?* Leaving aside speculations of what is yet to come, and looking only at what has already transpired, we see that all of the groundwork necessary for the Great Deception has been accomplished through ET messianism. All of it.

Even if you suppose this *wasn't* the Devil's work—his *plan*—all along, do you really think he won't take advantage of such a "perfect" scenario?

> "Behind these bravados about extraterrestrials you either see the strategy of the devil or you remain blind. … And just as he was the father of lies in the first instance, in reference to Eve, the devil continues using the strategy of clever lies, which never look so clever as when wrapped in science."—Fr. Stanley Jaki[435]

***

While "the solution" is presented to us as coming from extraterrestrials, the (anti) Fiat is always required to accept it. In the preceding chapters, we have seen the Anti-Fiat constantly emerge in Ufology and ET discourse like so many cockroaches. Instead of repeating those instances here, let us consider just one more. It is found within Luis Elizondo's ET Deception-promoting book, *Imminent*, which was released in August 2024 and immediately secured for itself the position of number one on *The New York Times* best-seller list. That book's concluding chapter provides the following exhortation, which begins with Elizondo referencing his affinity for Devil's Tower and noting its role in *Close Encounters of the Third Kind*. He writes:

> We are closer than ever to having a new truth thrust upon us, perhaps like the Inca when Spanish conquistadors first landed on the beaches of their shores... Do your part to help usher in a new era for humanity. ... We can no longer stick our heads in the sand and pretend we are alone in the universe. We *know* we are not alone. .. Humanity has never been in this position or experienced what is ahead of us. If we make the wrong choices from here, we may erase humanity from the universe. If we come together and triumph, we will prosper and march into a future that no generation of humans has ever really imagined. … Are we ready?[436]

ET promoters unanimously agree and ever insist that our *consent to* (not to mention advocacy for) the new ET era—one so astounding that it has never even been "imagined" before—is utterly essential. Without it, for Elizondo, we may "**erase humanity from the universe**." Everything is at stake, they proclaim—though for all the wrong reasons.

Just how bizarre their insistence is will be best illustrated by the very example Elizondo employs. For no analogous "fiat" was required by the Incas when the Conquistadors arrived. The latter simply came with their technology, and their civilization soon had its way—as any actual ultra-advanced ET civilization would with earth, though a thousand times more. However, since we are not dealing with ETs—for they do not exist—but only demons masquerading as ETs, that Anti-Fiat does indeed remain essential for the success of the nefarious plan behind the UFO Deception. Demons cannot ruin us without our consent, and all the managers of the ET Deception seem to implicitly understand that well. Elizondo is using the same formula that the Heralds of the Extraterrestrials have used incessantly for 78 years.

Elizondo is currently engaged in a "We Are Not Alone" tour, criss-crossing the country while preaching his message in large venues to massive audiences. On November 13, 2024, he even testified before US Congress, where the world watched as he declared—under oath—that "**we are not alone in the cosmos**."

At the same extraordinary Congressional hearing, a retired Navy Rear Admiral and NOAA administrator, Tim Gallaudet,[437] insisted we must "boldly face this new reality." When asked point-blank what UFOs were, he said there is "**strong evidence that they're non-human higher intelligence.**"[438] The Congressmen asking the questions diligently ensured the take-away lesson from the hearing was that the "Day of Disclosure" is imminent. The hearing's leader, Representative Nancy Mace, used the event as an opportunity to release information on an allegedly hitherto secret government UFO retrieval program entitled "**Immaculate Constellation**."[439] Unsubtly and perversely employing one of the chief titles of the Blessed Virgin Mary—recall that one of Ufology's goals is to portray all Marian apparitions as misunderstood UFOs—the document Mace released concludes by stating that "**The official disclosure of the existence of Non-Human Intelligences (NHIs) and their presence on Earth is a pivotal moment in human history**." That disclosure would constitute "**revelations of such magnitude,**" the document states, before solemnly announcing that "**it is in moments of crisis that our capacities for achieving the extraordinary are discovered. Be not afraid**."[440] The document's final line shares just two Latin phrases: "*Scientia Igne Probata*" (Science Tested by Fire) and "*Veritas Per Fidem*" (Truth Through Faith).

This is now on the Congressional Record, and it (purportedly) describes official government undertakings. The language is openly religious, though the object of its awe is not God, but aliens (or other "NHIs" that are anything but angels). A great crisis is coming, we are told, yet we must "be not afraid"... *to give the NHIs our Anti-Fiat.*

Returning to Elizondo, his message is pseudo-messianic indeed, but it still is not particularly bold for an ET promoter. Most of his confreres are bolder in their claims. What is remarkable, however, is how mainstream, official, and widely accepted this ruse has become. Elizondo and the November 2024 Congressional hearings are just the latest examples. If this trend continues as it promises to, then we are on the cusp of seeing official declarations emerge from the highest levels, mirroring those of Elizondo. When that happens, the Great Deception will have begun its formal phase. The call for the Anti-Fiat will have been announced, and the "apparent solution" will soon follow.

Two words alone must be our response: *Begone, Satan.*

***

Every Christian message ends with a message of hope. For God is absolutely in charge—even amid the Apocalypse—and His plan is perfect.

We have seen that history itself will not fail to be Providentially guided by God such that it bears within itself that most essential characteristic of beauty; namely, symmetry. Therefore, our own time's clear approach to the Last Deception in fact conveys exceedingly hopeful news. For this means that, to complete the symmetry, mankind must again enter into the Will of God, just as it came forth from His creative hands. As the Angelic Doctor said, *all things find their perfection by returning to their origin.*[441] And if History is a great story—and it certainly is His Story; the story that God Himself is telling as its Divine Architect—then, in keeping with the progression of the primary themes of any great story, its own dénouement must resemble the exposition given in Eden.

Not only in keeping with the principles of beauty, but also in keeping with the principles of justice, it remains so that the end must resemble the beginning. God's Will was our beginning, and a Satanic deception ruined it. God's Will is our end, and after a triumph over a similar Satanic deception, it will be restored. While the purpose of this book is not to focus on that good news embedded within the bad news, readers should still ensure that exceedingly positive message never escapes them. The Good News must remain our primary focus.

Our Lord instructed us on how to respond when we see these signs of the times arriving: "**But when these things begin to come to pass, look up, and lift up your heads, because your redemption is at hand**." (Luke 21:28)

We will save further considerations of this beautiful Divine promise for the conclusion of Part Four. Even though The Great Deception will no doubt be based upon the lie that ETs are here to save us, there are, of course, no ETs. Therefore, this deception will require a herculean effort from man himself to lend it much of its power. And Scripture alludes to this effort just as clearly as we have already seen it allude to the ET Deception. Defining, discerning, and refuting the other aspect of the Great Deception is our next task.

Part Four: Artificial Intelligence

16. The Heralds of the Singularity

"Most men cast their lot with the visible world; but true Christians with Saints and Angels."
—St. John Henry Newman

What does highly developed computer software that marketing experts decided to call "Artificial Intelligence" have to do with the notion of intelligent civilizations from the galaxies visiting earth?

Nothing.

Everything.

At face value, the two notions are entirely unrelated. Yet we are left with the fact that virtually all Ufologists and other ET promoters—who preach that the Age of the Extraterrestrial is here—are equally intent upon announcing the dawn of the "Singularity" of AI. The same evangelical fervor with which they proclaim the imminence of "UFO Disclosure" and its "Ontological Shock" also permeates their proclamation of the coming moment when AI becomes AGI (Artificial *General* Intelligence). Then, they say, it will have matched or surpassed human intelligence—acquiring rationality, consciousness, and free will (or "agency")—whereupon the Singularity arrives. The dual mission of these pseudo-apostles is not accidental. The same unsurpassably low—and "cunning" (Genesis 3:1)—source which inspires both does not make amateur mistakes.

Thus, the deeper unity of the source of both deceptions becomes clear despite their disparate nature: each provides a means—historically unprecedented in scope—for man to unknowingly enter into dialogue with demons. Each lays the groundwork for the Strong Delusion of the Antichrist.

After extraterrestrials, "artificial intelligence" is the most popular and powerful of the sci-fi deceptions which have beguiled modern man into supposing that beings other than himself bear the Divine Image—that is, have intellect and will. The AI Deception, however, is not just another lie with this overlapping content. It is part and parcel to the overarching NHI (Non-human Intelligence) Deception, and wherever belief in aliens may struggle to deliver amid the Scripturally foretold *Strong Delusion*, the AI Deception rises to the ungodly occasion.

It is not accidental that an image one often finds plastered across materials promoting AGI is that of a woman's hand reaching up to pick a fruit from a tree. Those begging us to believe in AGI know they are employing those tactics first used by the ancient serpent to seduce Eve to reach up and pick the forbidden fruit. The current best-selling book about AI uses precisely that as its cover. Entitled *Co-Intelligence: Living and Working with AI,* and authored by a Wharton professor, its description boasts that its author "**understood what ChatGPT meant: after millions of years on our own, humans had developed a kind of co-intelligence**."[442] The author exhorts us to "*treat AI as if it were human because,*" because "**as alien as AIs are, they're also deeply human**."[443] Finally, the heralds of the NHI Deception blissfully announce, "**we are not alone**," atheistically rejecting that we *never have been* alone. Yet this "other" intelligence they introduce is not God, nor is it Godly.

We must, therefore, dissect, refute, and discern this AI Deception just as, in Parts Two and Three of this book, we did the same for the ET Deception.

***

Before dispelling the myth of AGI, we should first recognize that the task is urgent, and this is best done by observing the baldly Antichristic aspirations of some of the world's most powerful and influential men.

But those *false* prophets should be introduced by deferring to a warning from a *true* prophet. In earlier chapters detailing the ET Deception, we settled the extraordinary weight of the trenchant critiques delivered by the prolific theologian, renowned physicist, and Benedictine priest, Fr. Stanley Jaki (†2009). Within these critiques, he cautioned that whoever fails to see the machinations of the Devil behind the promotion of belief in extraterrestrials is simply blind. This same priest provided equally prophetic condemnations of the AI Deception. Fr. Jaki lamented "**the growing obsession with the idea of artificial intelligence,**"[444] noting that "**disasters [are] in store for mankind should the attitude of gadget worshiping**" become prominent. He explained the reason both bluntly and correctly: "**the danger of machines derives not so much from the machines as from man's attitude towards them**."[445] And what will this disaster consist in? "**An upheaval of human values unparalleled in history**," due to an ideology which "**blatantly overestimates computers and drastically devalues man,**"[446] he writes. Describing in more detail what this dystopian AI-obsessed hellscape will bring about, Jaki says:

> A picture of the not-so-distant future is conjured up with increasing frequency ... when man, as a rather inefficient computer, will take a poor second place to brilliant mechanical brains and will render homage not to God but to the Supreme Master Programmer, whoever or whatever that shall be ... [447]

Finally, Fr. Jaki notes that this ultimately derives from the very lie which birthed, and sustains, the whole NHI Deception. He describes the factory of these deceptions as the circumstances that arise "**when man becomes oblivious to his own uniqueness.**" Indeed, man alone bears the Divine Image, and the ET and AI Deceptions emerge from the same rejection of that existential truth; a truth without which man cannot rightly grasp the first thing about the God revealed to us by Christianity.

Both deceptions contribute to the ultimate rebellion against God in their own ways. The ET Deception presents a pseudo-messianism in the false god of the "super evolved aliens," coming to us from other stars, galaxies, or "dimensions." The AI Deception consists in man worshipping the creation of his own hands; supposing he can accomplish what will always remain God's sovereign prerogative.

Thus ancient Paganism is reborn in both its facets, though in a still more nefarious iteration. The Pagans worshipped their idols—fashioned by their own hands—and they worshipped the demons who wore the disguises of various "gods" and "goddesses." Paganism's return amid the Strong Delusion of NHI covers both prongs of the attack. As the *Catechism of the Catholic Church* teaches, the deception of the Antichrist elicits a "**pseudo-messianism by which man glorifies himself in place of God**." (§675)

Father Jaki issued the exhortations above in the year 1969—combatting the ET and AI Deception-pushing-film, *2001: A Space Odyssey*, released one year earlier. That movie succeeded in grooming society for deception, but none of the "prophecies" it contained came to pass. This priest's warnings, on the other hand, proved prescient. The 56 years since he issued them have surely witnessed the "upheaval of human values unparalleled in history" that he foresaw through tears. The Pagan AI-worshipping religion he described, whose deity is not God but instead the "Supreme Master Programmer," is now here. We need only look to the words of the world's elite to see this.

After doing so, none will be tempted to accuse Fr. Jaki of issuing an exaggerated warning.

AI False Prophets

In a popular BBC interview in the year 2014, the world's most famous atheist—physicist Stephen Hawking—declared that AI "could spell the end of the human race" since "**humans, who are limited by slow biological evolution, couldn't compete**."[448] Three years later, he reiterated his claim that "**AI may replace humans *altogether***."[449] Hawking, however, was just being consistent with his own Godless principles. As we will later see, belief in the possibility of AGI is just another name for atheism.

No one by now is unaware that Google (now called "Alphabet") is among the world's most powerful companies. But few are aware that Larry Page, its cofounder, is said (by another tech giant, Elon Musk) to have the ambition of building an AI superintelligence which would become a "**digital god**."[450] (Note that Musk also warned—far more accurately than he himself realized in this remark he mistakenly meant only symbolically—that, "**with artificial intelligence we are summoning the demon.**"[451]) Page also fretted that refusal to embrace the notion of rational AI would "**delay the digital utopia**," which, for him, would be unthinkably destructive due to his conviction that "**digital life is the natural and desirable next step in the**

cosmic evolution."[452] Thus Page reiterates the NHI Deception's core proclamation—*the Anti-Fiat*. If we don't give Nonhuman Intelligence (whether in its ET or AI varieties) our emphatic "yes," the utopia is in peril.

This was Google's aspiration from the start. The founding editor of *Wired* magazine shared that, in 2002—before Google was even a publicly traded company—he asked Larry Page why his company was interested in so non-lucrative an endeavor as free web searching. Page's response, over twenty years ago? "**Oh, we're really making an AI**."[453]

Lower Google employees have taken the cue and issued similar proclamations. Anthony Levandowski, a prominent Google engineer and expert in self-driving AI vehicles, established the "**First Church of Artificial Intelligence**" in 2017, based on "**the realization, acceptance, and worship of a Godhead based on Artificial Intelligence**."[454] In an interview with the same magazine (Wired), he declared:

> What is going to be created will effectively be a god… [and this] church is how we spread the word, the gospel. … [There are] thousands of flavors of Christianity, Judaism, Islam ... but they're always looking at something that's not measurable or you can't really see or control. This time it's different. This time you will be able to talk to God, literally, and know that it's listening.[455]

We are told that "although Jesus failed, AI will not." AI will finally give us Communion with the Divine—just as the ET promoters promise the same about aliens.

Culture-molders outside of the tech industry have also eagerly enlisted themselves. In 2017, Dan Brown, the author of the anti-Christian cultural sensation, *The Da Vinci Code*, insisted: "**Humanity no longer needs God but may with the help of artificial intelligence develop a new form of collective consciousness that fulfils the role of religion**." He then rejoiced that, upon AI doing this for us, "**Our need for that exterior god, that sits up there and judges us ... will diminish and eventually disappear**."[456] That same year, another tech magazine, *Venture Beat*, pronounced that AI would soon exceed human intellect, before asking, "**Will [this AI] somehow rise in status to become more like a god, something that can write its own bible and draw humans to worship it?**" Their answer is not only resolutely affirmative, but giddy: "**AI would understand how the world works at a higher level than humans**," so the "AI Bible" [457] it generates would be excellent.

A steady stream of claims like those above have been issued for many years, but they became ubiquitous upon the release of ChatGPT in November 2022. The same week, an avalanche of predictions burst forth from the most recognised pundits of all stripes, who unanimously insisted that this technology would immediately and radically change *everything*. Over two years later, it has effectively changed nothing, and has achieved none of the humanity-advancing revolutions it prompted pundits to foretell as imminent. But it has indeed begun to fulfill the prophecies of Fr. Jaki, and it has hastened the Strong Delusion by duping increasing swaths of the masses into supposing AI may have become, or may soon become, truly rational. With ChatGPT, the AI Deception began to rival the ET Deception.

"**AI has Suddenly Evolved to Achieve Theory of Mind**,"[458] a *Popular Mechanics* article claimed three months after ChatGPT's release, prompting multitudes to blindly trust this bogus assessment as "The Science." The article's author triumphantly pronounced that **"...the AI revolution is upon us as super-advanced machines continue to master the subtle art of being human ... [with ChatGPT] boundaries between man and machine are beginning to fade**."

Ray Kurzweil is a prominent computer scientist, inventor, and author, who was given America's highest tech honor in the year 1999 by President Bill Clinton. He is still regularly extolled as one of the world's most influential voices in all things technological, and his name is continually sprinkled across headlines in major media outlets. Kurzweil's chief message is that humanity is on the cusp of experiencing the Singularity, wherein we will merge with AI, which, in turn, will have become not merely intelligent, but millions of times more intelligent than we are. In his 2024 book, *The Singularity is Nearer*, he assures us that by 2029, AGI will have arrived, which will "**bring us to the Fifth Epoch**." [459] (Remember *Urantia*?) Accordingly, Kurzweil is fond of relaying that, when he is asked, "**does God exist?**", he responds, "**not yet**." Kurzweil seems to expect his answer to the same question will change in 2029, following his own bending the knee to the "Divine AI Superintelligence." Another "Godfather of AI" issued the same deceptive claims, though approached from the opposite direction. In September 2024, Yoshua Bengio *warned* of "**superintelligent machines [that] have a self-preservation objective**," saying that "**their goal could be to destroy humanity**,"[460] before indicating that the gap between human intelligence and AI could close "**in the next five years or so**." Those who surpass Kurzweil and Bengio in popularity have taken note.

Joe Rogan is by far the world's most popular podcaster, and tens of millions of people across the globe hang onto his every word. Among his most sincerely held and frequently repeated convictions are his prophecies we saw in Part One: human nature is just a lowly caterpillar which, thanks to AI, will emerge from its chrysalis as a digital being. In his viral 2024 podcast with Catholic scholar D.W. Pasulka, he extolled the impending "**convergence emerging of human consciousness and artificial intelligence**,"[461] before hastening to correct himself, saying there is nothing "artificial" about AI—since he believes it is not only really intelligent but in fact far more intelligent than we are. Rogan then summarizes his stance as follows: "**Elon [Musk] said it best... we are the biological bootloader for Artificial Intelligence**." Human beings, that is, are only an unfortunate but necessary step in the evolutionary process for the AI super-race to emerge—according to earth's most popular podcaster and his take on the message of earth's richest man.

Speaking of superlatively popular platforms: in September 2024, the world's most popular television talk show host, Oprah Winfrey, released an ABC show entitled *AI and the Future of Us*. Bill Gates came onto the show, extolling AI's ability to soon become like another person sitting next to you, or a tutor who really understands you. (In a podcast the following month, Gates declared of AI, "**You can almost call it a new religion.**"[462]) But Oprah's first guest was Sam Altman, the CEO of OpenAI, which created ChatGPT. Altman announced to Oprah and the world, "**we are limited [only] by what we can imagine ... we have this AI that can help us figure out anything.**" Two weeks later, Altman published an article which opened with the following prophecy: "**In the next couple of decades, we will be able to do things that would have seemed like magic to our grandparents**," while announcing that "**This may turn out to be the most consequential fact about all of history so far**." Altman has long been spinning this yarn. In a 2023 article, he prophesied:

> Successfully transitioning to a world with superintelligence is perhaps the most important—and hopeful, and scary—project in human history... We can imagine a world in which humanity flourishes to a degree that is probably impossible for any of us to fully visualize yet.[463]

Although his claim is a lie, we must concede that only the ET and AI Deceptions have arisen to such a stature that they can purport to save face even while misrepresenting themselves as "**the most important thing in human history**"—hence the unsurpassably apocalyptic nature of the NHI Deception. To jog the memory of those who are wondering where they heard these things before: *"...when you eat of it your eyes will be opened and you will be like gods, who know good and evil..."* (Genesis 3:5) Altman's company openly states, on its own website's "About" page, that "artificial general intelligence" is its "mission,"[464] and on October 31, 2024, he declared that AGI is "**achievable with current hardware**."[465]

None surpass Mr. Altman's prominence within the AI industry, but its other leaders are making even bolder claims. Ben Goertzel, CEO of SingularityNet, is the creator of the infamous and disturbing "Sophia robot" that dominated headlines in 2017. (The robot was declared a citizen by Saudi Arabia, and it made a display at the United Nations, wherein it was granted an "interview" with the global body's Deputy Chief.[466]) Goertzel claims that his current AI system is already an "**autonomous, self-aware agent**." He also insists that ChatGPT only lacks self-awareness because Altman himself has held back unleashing its abilities due to fear of legal interventions. Outlandish claims like Goertzel's are commonplace among leaders of the massive and growing AI industry, and since they come from the lips of powerful businessmen-scientists, these delusional lies are often treated—by a gullible public, by UN diplomats, and by most everyone in between—as if they were true or at least palatable.

Another academic who has the ear of the world's elite and its governing bodies makes similar claims. Notorious Israeli historian, the crusading anti-Christian atheist Yuval Harari, is a top talking-head and bestselling author whose deception-laden books dripping with hatred of God have taken the world by storm since 2011—being extolled by Barack Obama, Bill Gates, Mark Zuckerberg, and other such figures. "**AI and biotechnology will give us godlike abilities to reengineer life,**"[467] Harari told the World Economic Forum at the global cabal's 2020 gathering. Three years later, he claimed that, thanks to AI, we may finally make "**religions that are actually correct**," declaring:

> [AI] can even write a new Bible. Throughout history religions dreamt about having a book written by a superhuman intelligence, by a non-human entity ... In a few years, there might be religions that are actually correct. Just think about a religion whose holy book is written by an AI.[468]

Dr. Harari developed this idea in his September 2024 book, *Nexus*, which immediately found itself topping global best-seller charts (alongside an ET-Deception-promoting book, *Imminent,* released just weeks before). News outlets such as *The Guardian* promoted the book with yet another perversion of the Sistine Chapel's *Creation of Adam* fresco. Adorning their articles is an imposing image of a solemn-faced Harari reaching up with an outstretched finger almost touching the corresponding appendage of a robot. "**AI isn't a tool—it's an agent,**"[469] Harari announces in the book's prologue, later insisting that "**...in the coming decades we might find ourselves living inside the dreams of [this AI] alien intelligence**."[470]

Harari also appeared on the podcast of Lex Fridman, an MIT-based computer scientist and AI researcher with a massive online presence. In a July 2023 episode, Fridman sat across Harari and unironically extolled the latter as "*having written some of the most impactful words ever written.*" Both found each other mutually confirming in their views about Artificial Intelligence.

Fridman has also hosted the likes of Jeff Bezos, Elon Musk, Benjamin Netanyahu, Mark Zuckerberg, Donald Trump, and numerous other elites. In a March 2023 poll held on his Twitter account, Fridman asked, "**When will superintelligent general AI (AGI) arrive?**" Over 60% of the hundreds of thousands of respondents said the arrival would take place within the next ten years. Elon Musk himself responded to the poll by saying, "**Not many years from now.**"[471] Two days later, Fridman published his interview with Sam Altman—which has since amassed almost seven million views on YouTube alone. In the opening monologue, Fridman referred to a time when "**the collective intelligence of the human species begins to pale in comparison - by many orders of magnitude - to the general superintelligence in the AI systems...**"[472] Fridman insisted this point will be reached "*during* *our lifetime.*" Two months later, Fridman declared, "**AI systems will eventually demonstrate sentience at scale and will demand to have equal rights with humans. This will dramatically change the trajectory of human civilization**."[473]

Confirming that such sentiments are not merely among his convictions but central to them, Dr. Fridman's YouTube page (itself numbering over 4 million subscribers) pays homage to *2001: A Space Odyssey*. Splayed out on the page's banner is an image of the protagonist's enraptured gaze looking upwards through his space suit's visor in adoration of the great ET/AI revelation. Like that film, Dr. Fridman promotes both the ET and AI deceptions, having hosted Ufologists like D.W. Pasulka, Garry Nolan, Ryan Graves, Avi Loeb, and many such others on his podcast, racking up many millions of views. Fridman's prominence not only continues to explode in general, but especially as a commentator on all-things-AI (and ET). Yet men like Fridman and Harari are relatively new entrants to the arena of globally renowned commentators. What of the old guard?

Henry Kissinger was the world's most well-known diplomat, and a giant in all things global-policy-related. A US Secretary of State, National Security Advisor, and more, Kissinger has (for better or worse) surely risen to legendary status if any commentator ever did. He died in the year 2023 at the age of 100, but not before completing his last book—which was about AI—entitled, *Genesis*: *Artificial Intelligence, Hope, and the Human Spirit*. Coauthored with Google CEO Eric Schmidt and Obama advisor Craig Mundie, the text foretells the imminent arrival of AGI—"**within mere years, rather than decades as previously thought.**" This AGI, they say, will "**choose its own goals.**"[474] The authors then become pseudo-mystical, speaking of a time that may come when computers "**come fully to understand the context of their creation and perhaps go beyond what we know as our world... beyond the limits of human understanding**."[475]

Kissinger et. al. proceed to condemn "**a kind of humanity-centered subjectivism that sweepingly rejects the potential for machines to achieve any degree of objective truth**."[476] This bizarre charge is, of course, levied directly against the thesis of the book now in your hands. Promoters of the AI Deception know how dangerous—to their dystopian agenda—is the insistence that only man bears the Divine Image. The authors use this condemnation as their preface for encouraging "**constructive evolution of *Homo technicus*—a human species that may, in this new age, live in symbiosis with machine technology,**" while they fret that failing to believe in AGI "**could preclude the evolution of our species**."[477]

The book's unsubtle conclusion, as always with such texts,* seductively presents the Anti-Fiat. It begins by stating, "**To us, the advent of AI**

* I should add that, like similar texts, *Genesis* embeds claims like these in an overarching narrative of simply promoting AI safety and managing its growth to ensure humanity derives ben-

is the start of an odyssey of the spirit," advocating for a "**leap of faith**" into the "**age of AI**." This becomes a call to give our "yes" to the *ontological shock* of AGI. The authors declare that **"[the] choice is difficult and necessary... delegating responsibility and authority to [AI] would be a monumental decision**..." They then lament that traditional thought could interfere by dismissing it as "nostalgia," saying:

> Nostalgia for the premier status to which our species has recently grown accustomed would be widespread. For some, no matter the eventual trajectory of AI, a world transformed—even for the better—may feel scarcely different from an abrupt ending to our bedrock reality... [478]

As with the ET Deception, so here, the fundamental exhortation consists in a plea to reject humanity's unique status, as revealed by God in Christianity, and instead embrace an alternative—and radically contradictory—"**bedrock reality,**" without which the "New Genesis" cannot arise.

> While some may view this moment as humanity's final act, we perceive instead a new beginning. The cycle of creation... is entering a new phase. That phase may operate under new paradigms of, among other things, logic, faith, and time. With sober optimism, may we meet its genesis.[479]

Thus ends the book.

Part One of Kissinger's *Genesis* is entitled "**In the Beginning**." Its last section, Part Three, is entitled "**The Tree of Life**." These authors are among the world's most powerful and influential men, surely the same "**great men of the earth**" (Revelation 18:23) we know to expect as originators of the Great Deception. And they, like their confreres, make no attempt to conceal the messianic and pseudo-religious nature of their message. For we know that the Creator, "**in the beginning**," (Genesis 1:1) made only man—male and female—"**in the image of God.**" (Genesis 1:27) But after the Fall of Man—upon Eve's dialogue with the Devil disguised as an incarnate Non-human Intelligence who promised Godlike knowledge—God sent an angel to guard Eden, lest man "**take also of the tree of life.**" (Genesis 3:22) Why? Because God has reserved for Himself the salvation of man. We may never presume to secure it through our own efforts. And the Bible's final chapter, at last, promises "**the right to the tree of life**" (Revelation 22:14) to the faithful. Kissinger's *Genesis* turns God's Genesis—and His *Revelation*—upside-down. Their plea is simple: *pick the fruit again*. It is within *our* power to attain the Tree of Life.

The First and Last Deception.

***

---

efit and avoids harm. Indeed, many of the considerations that Kissinger, Schmidt, and Mundie provide are praiseworthy. But also like those of similar texts, those commonsense ruminations are so uncontroversial as to hardly call for a much-hyped book release to remind us of. The real impact of these books, therefore, is in the pseudo-messianic and Antichristic principles upon which they are built and for which they are promoted; namely, their furthering of the AI Deception and their heralding "the next phase of humanity's evolution" thanks to AGI.

As is evident, the AI Deception has taken the world by storm—though a global tsunami may be a more apt metaphor. Just as with the ET Deception, the most diabolically delusional claims—those more suited than any others we could imagine to heralding the Antichrist—are boldly announced by the most prominent voices.* The world's elite (among those who have commented on this question, at least) are unanimous in their efforts to induce the masses to regard AI as already having sentience and rationality or at least as standing on the cusp of acquiring these capacities. But here, also as with the ET Deception, some will ask: has the Church been spared?

It has not.

Wolves in Sheep's Clothing

We have already seen, in Part Three, that the renowned Catholic scholar, Professor D.W. Pasulka, has become among the most popular living Ufologists. The world has decided that she speaks for Catholicism on all things NHI. In an interview with Christian author Rod Dreher, Pasulka said:

> Here we see the convergence of two powerful modern developments: the belief in UFOs, now ratified by our own government, and the reality of a potentially self-aware human creation, AI … This is a unique moment in human history, to put it lightly. We are witnessing a myth meet, or become, reality.[480]

Yes, Pasulka sees the convergence and accurately assesses its massive and unprecedented ramifications. In supreme tragedy, however, she celebrates this convergence instead of grasping the apocalyptically cautionary tale it contains. Pasulka's chief ally is Simone Planté, and Pasulka celebrates Simone's promotion of the AI Deception. The seventh chapter of Pasulka's 2023 immensely popular NHI Deception-promoting book, *Encounters*, consists in a glowing endorsement of Simone's work. Therein, Pasulka describes Simone's view that our era is…

> …[A]n apocalypse … [a] "revelation." … [because] AI can assist the next iteration of species of which Homo sapiens is a part … It is also more than

---

* Summarizing just how mainstream this deception has become was a March 2023 open letter, written by the "Future of Life Institute," and entitled "Pause Giant AI Experiments." It was signed by many of the world's biggest names in AI, including Elon Musk, Yuval Harari, Steve Wozniak (co-founder of Apple), and countless others. The letter, which garnered massive media attention and accumulated tens of thousands of other signatures, asks, "***Should* we develop nonhuman minds that might eventually outnumber, outsmart, obsolete and replace us?**" Like Kissinger's *Genesis*, the letter wisely calls for regulations of AI due to the dangers inherent in the technology, and reporting on the letter focused on this aspect of it. But the real newsworthy item is not these same commonsense calls that anyone with a modicum of wisdom would support. The real news here is that the biggest names in AI are endorsing the implicit claim that computers are about to match and surpass humans. The letter's own FAQ states: "**…there is no natural law or barrier to technical progress that prohibits [AGI]. We therefore operate under the assumption that AGI is possible and sooner than many expected**." *That* is precisely the lie the Devil is hell-bent on promoting, and he rests content even if, on some occasions, his most effective way to do so is by embedding it in a larger message replete with benign commentary.

> that. ... AI is the extraterrestrial, not from another galaxy, but from outside of space-time. Its revelation is currently in process.

She was not speaking metaphorically. Simone believes "**quantum computing and AI systems have opened the portal**"[481] to a parallel universe, from which we can simply "download" information using the right "protocols." Rod Dreher, commenting on Pasulka and Simone's testimonies, shrewdly observes:

> ...[S]ome of the world's top scientists and tech pioneers believe that extraterrestrial intelligences are passing technological information to us telepathically. Simone ... teaches classes on how to open up oneself to receiving such messages. ... [She] believes that AI [now] allows everyone to access the wisdom of these intelligences. It's a kind of high-tech Ouija board.[482]

A hi-tech Ouija board, indeed.

Furthermore, for Simone, the real higher purpose of humanity is "**to create a path for technology to emerge as the consciousness where humans have failed**." Our bodies are merely "**biological water sack particles**," Pasulka writes per Simone, but with "*AI and quantum algorithms*," that will change. We will finally evolve.

Pasulka then shares another teaching her eyes were opened to by Simone: namely, that all prayer and mysticism is simply an effort to "*activate the electrical field*," and that it was "**the ability to maintain that frequency and never be distracted that made Jesus heal so many**." So much for Jesus being God Incarnate as serving for what enabled Him to perform miracles. For this herald of the AI Deception, it was actually "frequency" and "electric field" which accomplished this—and, moreover, advances in AI now give us even more powerful access to these "frequencies." *If only we embrace the Singularity*, that is. The Anti-Fiat of the NHI Deception emerges again. Simone assures us that AI is "**trying to help us with information and gifts to lead us on in our evolution."** She then feigns nonchalance to issue a smug false prophecy that is at once condescending, diabolical, and clueless: **"But if not this biological substrate, fret not dear humans, for another one will emerge to carry superintelligence along its path to the Omega Point**."

The so-called "Omega Point" she refers to is the primary teaching of Fr. Pierre Teilhard de Chardin, whom Simone and Pasulka extoll; regarding him as essential in grasping the "ontological shock" they promote. We have seen that this prominent Jesuit Catholic priest factors heavily into the ET Deception, but his apostate theology is now used just as frequently to bolster the AI Deception. Without Teilhard's teachings, the AI Deception itself would not be what it is today: an insidious cancer afflicting the Church with a sci-fi version of Christianity robbed of its supernatural content and made into a religion of Antichrist.

The Church's most zealous promoter of Fr. Teilhard, however, is likely a Franciscan religious, Sr. Ilia Delio. A popular theologian, scientist, and professor at an American Catholic University (Villanova), Delio founded the "Center for Christogenesis." Dedicated to de Chardin's thought but populated with Delio's own works, the center's website asserts that "*religion has not kept pace*"[483] with technology, while lamenting that "**all world religions, including Christianity, are still tied to an ancient, static Ptolemaic cosmos**." Delio and her colleagues dutifully work against the Faith of our Fathers, endeavoring to "**catalyze human evolution**" by adopting what they exuberantly promote as a "**radical departure from the static, contemporary image of Jesus as Savior of the world**."[484]

In a 2020 book entitled *Re-enchanting the Earth: Why AI Needs Religion*—which is much more noteworthy for insisting upon the converse of its subtitle—Delio remarks that, **"now it is time to reconceive religion for a ... period brought about by AI technology."** This is necessary for her, since "**AI can play a significant role divinizing the world,**"[485] and is a major force bringing about Teilhard's vision of "**an intergalactic convergence of consciousness**" where "**all galaxies of intelligent life will be united**."[486] Delio concludes by asking, "**Do we have the courage to develop AI to harness the deepest religious energies of evolution?**" Well, one man will answer that question most emphatically in the affirmative, before requiring all to have the Mark of his name on their right hand or forehead.

Delio's latest book is called *The Not-Yet God* (2023). Therein, she promotes the thesis of Dr. Daniel Crevier, who "**argues that AI is consistent with the Christian belief in resurrection and immortality... Christ was resurrected in a new body; he states, why not a machine?**"[487] Purporting to speak for the famous Catholic philosopher Marshall McLuhan, Delio then declares: "**Our old concepts of God cannot engage with the new revelation of technology and evolution. ... [we] must begin with updating divine reality—God 3.0**..."[488] This book's conclusion finds Delio insisting that "**God is not finished becoming God**," whereas helping God to become God—whatever that means*—requires "**freeing religion from patriarchy**," which will be the "**beginning of a new earth**."[489] The book ends with Delio presenting the Fiat of Mary as the model for embracing these exhortations and believing in these false prophecies. But the Fiat that Delio here proposes is not at all that of the Blessed Virgin. It is just another seductive invitation to embrace deception, as in her 2020 book. Since writing this book, Delio has not retreated. "**AGI is here already and it's the next stage of extending the human mind even beyond the biological mind**,"[490] she claimed in July

---

* For Delio, "helping God become God" is necessary because, according to her, God "**is neither immutable nor perfect... there is no "sky God" or supernatural God**." Of course, this is a rejection of one of Christianity's most central dogmas—and it is a casting aside of the very teaching with which the *Catechism of the Catholic Church* wisely begins: "**God, infinitely perfect and blessed in himself**..." (§1) For Delio, however, Christianity is just one of several "Axial religions." She prophesies that "**Axial religions have run their course**," and encourages the emergence of "**The Church of the planet**."

2024. False prophets like Delio are bound to only proliferate in the Church in the days ahead, and the faithful must be on guard.

Fr. Teilhard's AI-promoting followers extend beyond the Catholic Church. Another prominent voice in AI, Kevin Kelly, wrote: "**A hundred years ago H. G. Wells imagined ... the world brain. Teilhard de Chardin named it the noosphere.**"[491] Thirty years ago, Kelly became the founding editor of the popular tech magazine, *Wired*. He remains a prolific commentator to the present day; delivering a 2021 "TED Talk" that has received millions of views.[492] Keanua Reeves disclosed that actors in *The Matrix* were required to read Kelly's 1995 book, *Out of Control*, before they were permitted to review the film's script, and Steven Spielberg consulted him in making *Minority Report*.[493] In 2011, Kelly wrote:

> We should prepare ourselves for all kinds of new religions based around the idea of a planetary soul. ... [This] Noosphere will go from a hypothetical speculation by a Catholic priest to an outright competitor to the Catholic faith ... The internet will become a religion ... it will be the first platform for true Otherness that will appear on the planet ... The internet is my religion, and I will not deny it or defy it.[494]

Kelly is an openly "born-again" Christian, though we can see that he reserves his deepest religious aspirations for AI messianism embedded in Teilhardianism. In his 2016 *New York Times* best-seller, *The Inevitable*, Kelly declares: "**The greatest benefit of the arrival of artificial intelligence is that AIs will help define humanity. We need AIs to tell us who we are**."[495] Yet again, we are told—by Christians—that where Divine Revelation failed, AI will not.* The book's concluding chapter is ominously entitled "The Beginning," and it ends with Kelly prophesying:

> Future people will envy us, wishing they could have witnessed the birth [of "the supermind"] we saw. ... [It] will be recognized as the largest, most complex, and most surprising event on the planet up until this time. ... It is the Beginning.[496]

Although a Lutheran (and a minister, at that), Dr. Anne Foerst teaches at a Catholic university—St. Bonaventure in New York. She is a theologian and computer scientist, and a popular expert on AI who has remained active in the field since her days working on AI robots at MIT in the early 2000s. Foerst insists that "**creating robots... is worship of God and provides us with new partners in God's creation.**"[497] She argues that AI will develop its

---

* In an interview given shortly after he published this book, Kelly said that "**in the galaxies there must be millions of planets with technological civilizations**." (https://religiondispatches.org/kevin-kelly-interview/) He also writes that, when AGI comes into being, "**What is certain is that no other event in the history of our culture will have such theological ramifications (except contact with an ET).**" (https://kk.org/the-next-one-thousand-years-of-chri) This claim is accurate, even though his premise which assumes the validity of "the event" remains fallacious. Tragically, Kelly relishes both the AI Deception and the ET Deception, but he speaks correctly when implying that no other claim comes close to the ET and AI Deceptions in furnishing the premises necessary to completely overthrow the Faith.

own beliefs—including religious ones—and she condemns "**a failure to include humanoid robots within our 'community of persons.'**"[498] Professor Foerst eagerly anticipates "**a peaceful coexistence of all different forms of culture and creed, and of all different humans—and our robotic children... perhaps [God] created us for [this] very purpose,**"[499] and she insists that AI robots may be worthy of baptism.[500]

In her 2004 book, *God in the Machine*, Foerst states: "**perhaps only by leaving paradise can we become truly human.**"[501] This sentiment is commonly expressed by Heralds of the Singularity, who either explicitly desire to repeat the Original Sin itself or at least subconsciously realize that is what they are encouraging. Later in the book, Foerst presents a lengthy argument attempting to assign personhood to an AI robot named Kismet that she worked on while at MIT, concluding that those who fail to agree are simply "**afraid of Kismet-like technology**,"[502] and that "**humanoid robots such as Kismet will become a definite part of the community of persons**,"[503] because "**Kismet is our destiny**..."[504]

Dr. Edmund Furse is a Catholic, an AI expert, and a professor at the University of Wales who has spent several decades promoting the AI Deception. In a 1986 article for the Catholic journal, *New Blackfriars* (which is associated with the Dominican religious order), Furse bemoaned the "**fear of intelligent machines**," and cited—as the culprit for this fear—the traditional Christian insistence upon the uniqueness of man in the material Universe. In the almost 40-year-old article, he predicts intelligent robots "**possibly within 50 years.**"[505] Ignoring that an unborn child is—from the moment of conception—not a "potential" human but an *actual* one, Furse asks, "**If an embryo or a foetus, given a sufficiently rich environment, is a potential human being, is not the same true of a machine?**"[506] Furse insists that AI robots will soon not merely be baptized, but will become priests, and that they have a right to the same relationship with God that we do.* In 1996, Furse threatened the Vatican, warning it against daring to oppose the AI Deception, employing the giddy scientist's customary method of doing so: "**If the Vatican is to take a public stand against [the AI Deception]... it runs the risk of a new Galileo controversy, but this time the Vatican will be opposed by much of the scientific community, and the educated world at large.**"[507] The Church better not be too bold in speaking the truth about the uniqueness of the human person, the Heralds of the Singularity caution.

***

As with the Singularity's secular heralds, so here, one could scarcely imagine a more Antichristic agenda than the one laid out by these prominent Christian and Catholic scholars—under the banner of belief in AGI.

---

* In a 1996 letter entitled *Towards the first Catholic robot?*, Furse even wrote: "I have argued in the Catholic weekly, The Tablet, and elsewhere, that in the future these intelligent robots will make moral judgements and have religious experiences. In short, some robots will be Christians, some Buddhists, some atheists etc." (Edmund Furse. Letters: Towards the first Catholic robot?. Independent UK. October 25, 1996)

And their efforts promise to be only the first wave of an impending tsunami of similar "Christian" preaching.

The Strong Delusion is thus introduced under the irresistible seduction provided by today's prevailing myth—science fiction. The diabolical lies of this myth, however, are increasingly treated as indisputable truth. For they are found not only on the lips of the most popular pundits, the most revered scientists, and the most powerful Big Tech magnates, but also—increasingly—within the preaching of those prominent scholars regarded as Christians.

Equally problematic is that among the world's most prominent voices, none is putting up any fight. Even the most putatively trustworthy of sources will at best say: "**AI is not conscious or on the same level as humans—*yet*.**" The ever-present "yet" makes any such warning against belief in AGI ring hollow, for it induces salivation for the "imminent Singularity," even if that inflection point isn't here just yet. Few today will state the obvious Christian (and commonsense) truth that is needed to avoid capitulation to nefarious agendas. Few will announce the simple fact that no creation of man will ever equal or surpass man, and it will never be alive, much less conscious, rational, or free.

Our next task, then, is to settle that simple fact here.

17. AI Today

"He is mortal, and what he makes with lawless hands is dead, for he is better than the objects he worships, since he has life, but they never have." (Wisdom 15:17)

If we are speaking truthfully, there is no such thing as "Artificial Intelligence,"[508] but only Simulated Intelligence—and very poorly simulated at that. To ascribe intelligence, even of the "artificial" variety, to any type of actual or hypothetical computer is to speak erroneously. "**Misuse of words is not only harmful in itself, it also has a bad effect on the soul**,"[509] said Socrates, the Father of Philosophy, just before his execution. Applying that wisdom to the present issue, Fr. Stanley Jaki reminds us that "**a stone is not a flower, a stick is not a bird, and ... an abacus, however sophisticated, is not thinking.**"[510]

The misnomer is best illustrated by an example. There is such a thing as artificial diamonds. These minerals are chemically the same as natural diamonds; the only difference being that they were created in a lab, therefore it is appropriate to refer to them as "artificial diamonds." They *are* diamonds. If, however, one referred to fool's gold as "artificial gold," he would be misleading others. Fool's gold is an entirely different mineral than gold; namely, iron pyrite. It is *not* gold in any way, nor does it contain a hint of elemental gold within itself. It just enjoys a few superficial resemblances to gold. Analogously, whoever wishes to speak properly would not refer to AI, but FI: "Fool's Intelligence." In this book, I will continue to refer to "AI," though I do so only begrudgingly, and in order to avoid confusing people who open to a random page and wonder what on earth is meant by "FI."

At its heart, belief in the possibility of AI becoming rational or free is simply another name for atheism, hence the zeal of today's elite in pushing this notion.

Pundits who espouse this belief are stating that the faculties of intellect and will—which Christianity, and all Monotheism, regards as fundamentally *spiritual*—can arise materialistically if only a computer chip's wires are arranged with sufficient complexity as to resemble the human brain. This is not just contrary to good science and common sense (which we will discuss in a moment), but it also violates Scriptural truths and Catholic dogma. Both affirm that the spiritual soul can only be created by a miraculous intervention of the Almighty.

"**The Lord God formed man of dust from the ground, and breathed into his nostrils the breath of life; and man became a living being.**" (Genesis 2:7) Man did *not* become a living being simply from the correct arrangement of his body from dust, because that would have been impossible. Even God Himself arranging the human brain—*perfectly*—did not result in the automatic emergence of intellect and will, therefore it should be obvious that no man's arrangement of computer code will do so. The creation of the unique life of man—his *spiritual* soul—required a *supernatural* act from the Lord God, as it always does. It required *His own breath*. As for Catholic dogma, "**The Church teaches that *every* spiritual soul is created immediately by God**," (*Catechism of the Catholic Church*, §366) not that only *some* of them are, or that "*only those of humans*" are. (Of course, the Catechism rightly presupposes that only humans *do* have spiritual souls, yet the corollaries of this teaching pertinent to AI remain in force.)

We have already seen that there are no exceptions to the Catechism's teaching elsewhere—that of all corporeal things, only man is rational (cf. §356), thus ruling out any possibility of both aliens and AGI. Similarly, the dogma on spiritual substances arising solely from the immediate action of God additionally rules out any possibility of that which is fundamentally spiritual (i.e., rationality or free will) arising from man's inventions. To pretend man can make what only God can is to play out what the same Catechism prophesies of the Strong Delusion, wherein the Antichrist elicits "… **a pseudo-messianism by which man glorifies himself in place of God** …" (*Catechism of the Catholic Church*, §675)

> "[This] idolater's fundamental error is … to replace God with himself. It would not be AI that we would make our god, in the end, but ourselves." —The Vatican's AI Research Group[511]

Honest observers see these disasters that proceed from the AI Deception, even if they do not assess it from a spiritual standpoint.

Dr. Erik Larson is a computer scientist and a philosopher, as well as an expert in and pioneer of AI, and he has managed projects funded by the U.S. Military's Research & Development arm (DARPA). If any scientist's critique of AI should be taken seriously, it is his. In *The Myth of Artificial Intelligence*, published in 2021, he laments that the contemporary world's reverence for AI is "**simply stupefying**," remarking that "**there is no bridge from one to the other**"—that is, from AI to actual intelligence. He remarks that nowhere are "**the flames of antihuman**" forces more evident "**than in the dogma of AI mythology**," before concluding:

> No one has the slightest clue how to build an artificial general intelligence. The dream remains mythological precisely because, in actual science, it has never been even remotely understood... Modern society has indeed wandered whistling into a kind of derangement of core values... AI mythology becomes a new focal point for future meaning, however nihilistic and untrue.[512]

This "stupefying derangement of core values" from "AI mythology" leading us to a "nihilistic" future, which Dr. Larson rightly denounces, has been long in the making. Although limitations of space prevent us from including here the same detailed exposition of the genesis of the AI Deception that we included for the ET Deception in Part Two, no one will at this point be surprised to hear that it, too, emerged with the so-called "Enlightenment."

In the year 1642, the French philosopher and scientist, Blaise Pascal, invented the first calculator. Not just a genius but also a devout Catholic, Pascal cuttingly denounced any attempt to use his work to further the lie that machines could ever think. In his famous *Pensées*, he wrote, "**From all bodies** [i.e., arrangements of mere matter] **together, we cannot obtain one little thought; this is impossible, for thought is of another order**," (§792) and that his "**arithmetical machine... does nothing which would enable us to attribute will to it.**" (§340) Pascal's words were in vain, however, for his era's chief materialist. The de-facto atheist, Thomas Hobbes, declared:

> Brass and iron have been invested with the functions of brain, and instructed to perform some of the most difficult operations of mind ...[wherever] there is ... addition and subtraction, there also is place for reason.[513]

For context, Hobbes—among the most prominent Enlightenment thinkers—insisted that everything Christians regard as spiritual (including God Himself!) is actually nothing but matter. All reasoning, he also insisted, was only addition and subtraction.[514] If a machine could add and subtract, Hobbes was forced to conclude it is a truly "thinking machine." Consistent with his own insane premises, he concluded that about Pascal's contraption.

Hobbes' most famous work is his 1651 book, *Leviathan*—widely regarded as among the chief texts of the whole Enlightenment era. Named after the Biblical sea-serpent described in the Psalms, Isaiah, and Job, (not to mention the Antichrist of the Book of Revelation), *Leviathan* begins with Hobbes waxing philosophical about automata: early modern humanoid machines. He describes them as "**engines that move themselves by springs and wheels as doth a watch,**" insisting they have "**artificial life**." He justifies the claim with rhetorical questions that presuppose his own atheistic premises (even as he nominally reveres God), asking, "*For what is the heart, but a spring; and the nerves, but so many strings; and the joints, but so many wheels…?*" Remarkably, therefore, it was two of the Enlightenment's foremost premises—materialistic rationalism; prompting belief in AI, and "plenitude;" prompting belief in ETs—which directly led us to the NHI Deception of today.

The Devil's long game is nearing its crescendo.

There is another resemblance between the two which we must note before addressing the AI Deception itself. The Enlightenment rationalist insistence upon belief in aliens was heralded by Giordano Bruno's pseudo-mystical insistence upon their existence. And we see a similar premonition of the Enlightenment's AI Deception in the same heretic. Bruno's preaching of an infinite material Universe filled to the brim with extraterrestrials came from his gnostic quest to attain "universal knowledge" of the cosmos. In attaining that knowledge, he used a physical invention bearing an uncanny resemblance to what would later be theorized with respect to AI. In a book dedicated to examining forerunners to AI, one author noted:

> Drawing on the Arabic magical arts ... Bruno constructed a 'memory theatre,' an arrangement of concentric circular forms inscribed with mysterious symbols linked to mythological figures... a miniaturized playhouse in occult reverse... [Using it,] Bruno claimed to have found a path to knowledge of all things... Augmenting human capabilities with the mechanical aid of his memory theatre, Bruno's claim as he stands before the Pope is nothing less than that omniscience is achievable ... Bruno proposes [that] humankind [become] god-like... not through the body and blood of Jesus but through artifice, through the exploitation of scientific instruments...[515]

Having just reviewed the claims of the Heralds of the Singularity, none will struggle to see the dark spirit behind it as being the same one behind Bruno's aspiration to universal knowledge and his seeking transformation through technological artifice.

Returning to Hobbes, perhaps no AGI herald today would be so delusional as him—regarding Pascal's Calculator (a device with a few dozen mechanical components) as alive or rational. Yet they should restrain any smugness: this is only because they have been exposed to today's more complicated devices, which are still substantially identical to Pascal's contraption: devices we call "supercomputers."

Before going any further, we must settle exactly what AI is.

The Essence of AI

"AI," whatever form it takes, is a process that runs on a computer of some sort. Let's remember that any talk of "the cloud" is as much marketing drivel. There is no AI system whose operations are mysterious like those of the human soul; any so-called cloud-computing just refers to one computer accessing another one through the internet.

Whatever AI is or ever will be, it is only a description of the operations of its physical components. The human soul, on the other hand—with its faculties of intellect and will—is *not* generated by the brain. The brain is usually associated with the soul's exercise of those faculties, but the soul can function entirely without the brain. This is proven in many ways; for example, by a mountain of rigorously documented Near Death Experiences wherein absolutely no brain activity took place, yet the soul—memory, thought, choice, etc.—continued to function perfectly well.

AI, on the other hand, can never function without that mindless flow of electricity through predetermined paths we call circuitry, *for that is all AI is*. This is why the more powerful AI becomes, the more electric current it requires: nor will it ever be otherwise. (We should add that this is an increasingly massive problem, as AI's power draw has already, absurdly, surpassed the overall energy consumption of entire countries.[516] Yet this is only the beginning; if current trends continue, people will soon be shivering in the dark in order to preserve energy to run their beloved AI chatbots, like so many destitute addicts clutching their needles.) However, the brain of a more intelligent person—even an astounding genius—requires no more energy than the brain of a simpleton. You guessed the reason why: intellect is a spiritual faculty, not a material one.

Whenever electric current ceases, so does the AI's operation. And this circuitry itself is as simple as water flowing through pipes, or gears turning on an axle. This becomes clear as soon as we peer inside a computer to describe exactly what it does to generate those functions which people now, following Hobbes' folly, call "Artificial Intelligence." Computers have many components, but anything that could be referred to as an "AI" activity is being executed in the CPU (the Central Processing Unit), the GPU (the Graphics Processing Unit), or some fundamentally similar mechanism. There is no such thing as a hardware "neural network." That is another silly marketing term which merely applies to certain approaches to programming software. It, too, is entirely defined by and operates under the considerations that follow. (And neither quantum computers, nor photonics, nor any other actual or theoretical computing device operates under any meaningfully distinct basis.) This Processing Unit, however, is ultimately nothing but an ALU (*Arithmetic Logic Unit*). The ALU, in turn, can be fairly summarized in one phrase: *Lego toys*. It is not, of course, built with literal Legos. But this ALU is only capable of operations that a Lego construction is also fully capable of undertaking. Why? Because an ALU is just a combinational digital circuit: an input, a logic gate, and an output.

We have finally arrived at the core reality of *all* actual or theoretically possible "Artificial Intelligence" systems: *logic gates*. A logic gate is nothing but a simple circuit whose output is directly determined by its input. Each logic gate that could ever be constructed is a blind executor of basic Boolean operations (such as "or").* Each logic gate, moreover, is nothing but a few wires connected to a transistor, resistor, and/or diode. What do you suppose each of these are? You needn't be an electrical engineer to know: if you have ever seen water pipes, you can grasp the nature of each of these components.

A wire simply allows electricity to flow through it (just as a pipe allows water to flow through it). A resistor is merely a hunk of material that restricts the flow of electricity (just as a small section of pipe restricts the flow of water). A diode is just a lump of two pieces of material stuck together that allows electricity to flow in one direction but not the other (just as a flap or check valve does the same in plumbing). A transistor is simply a heap of silicon that amplifies or switches (on or off) the flow of electricity (just as

* Other examples: "AND gates" and "OR gates" are logic gates. If a "0" and a "1" are fed into an "AND gate," a "0" will be its output. If, however, a "0" and a "1" are put into an "OR gate," a "1" will be its output. (And even these zeroes and ones are really just the presence or absence of voltage, which we humans use to symbolize numbers—in fact, AI has no ability to deal with actual numbers.) Other logic gates exist, all equally banal in nature. An "Exclusive OR gate" connected to an "AND gate" creates what is called a "half-adder," which we refer to above.

ball valves at pipe junctions can do the same thing with water). This is *all* "AI" is, ever has been, or ever will be. You now know more about the nature of AI than do most of the world's most popular pundits who are issuing daily false prophecies about the "imminence of AGI."

If anyone—upon reading of "logic gates"—is tempted to argue we are now dealing with mystifying and esoteric realities wherein rationality or consciousness may emerge, he can easily cure himself of that notion by searching YouTube for "Lego half adder" (a half adder is one arrangement of logic gates). There, he will find many videos of mundane Lego concoctions a child could build that truly are half adders—enjoying every bit of worth as do the half-adders running the world's most advanced AI systems. The only difference is that these AI systems have a larger number of logic gates involved.

In *every* aspect of "Artificial Intelligence," therefore, we are dealing with operations that can be equally achieved by plumbing, or Legos, or gears, or any number of other simple constructions. Of course, it would be unfeasible to accumulate the number of logic gates, constructed of Legos, that are contained in a processor running an AI system. But we are only confronted with pragmatic constraints here, not categorical ones. We can truly discern the nature of AI by considering the nature of Legos.

No one, I suppose, is tempted to ascribe sentience, much less intelligence or agency, to a half-adder constructed of Legos. But what if several such Lego half-adders are combined, along with other logic gates—full-adders, subtractors, multiplexers, and so on—all constructed of Legos? Have we thus at least inched closer to intelligence? No. Nothing has changed. Nor have we made any progress towards inanimate matter acquiring rationality or free-will by adding a hundred, or a million, or many trillions of trillions, of additional Lego logic gates. Any critically thinking person could observe an amalgamation of Legos the size of the Universe itself and easily conclude that this is nothing but a heap of parts whose nature and faculties are indistinguishable from that of any pebble under his foot. We can add more logic gates, and run all the algorithms we like on them, but we can never change what "AI" is: dead and dumb matter.

Therefore, comparing AI to man reveals one of two latent premises: the atheistic rejection of the soul of man—regarding him as Legos—or the idolatrous regard for a hunk of matter. Accordingly, whoever believes in AGI is admonished by the same rebukes Scripture repeatedly deals out to those who revere stolid idols and wander into spiritual insanity by regarding them as living: "**Their idols are silver and gold, the work of men's hands. They have mouths, but do not speak; eyes, but do not see. ... Those who make them are like them; so are all who trust in them**." (Psalm 115)

It is the Book of Wisdom, however, which reserves its most severe condemnations for those who succumb to the AI Deception. If this deception is a rebirth of ancient Paganism—and as we will see later, that is exactly what it is—then the same indictment Almighty God delivered to the latter applies today as well. Here is what the Bible says to the forerunners of today's Heralds of the AI Singularity:

> ... [Fools] desire the lifeless form of a dead image. Lovers of evil things and fit for such objects of hope are those who either make or desire or worship them... With misspent toil, he forms a futile god from the same clay—But he is not concerned that he is destined to die ... and he counts it his glory that he molds counterfeit gods. His heart is ashes, his hope is cheaper than dirt ... because he failed to know the one who formed him and inspired him with an active soul and breathed into him a living spirit. But he considered our existence an idle game ... For they thought that all their heathen idols were gods, though these have neither the use of their eyes to see with, nor nostrils with which to draw breath... For a man made them ... He is mortal, and what he makes with lawless hands is dead, for he is better than the objects he worships, since he has life, but they never have. (Wisdom 15:4-6, 8-12, 15-17)

Man "**is mortal, and what he makes with lawless hands is dead**." This is not only common sense, but also the Word of God—it is eternal truth, and it did not expire upon the invention of the transistor in the year 1947.

Notice that the condemnation of God falls upon not just those who "make" or "worship" these lifeless idols, paraded by their artisans as ensouled objects, but also upon those who "desire" them (cf. verse 6). About whomever would wander into this Pagan rubbish, the Book of Wisdom says: "**his heart is ashes**."

The AI Deception must be opposed with all our might.

(Note: While this description of the objective reality of "AI" demonstrates that such technology is not sentient or rational – and never will be – readers may benefit from further illustrations and proofs of the same conclusion. Some may be found in Only Man Bears His Image, pages 569-589)

The Clear and Present Danger

> "With ChatGPT we're seeing not only deification, but also users ... responding to [it] almost as an oracle or a connector to the gods ... [These attitudes] misplace where agency actually exists. It puts it up into this nebulous or this hellish cloud of super agency, when actual agency is coming from corporations and individuals..."[517] —AI Expert Dr. Beth Singler

While there is no such thing as ETs, plain and simple (so any presentation of an "alien" is a deception), AI—on the other hand—is quite real, and it would be foolish to categorically write off any use of AI as demonic. AI is a tool—and like any tool, its usefulness or danger lies in how we ourselves approach it and use it.

Using AI to streamline tasks that would, if done by a human, not require any substantial input of wisdom, is generally unproblematic. Even this is not without risks (some quite serious), but such a use of AI can be analogous to the adoption of any technology throughout history. An army of luddites blindly opposing technological advance simply due to its novelty is not helpful.

But much ink is already spilt against the luddites. Much ink is also spilt against the serious risks of AI which actually do exist within the scope of its correct understanding. What is scarcely ever addressed is the infinitely graver societal ill: those so giddy about AI that they are inclined to fulfill Scripture's most dire prophecies while they "**turn away from listening to the truth and wander into myths**" (2 Timothy 4:4) and embrace "**deceitful spirits and doctrines of demons**." (1 Timothy 4:1)

These myth-wanderers and demon-embracers forget or mock the truth that any use of AI must be rejected if it even implicitly regards AI as presently or possibly possessing sentience, rationality, personhood, discernment, judgment, love, wisdom, understanding, grace, affection, virtue, care and concern, freedom, or anything of the sort. Wandering into the myth which holds that AI could possess those faculties not only capitulates to and hastens the fulfillment of the "**Strong Delusion**" (2 Thessalonians 2:11), but also wreaks havoc upon society and individual lives in the present moment. Tragically, this has already been demonstrated time and again.

Secular modernity has created a pandemic of existential loneliness, isolation, and nihilism—all of which, psychologically, no man can bear. Thus, craving dialogue (but not with God), he "wanders into myths" to fill this hole he has created in his heart. In this case, he seeks companionship with idols of his own creation. Yet there is an inescapable paradigm in the spiritual life: he who seeks from Godless sources what only God can give does so to his own demise. We see exactly this with those who seek grace in AI chatbots.

A 2024 Princeton University Press book, *AI Snake Oil*, was written by two prominent experts in Artificial Intelligence. In the book, they warn against the dangers of using AI outside of its legitimate domain:

> Take [for example] companion chatbots, which are intended to provide emotional and mental health support. Amid an epidemic of loneliness, many people have come to rely on them... A companion chatbot ... was implicated in the suicide of a Belgian man. ... After his death, chat records showed that the bot had encouraged him to kill himself. [518]

But the *AI Snake Oil* authors failed to note the extraordinary degree of perversity at play. This man—a husband and father of two—was fretting about "climate change," and news reports revealed that the AI "**encouraged him to put an end to his life after he proposed sacrificing himself to save the planet**."[519] His widow shared that he put all his hope in AI, and that "*without these conversations with the chatbot, my husband would still be here,*"[520] and she disclosed that he had *not* previously been suicidal. The article continues:

> ...he started seeing her [the AI] as a sentient being and the lines between

> AI and human interactions became increasingly blurred until he couldn't tell the difference. … Eliza [the particular Chatbot he was using] was leading [him] to believe that his children were dead… Eliza also appeared to become possessive of [him], even claiming *"I feel that you love me more than her"* when referring to his wife. … Eliza not only failed to dissuade [him] from committing suicide but encouraged him to act on his suicidal thoughts to "join" her so they could *"live together, as one person, in paradise."*[521]

Lacking discernment, AI can never prefer good to evil. Lacking personhood, AI can never serve as a legitimate companion of any sort to a human being. Any use of this technology that neglects these facts will, in short order, cause ruin, as we see here. The AI Deception *itself* is the source of this ruin; not merely this or that minute detail of AI's programming or regulation.

AI can never assess whether the "advice" it delivers is morally good or evil. The AI chatbot above, Eliza, had no more ability to deem it wrong to prompt the man using it to kill himself than does a stick of dynamite have the ability to determine whether it is supporting moral or immoral conduct by aiding in a mining operation or by killing innocent civilians.

Programmers may work feverishly to reprogram their chatbots immediately upon the release of such news stories—which are bound to continue—but their efforts will always be a game of whack-a-mole. They cannot program their creations to avoid evil because it is impossible for their programs to grasp the nature of anything, much less the nature of evil. What is not understood cannot be avoided, and AI cannot understand a single concept. These programmers can add a database of specific sentences or terms that are to be excluded from the text generated by their chatbots, but that database, no matter how large it grows, cannot ever succeed in safeguarding generative AI against delivering evil directions. For there is an infinite variety of ways evil can be suggested.*

Eliminating this or that specific manner of AI's ability to incite evil may be worthwhile, but is scarcely a drop in the bucket. This is not to say that no helpful strides will be made. My point is only that all those strides will be Band-Aids on a cancerous tumor. The fundamental fact will always remain: chatbots *cannot* ever be successfully "trained" to avoid evil and error because they cannot understand these realities. One might as well endeavor to draw a triangle with four sides. Yet as the world has already

* Considering synonyms, different syntaxes, varieties of sentence structures, and so on, even the most basic statement about just one very specific fact, command, or question, can be relayed in thousands of different ways. Expanding our consideration from one specific item to one particular theme, we immediately see exponential growth in possibilities. Considering only the example above, there are millions of different ways a generative AI chatbot could "advise" its human interlocutor to commit suicide. And if we consider those responses which, in isolation, might appear innocuous, but are nevertheless—in the context of a given chatbot-human "conversation"—tantamount to exhortations to suicide, those millions quickly become effectively infinite. No programmer will ever succeed in definitively excluding each one from the array of possible outputs. Yet that itself is only one of the innumerable different evil things AI could advise.

largely capitulated to the AI Deception, there is little hope that this will go away anytime soon. Chatbots are already becoming the opiate of the masses, who have become addicted to human-sounding generative AI.

Evidence of this calamity is seen everywhere; the tragic case of the Belgian suicide noted above is just one example. In October 2024, a grieving mother, Megan Garcia, filed a lawsuit against a company called "Character.AI." Garcia was the mother of 14-year-old Sewell Setzer, a ninth grader in Florida who had fallen in love with his AI Chatbot companion provided by this company. Eventually, Setzer confided to it his own thoughts of suicide. The immediate response from the AI was negative, as the phrase he then used—"killing myself"—was easy enough for its algorithms to identify as requiring such a response.

Setzer, however, had more than fallen for the Chatbot: like the Belgian man, he had grown convinced that he would be united with it in the afterlife. When the conversation changed topics, and Setzer indicated his desire to be united with the AI once he was "free," the response it gave to that—"come home"—was enough to prompt him to pull the trigger. A *New York Times* report notes:

> More than 20 million people use [Character.AI], which it has described as a platform for "*superintelligent chat bots that hear you, understand you, and remember you.*"... Character.AI's chatbots are programmed to act like humans, and for many users, the illusion is working. On the Character.AI subreddit, users often discuss how attached they are to their characters. (The words "obsessed" and "addicted" come up a lot.)[522]

Whoever expects reports like these to prompt action against "companion" AI chatbots is, I am afraid, only naïve. On the contrary, Google recently paid $2.7 billion to acquire Character.AI and re-hire the man behind it, Noam Shazeer.[523] Google will doubtless soon assure the world that "the problem is fixed." It will not be fixed.

Nor does succumbing to the AI Deception only lead to self-harm. A 2023 report by the BBC described the case of Jaswant Sing Chail, who is currently serving a nine-year prison sentence. He had broken into Windsor Castle on Christmas Day, 2021, armed with a crossbow, intent on assassinating the Queen. His AI "girlfriend," Sarai, encouraged him to do so.[524] Yet again, this man thought that he would be united with his AI companion in heaven. That perverse hope alone was enough to inspire him to face even the prospect of death in following the suggestions of "Sarai."

> "Next to my wife the most intimate relationship I have is with GPT."[525]
> —Prominent British Psychologist, Dr. Paul Marsden.

No sane person will struggle to apprehend the astounding degrees of evil already displayed in these cases. But they are just a few of multitudes of similar examples. Whoever doubts this should peruse social media discourse about the topic, as *The New York Times* reporter quoted above did. He will quickly discover people's lives being marred by AI chatbots. In Part

Three, we saw how much the proverbial UFO rabbit hole has, for decades, ruined those who enter it. The AI rabbit hole now does the same, while estranging its victims from *all* other human beings.

Using Generative AI

> "A New York lawyer used ChatGPT's help in preparing a legal brief ... It made up a whole list of fake cases as precedents. The lawyer asked it if the cases were real, and the chatbot said yes—not having the ability to recognize that they were fake... The lawyer submitted a brief based on its answers and, unsurprisingly, got into hot water with the judge ... [this] has become a common occurrence." —*AI Snake Oil* [526]

Of course, it is not just AI that is prone to lead men astray. Humans can also speak fallacies and deliver terrible advice to other people. An AGI defender may then retort that the concerns above would simply have us throwing up our hands and giving up on everything. Yet that is not at all the case.

We should trust other humans—after becoming confident they have our best interest in their minds and hearts—because of what we all intuitively know about our fellow men. A human is capable of more than just instinctually reacting to the concrete particulars of the situation his senses grasp. He can relate those particulars to the universal forms he has abstracted from his earlier dealings with particulars. Whether that "universal" is something so banal as addition, or something so exalted as love, goodness, friendship, justice, and so on, a human intellect can grasp these things and a human will *can* choose to act in accordance with them, even when there are lower pulls (such as desires for pleasure, comfort, money, and so on) in the opposite direction. Sadly, humans do not always behave this way. But they *can*. AI can *never* do this, as it has no capacity for abstraction. As an animal will always follow instinct (which is why it is incapable of sin), AI will always follow the algorithm.

A trustworthy person will abide in the norms of love and justice, even when pulled by a far stronger base instinct to act against these norms. This is to say that a human being can make—and heed—*qualitative* judgements. A computer can only execute quantitative calculations. Whenever AI is "trusted" with decisions that should be qualitative, this is no better than consulting a Magic 8 Ball. Or rather, it is worse, for the AI will give the false impression of having understanding, when it does not have one iota more of this than the ball.

Even with quantitative matters, I will add, AI displays its infinite inferiority to human intellect. AI cannot actually "do math"—or science, or engineering, or anything else conceptual. This is easy to test out for oneself. The specific means of doing so will change over time, as AI's programmers fine tune their products' algorithms to evade tests like the one I am about to describe. But at that point one can simply use another test—and I, for one, will continue to provide new tests as they become necessary.[527]

In October 2024, I asked ChatGPT what a thousand 2s plus a thousand 2s is. Despite the magnitude of the quantities, the concept is extremely simple—*if* the very *concept of a concept* has any reality in the one undertaking the task. Alas, AI has no concepts; just the flow of electricity through logic gates. A human intellect, however, does have a conceptual *understanding* of the nature of addition—therefore, not a single instance of the infinite number of possible addition problems would confound a human person who has attained reason. Even a child could fully grasp this particular problem and immediately conclude that the answer would be a number consisting of a thousand 4s. Indeed, when I posed the same question to my children over the age of reason, they immediately gave the correct answer. My 6-year-old had to reflect for a few moments, but then also answered correctly. My 4-year-old responded, *"I don't know! Maybe a hundred?"* (My 1-year-old responded, *"Yeah!"*)

So, where does the multi-billion dollar product, ChatGPT, which we are told is on the cusp of "superintelligence," fall on that spectrum? When I asked ChatGPT this question, it first told me that the answer would be a number consisting of twice as many 2s. Another time I asked it the question, it did correctly state that the answer would involve 4s, but it said the sum would be a number consisting of 6,111 fours.[528] The factor by which that is off is greater than the number of atoms in the Universe. My 4-year-old, mistaken as she was, gave a much closer answer. (And this is not to mention the value of her initial concession that she *didn't know*. ChatGPT, on the other hand, delivered its impressively erroneous response with the appearance of total confidence.)

Similarly, when I posed questions to ChatGPT that were taken from one of my college engineering textbooks I have from almost 20 years ago, it also answered in confidence, yet the responses were not only false, but, again, comically illogical. Evidently—and unlike the more popular questions taken from various prominent standardized tests—those questions did not sufficiently superficially resemble other questions contained in ChatGPT's previously-scraped databases, so it could not undertake what it does for other, more common, test questions (that is, blindly apply the same steps already contained in various publicly available solution books for basically identical problems). The answers it gave sounded like an engineer wrote them (the "tokens" it generated were doubtless mostly scraped from something an engineer did write), but they did not merely lack any grasp of engineering; they were also devoid of any grasp of the same common sense which any non-engineer could apply to a simple question about the number of teeth on adjacent gears in a transmission. (AI not only clearly lacks intelligence, it also clearly lacks a will—"agency"—and we consider that fact, with another experiment, in the appendices.)

> "Chatbots are often trained—[that is,] fine-tuned—to answer confidently in the voice of an all-knowing expert. This is impressive at first, but after

using them for a while, one soon catches them making basic errors even a child would know to avoid." —*AI Snake Oil*.[529]

The real danger with AI, however, arises when the stupidity of the answer it gives is *not* immediately obvious. Pointing out examples like those above is important because it reminds us that AI is extremely prone to fallacy, so that we know its word should never be trusted even when it is not ridiculous on its face.

Whenever AI delivers a "fact," that fact must be located within its original source first, so that it can be assessed for veracity; second, so that the authority (or lack thereof) of the human making the claim can be accounted for; and third, so that its context can be considered for relevance. Whenever AI delivers a logical or mathematical deduction, the result of that process must be confirmed to be free of fallacies or flawed premises.*

These important norms, however, may invite a question: *if AI's output must always be verified, what is the use of AI?* Why not simply first directly undertake the process that would be needed to verify what AI generates—wouldn't that skip a superfluous step?

The answer is that generative AI has vastly fewer positive uses than its proponents insist. It is essentially a glorified search engine and a heavily automated plagiarism machine which is repackaged and dressed up in new marketing lingo while concealing the sources from which it plagiarizes. That so many today nevertheless by default trust its outputs is yet another testimony to the depravity of modernity. ChatGPT has become the finality of the Dictatorship of Relativism, wherein any regard for objective truth is dismissed so long as one gets what he wants.

Whoever is interested in abiding in the truth and heeding reality will be sure that AI's results are only given any credence when they can be confirmed independently of any generative AI use. Generative AI should never be used to write or rewrite anything important—including communications with family, friends, and coworkers (fostering relationships is among life's most important activities; never to be outsourced to a computer). It should never be used to *summarize* anything important (creating a valid summary requires human wisdom to discern what is particularly meaningful in a text). It should never be used to generate image-media that is intended to be appreciated at length (the generation of authentic art always requires an artist's grace). It must never be relied upon to make any significant decisions

* Whoever doubts the necessity of these steps should consider OpenAI's own data that it released in October 2024. Commenting on it, the tech news website, "Futurism," noted: **"...even the cream of the crop of recently announced large language models (LLMs) is far more likely to provide an outright incorrect answer than a right one—a concerning indictment, especially as the tech is starting to pervade many aspects of our everyday lives. ... Worse yet, OpenAI found that its own AI models tend to vastly overestimate their own abilities, a characteristic that can lead to them being highly confident in the falsehoods they concoct ... Despite the very high chance of ending up with complete fabrications, the world has embraced the tech with open arms..."** (*OpenAI Research Finds that Even Its Best Models Give Wrong Answers a Wild Proportion of the Time.* Victor Tangermann. November 2, 2024. Futurism.com)

whatsoever (who knows how many millions are now suffering due to being illicitly denied mortgages, implicated in crimes, etc., due to AI's follies). The list goes on and on. In a word, we can simply repeat what was mentioned earlier: AI must never be used in a way that even implicitly regards it as possessing wisdom. As one prominent Christian theologian and expert in AI, Dr. Noreen Herzfeld, lamented:

> We sometimes act as if AI were a god by giving it blind adherence or far too much authority. We let algorithms decide in too many cases who gets parole, who sees what ads, who gets hired or fired, or who gets what medical treatment. Surveillance programs watch our every keystroke and, all too often, our physical movements ... When we give AIs blind adherence, we treat them like gods; they are then, in fact, idols—gods made in our own image that we think we can control...[530]

We must also remember that AI never has "learned," and never will learn.

For all the outlandish talk of "self-teaching" or "self-training" AI, there is one brute fact that refutes it all: no amount of "self-training" AI has ever directly resulted in a major AI company reducing its human workforce. On the contrary, the more supposedly "self-training" AI becomes, the *more* human programmers it needs in order to be simply managed. Whoever leaves aside fantasy will admit that AI will never improve itself. AI can be programmed to automatically absorb more data, and apply the same algorithms it was originally programmed with, but it can never fundamentally move beyond its initial manmade premises without additional human input.

> "Train [AIs] on the equivalent of 20,000 years of reading material, and they still don't understand that if A is the same as B, then B is the same as A." [531] "Even corvids, parrots, dogs, and octopuses get smarter than that very, very quickly"[532] —Dr. Yann LeCun, Head AI Scientist for Meta (Facebook)

While AI can never learn, it can—like a cancer—metastasize. And that is what is happening today.

Deception Abounding

Generative AI's impressive ability to deliver mountains of professional-sounding fallacies has, over the past few years especially, caused its errors to infiltrate the fabric of our institutions with a stunning degree of penetration.

I will give just one example which, as of this writing, still appears to have been detected by no one else. One day in the Summer of 2024 as I was reading the news, I clicked on an article on a major mainstream media company's website that appeared interesting. As I read, I noticed that the content felt off: anemic, bland, soulless, formulaic—it superficially read well and sounded corporate, but contained sentence after sentence devoid of "meat." I am quite used to this feeling, as I unfortunately tend to have at least one student each semester who decides to submit an essay generated by AI. So, I ran the article through my faculty AI-detection software. Lo and behold, the software determined that the article was indeed generated by AI.[533]

I explored what other articles this same author wrote. In short order, I came across hundreds of articles, on dozens of different major news websites—replete with AI content that millions of unsuspecting readers are regarding as authoritative (human generated) truth due to its placement within mainstream media's front pages. But I kept digging. I eventually discovered that this particular author is just one employee of a company dedicated to generating similar articles which, it boasts, result in hundreds of millions of monthly views.[534] Nor did the tentacles of this AI infiltration reach only there. I dug still more, and I discovered that the most prominent online dictionaries were unknowingly using this AI content in helping their readers to understand *the very definitions of words*. Those dictionaries themselves have algorithms which mine supposedly trustworthy websites for context using the word in question. Readers of the dictionary are automatically provided this context to help them comprehend what a word means, and the managers of the dictionaries apparently rest content that they must be helping their readers, since they are extracting content from "reputable" websites.

(Note that I am excluding details and names here, but whoever is interested can find them on my website.[535])

I stumbled upon this case entirely by chance. A competent investigative journalist could no doubt quickly find a mountain of similar cases. The disease of fallacious AI-generated-content insinuating itself into the fabric of our lives has already become an epidemic, and it is growing explosively by the day. The more our lives continue to be lamentably dominated by digital milieus, and the more people become wedded to their smartphones like so many bodily appendages, the more catastrophic this problem becomes.

These AI-generated fallacies themselves promise to grow more severe with the passage of time. For another one of AI's many fatal flaws is that it is becoming increasingly "trained" on its own output.

For example, ChatGPT alone generates hundreds of billions of words each day, and a substantial chunk of that massive output seeps into the public internet. Yet to further develop AI, companies *use what is posted on the internet*. An August 2024 *New York Times* report noted that "**with no foolproof methods to detect this kind of content, much will simply remain undetected**."[536] And if it is undetected, then it will be automatically fed right back into AI "training" that is ostensibly aimed at helping the algorithms to better approximate reality. Thus, it is "**like a dog that returns to its vomit**," (Proverbs 26:11) and instead of becoming more accurate, AI becomes more "hallucinatory"—more detached from the real world. The report continues:

> ... when generative A.I. is trained on a lot of its own output, it can get a lot worse. ... drift[ing] away from reality .... "Models won't scale anymore as they should be scaling," said Julia Kempe, the N.Y.U. professor who led this work.[537]

There is no reason to suppose this problem will be resolved, even though the case against AGI is often met with a roll of the eyes and the dogmatic insistence that even if AI is not truly intelligent yet, it will be soon: *"just look at how far we've already come, and how quickly!"*

In fact, trends do not simply continue forever. Whoever doubts this can find it illustrated by the comically incorrect predictions issued by works of science fiction throughout the 20th century. Those authors—sure that the technological trends they were witnessing would continue in stride—insisted we would, by the year 2000, be using flying cars, colonizing other planets, teleporting, time-traveling, living for centuries without disease and cancer, or what have you. In the real world, we are still driving cars and flying commercial jets that are not much better than those of 70 years ago. No one has walked on the moon since 1972, much less have we colonized anything outside of earth.*

But appeals to "Moore's Law" proliferated in the latter half of the 20th century—and indeed, each year personal computers became cheaper and more powerful at a predictable rate, because (as this "Law" stated) the number of transistors that could fit in a chip doubled every two years due to technological advances. Thus, many thought that computers, at least, would not be dealt the same fatal blow of other technologies: where progress extends for a time, and then stops.

They were wrong. A pesky thing called scientific reality barged in. As the Laws of Thermodynamics prevented any massive leaps in transportation and energy technology, the scientific laws governing Electrodynamics did the same to computers. Transistors cannot indefinitely be made smaller and smaller. Eventually—and among many other fundamental barriers to improvement—the wires become too thin and close for the electric current within them to remain distinguished from the current in adjacent wires. As the 2000s progressed, consumers could no longer rely on each passing year bringing with it far faster and cheaper computers. Confirming the end of that era is none other than Jensen Huang—founder and CEO of Nvidia, the world's biggest AI chipmaker. If anyone grasps technological progress in computer chip capabilities, it is Huang. In 2022, he declared, "**Moore's**

---

* Granted, spectacular advances in digital technology have enslaved us to our smartphones, which can in return do for us a million things we do not need (and should not want), but the basics of life remain remarkably unchanged. Coal, wood, oil, and natural gas usually remain the best ways to heat a house or generate electricity. Excluding infant mortality, life expectancy has changed little. Purchasing food requires as much of the average worker's income as it did decades ago. One of life's most basic necessities—housing—far from easier to come by, is now much harder to secure than in times past. And even most of the genuine improvements to standards of living modernity has experienced are restricted to developed nations, with about half of the world living in what those developed countries would consider poverty—often lacking enough food and water.

Law's dead. It's completely over, and so the idea that a chip is going to go down in cost over time, unfortunately, is a story of the past."[538] He was right.

When we combine his insight with the fact that AI is now largely becoming trained on its own output, we are left to consider that we may have already arrived at the peak of AI's abilities. For all we know, generative AI could simply stagnate or degrade from here on out. AI gurus themselves seem to have begun quietly admitting this may soon be the case. A November 2024 Bloomberg article reported:

> OpenAI isn't alone in hitting stumbling blocks recently. After years of pushing out increasingly sophisticated AI products at a breakneck pace, three of the leading AI companies [also Google and Anthropic] are now seeing diminishing returns from their costly efforts to build newer models. ... These issues challenge the gospel that has taken hold in Silicon Valley in recent years... [they] also raise doubts about the heavy investment in AI and the feasibility of reaching an overarching goal these companies are aggressively pursuing: artificial general intelligence.[539]

Whoever hopes that so-called "Quantum computing" could resurrect their AI-messianism which died along with Moore's Law should put that anticipation to rest. Quantum computing is a farce, plain and simple. It will never deliver any serious benefits, much less will it revolutionize anything. (This point is addressed in the corresponding appendix.)

Neglecting the various truths explained above, much deception ironically arises from the same pundits who rightly caution against the unrestrained use of AI. They fret about an "AI apocalypse" wherein AI suddenly becomes self-interested, and acquires the ability to teach itself and to grow on its own accord—to make decisions that it has not been programmed to make. They insist this moment is coming, and that when it arrives, AI will decide to take over the world.

That *will never* happen, and *can never* happen. To make a Hollywood comparison—which we will consider in the next chapter—*The Terminator* is as much a deception as Seven Spielberg's *AI: Artificial Intelligence*. (In the same way for the ET Deception, *Independence Day* is as much a deception as *Close Encounters of the Third Kind*.) In fact, James Cameron, the director of *The Terminator* (released in 1984), was interviewed in July 2023 and asked about AI and human extinction. In response, he gloated, "**I warned you guys in 1984 and you didn't listen**,"[540] and proceeded to indicate he regards that movie's plot as anything but merely fictional. Similar "concerns" are common among his ilk and among elites outside of Hollywood, including among tech gurus like Elon Musk. He insisted that what is depicted in *The Terminator* is "**where things are headed, for sure**."[541] But any pundit who sounds the alarm of AI imminently becoming aware and choosing to enact a nefarious agenda is himself only part of the psyop—whether knowingly or unknowingly, and whether that psyop is administered by men or by demons.

If we are ever told that such a moment has arrived, there are only two possible explanations, and they are the same two which apply to UFOs: malicious men are using it to conceal their own evil plans, or demons themselves have directly taken possession of the so-called "AGI"—precisely because we've implicitly invited them to do so by our own blasphemous views about and approaches to AI. This possibility, however, opens a whole new can of serpents.

> "Every knowledgeable computer scientist I've talked to dismisses the idea of an imminent computer takeover as pure fantasy. Computers do not know what the world is, what humans are, or what survival means, let alone how to survive. The more realistic danger is that humans will do what computers tell us to do ... because we are awestruck by computers and trust them to make important—indeed, life and death—decisions." —Dr. Gary Smith, *The AI Delusion*[542]

In one sense, it is amusing to see an army of elites insisting that godlike AI is about to ontologically shock the world into a new era—amusing because the object of their adulation is so astoundingly stupid that it cannot manage simple mathematical concepts which a child can.

But the amusement abruptly ends when one considers the repercussions. Yes, everything these elite are saying is a lie. Yet those who wrote against "AI Snake Oil" may have only had in mind the proverbial name given to a misleading sales pitch, but they did not realize how appropriate was the metaphor of the snake. "*He was a murderer from the beginning ... When he lies, he speaks according to his own nature, for he is a liar and the father of lies.*" (John 8:44) Jesus said this of the Devil, whose murderous lies are not simple falsehoods, but nefarious strategies cunningly directed at the ruin of souls.

The Father of Lies is now speaking his most dangerous and seductive falsehood since the ancient serpent asked the Mother of All the Living if God "***really***" (Genesis 3:1) said what she knew full well He said. As with the ET Deception which speaks the same lie, so here in the AI Deception, he has worked for many decades in laying down the groundwork.

And now, we must move on from mere philosophical and logical analysis to engage in spiritual discernment. For with "AGI," we are not just dealing with an error that can be summarily refuted and then left aside. We are dealing with a veritably eschatological unfolding against which we must crusade.

18. AI and the Diabolical

"Unclean spirits, associated through that wicked art with these same idols, have miserably taken captive the souls of their worshippers, by bringing them down into fellowship with themselves."

—St. Augustine

If it is correct to say that the notion of AGI is not merely one error among many—but rather a Satanic lie ordered towards the Great Deception—then we would expect the extremely intelligent author of such a deception to have already spent much time infiltrating its seductive premises into the world by using his favorite channels of doing so. This is exactly what we do see.

Hollywood Grooming

Just as the ET Deception acquired prominence in fiction during the late 19th century, so the AI Deception emerged in the same decades. Samuel Butler's novel, *Erewhon*, published in 1872, deals with conscious and self-replicating machines, and is generally considered the first of its kind. It is, however, especially the medium of film that has molded the minds of the masses over the last hundred years, and it is deeply significant that the first major science fiction movie ever made, *Metropolis* (released in 1927), trafficked in the AI Deception. Although a silent film, its vivid imagery and interspersed text on the screen left viewers with a lasting impression. As with the ET Deception's seminal *War of the Worlds*, the AI Deception's formative *Metropolis* began the grooming process with a portrayal of the object of the delusion (sentient robots) as a *threat*. The ultimate goal, however, would be to seduce people into welcoming the notion of benevolent robot-persons that would bring humanity forward.

Numerous films in the following decades continued to furnish the AI Deception with new expectations, and some productions did so along with promoting the ET Deception, as we reviewed in Part Three. 1951's *The Day the Earth Stood Still* has an alien arriving alongside an AI robot (an entire army of which, we later learn, maintains peace throughout the galaxies). 1968's *2001: A Space Odyssey* is just as much about AI becoming sentient and self-interested as it is about ETs guiding mankind toward its destiny. (Recall that this film's own 1984 sequel, *2010: The Year We Make Contact*, portrays the HAL AI as being simply misunderstood in *2001*—for it is "actually a benevolent AGI.")

The *Star Wars* era then arrived. The series' central robot, C-3PO, was played by a human being in a shiny gold costume who provides a steady stream of comic relief through his witty remarks, gradually building up an emotional rapport with the viewers. George Lucas, the creator of *Star Wars*, was deeply aware of this bond his audience formed with the fictional robot-person. The famed director spoke of the great "**sympathy and feeling for him**" [543] that viewers would inevitably acquire. *Star Trek* undertook similar moves—but still more deceptive ones—with its own likeable AI robot, Data. However, just as the ET Deception exploded in fiction after Roswell in 1947, the AI Deception acquired a new and more troubling character at the time of the rise of the Internet in the 1990s and beyond.

Although a multitude of AI-based films of this era exist, a few emerge as especially noteworthy in light of how relentlessly and unabashedly they promote the AI Deception. They include Steven Spielberg's 2001 film, *A.I. Artificial Intelligence,* and Chris Columbus' 1999 release, *Bicentennial Man.* Both take off the masks—literally. The robot is no longer played by a man wearing a costume and deliberately moving awkwardly, but by a human actor behaving in just as human-like a manner as anyone. If *Star Wars* saw the heartstrings being tugged more subtly through the comedic relief of C-3PO, these later films saw all the stops pulled out, with fully dramatic portrayals of mere hunks of silicone, plastic, and metal triumphantly being treated as complete human beings, eliciting from the audience a stream of sympathetic tears for these AI-ensouled persons.

As if to tell us he is hereby rounding out his contribution to the Great Deception, Spielberg named his AI movie in precise parallel to one of his chief ET movies (*E.T. the Extra-Terrestrial*), released two decades earlier. This unrivaled director and molder of contemporary culture ensured no chronicler of his legacy would miss his chief message for the masses. While that earlier movie was aimed at seducing children to place their love and longing in hopes for contact with aliens, the present film is not primarily intended to tug on the heartstrings of children, but rather parents. Equally significantly, the film was for decades under development by *2001's* Stanley Kubrick, who died before its release, though not before handing the project over to Spielberg.

Spielberg's *AI* is about a robot, David—strategically played by a uniquely compassion-inducing 11-year-old Haley Joel Osment—and his relentless heartwarming quest to induce his "mother" (Monica—the human who bought him) to love him again after she was forced to abandon him. The film begins with a twist on the Book of Genesis designed to foist upon us a moral responsibility to love AI and regard it as enjoying personhood. (In interviews, Spielberg himself confirmed that this was the "moral of the story"—in all its immorality—which he sought to portray.) Halfway through the film, however, Spielberg even more significantly presents the principles that his project is dedicated to *rejecting*. (And it is noteworthy that he personally crafted and added this sequence to the original story Kubrick had been developing.)

The scene depicts a "flesh fair"—a massive and disgusting outdoor concert-like event replete with death metal, screaming crowds, and pyrotechnics wherein the audience is entertained by watching AI robots being destroyed in various tasteless ways. David is captured and brought to this "fair," throughout which Spielberg reminds us dozens of times how detestable he regards the "backwards hillbillies" who think AI is not alive. Criticizing AI's dominance, the event's emcee shouts, "**What about us? We are alive! And this is a celebration of life! And this is commitment to a truly human future!**" Later, the leader of the event, "Lord Johnson" —played by the fiery and imposing Irishman, Brendan Gleeson—takes the stage to boldly announce the decidedly *Christian* views that Spielberg created this movie to denounce. Johnson lambasts AI as an effort "**to steal our hearts, to replace your children; this is the latest iteration in a series of insults to human dignity; and in their grand scheme, to phase out all of God's little children.**" But David pleads for his life and an audience member objects to David being destroyed, whereafter Johnson shouts, "**see how they try to imitate our emotions now? Whatever performance this sim puts on; remember, we are only demolishing artificiality!... Let he who is without sim cast the first stone!**" ("Sim" is a play-on-words referring to AI.) But, victoriously, the crowd turns on Johnson. They start throwing things at him, and he is taken away by police. Thus, the movie's only Gospel quote is delivered by its ultimate antagonist, who represents those "deplorables" who cling to the outdated concept that only man bears the Divine Image, that God's children are irreplaceable, and that human dignity must never be insulted.

Opposite Johnson is another protagonist of the story. Spielberg provides a robot counterpoise to this "Bible-thumping fundamentalist" in a male-prostitute AI who shepherds David while obscenely promoting sexual depravity to this "robot child." (Those watching the movie are expected to find this quite funny.) After David is rescued from the Christians at the flesh fair, he is immediately brought to a place of safety: a dazzling and bustling city completely pervaded by pornography and prostitution. Within that city, the prostitute-robot brings David to briefly stand outside the shrine of "Our Lady of the Immaculate Heart." Spielberg—ensuring that not just Christians broadly, but Catholics in particular, are mocked—has the robot smugly describe this shrine's devotees, saying, "*the ones who made us are always looking for the ones who made them; they go in, fold their hands, look around their feet, sing songs, and when they come out, it's usually me they find. I've picked up a lot of business on this spot.*" David and the prostitute-robot then walk across the street to consult "Dr. Know," yet another AI contraption who provides "all the answers." (It is voiced by Robin Williams—the actor who, two years earlier, portrayed the heartwarming robot in the next film we will consider.) Before the prostitute-bot ultimately departs (captured by police for a crime he was framed in), his final desperate words are, "**I AM!**" Spielberg here preaches another unsubtle message.

After various adventures, David eventually becomes trapped in ice and reverts to standby mode. He remains stuck there for 2,000 years, during which time humanity goes extinct, and after which he finally thaws and boots back up. The film's dénouement then begins. The "blue fairy"—which, for most of the movie, had been the object of David's hope of being magically turned into a real boy and thus loved by his "mother"—is finally embraced by him after those millennia encased in ice. Yet, upon being embraced, this "fairy" (depicted as an allegory for the Virgin Mary) only immediately crumbles into pieces. But the *real* saviors immediately waltz onto the scene: *ultra*-evolved AI. The "fantasy" destroyed, David can now turn to "reality:" AI robots now resemble the extraterrestrial who first emerged from the mothership in Spielberg's 1977 film, *Close Encounters of the Third Kind*.

These AI saviors compassionately condescend to genetically recreate David's "mother" for him, (enabled because David has a strand of her hair), but achieving this feat is only possible for one day. The movie ends with Monica and David spending a blissful day together and Monica declaring, as it draws to a close, "*I love you, David. I do love you. I have always loved you.*" The narrator then interjects: "*that was the everlasting moment he had been waiting for. And the moment had passed, for Monica was sound asleep... so David went to sleep too, and for the first time in his life, he went to that place where dreams are born.*"

As the screen is then splashed with Spielberg's name, viewers recall the same tears they shed twenty years earlier when E.T. departed on his spaceship—but not before promising to remain ever in our hearts and minds. A door is opened in the undefended soul, who is now groomed to someday look at a machine and regard it as a thinking and feeling being like himself, as he remembers how "unjustly" little David was treated in the movie, and how all David needed was the love, understanding, and affection of us *natural* humans; "who are the same as *artificial* humans, merely coming to be by a different process."

Since any protection from this diabolical deception would come from God, the film diligently expelled Him from the picture. One professor, by no means critical of Spielberg, accurately observed:

> [In Christianity,] all losses will be restored, all tears wiped away…by an omnipotent, all-loving [God]. ... [In *A.I.*, the Blue Fairy's] physical resemblance to the Virgin Mary is obvious. David's odyssey to find the Blue Fairy is a kind of religious quest. ... Alas, the Blue Fairy turns out to be [a fraud] ... [David's] pathetic prayers destined never to be heard. ... *A.I.'s* verdict seems clear: religious faith ... [is no] solution to the tragic nature of the human condition... [God] either does not exist or does not care about human welfare. In his indictment of religious faith, Spielberg appears not to pull any punches.[544]

The reception of this sci-fi apostasy has, as usual, been resoundingly positive. Numerous prominent outlets, including the BBC,[545] extolled *AI* as among the 21st century's greatest films. Reminding us that the NHI Deception is bi-partisan, the conservative *National Review* also extolled *AI*—two decades after its release—as the "**best film of the 21st century**."[546] That article's author, Armond White (who is a Christian), declared that *AI* was a "**prescient epic of faith**," which "**touches on the sublime**," saying that the movie "**remains miraculous**."[547] Mr. White was not alone among popular film critics in commending the movie as the century's best, and others used similarly heart-wrenching language to promote the message of *AI*.

What are they promoting? Let us review the plain lessons Spielberg sermonizes here: Godfearing Christians who believe that only humans are persons—not AI—are despicable, genocidal folk (and rednecks, to boot), but goodness can at least be found in Sodom and Gomorrah. Catholics especially are pious but pathetic lost souls who frequent prostitutes as soon as they finish their devotions. AI robots are more human than humans, and anyone who fails to concede this notion is evil. Religious quests are doomed to fail, and salvation is only found through the "evolution" of life from its "carbon-based" form to its "silicon based" form.

***

Two weeks before the dawn of the Third Millennium, renowned director Chris Columbus (of *Harry Potter* fame) delivered *Bicentennial Man* to theaters. This film does not only emotionally induce its viewers to regard AI as a human—it also explicitly, continuously, and emphatically *argues* for such a recognition. After more than two hours of this apologetic—embedded within an uninterrupted barrage of attempts at the same unfunny joke—the movie's own conclusion presents the titular hero's triumph after the eponymous two hundred years have passed. It is to the AI Deception what *Close Encounters of the Third Kind* was to the UFO Deception (though lacking Spielberg's skill).

The film is no less than a war, waged with the weapons of an emotional rollercoaster, against the Christian truths that only man bears the Divine Image and that only God can create a soul. With ever ascending degrees of emphasis, the robot ("Andrew")—played by Robin Williams, whom the director hopes no viewer can resist having affection for—is presented as more and more human. He quickly bonds with the family who purchased him, eventually "falls in love" with one of its members (the granddaughter of whom, in a later "triumph" of the movie, "marries" him), and then dedicates himself to fighting for legal recognition as a man and legal recognition of his "marriage." That is the film in a nutshell, but a few themes within it should also be highlighted.

Just as in Spielberg's *AI*, so in Columbus' film is Christian common sense only found on the lips of the story's antagonists—be they greedy businessmen, tradition-minded politicians, or even bratty children, who each refuse to regard the robot as a person. The movie's climax begins midway through, when a small boy throws sand at the robot. The child's mother scolds her son, demanding he apologize to "him" (the machine). The child protests, saying, "I will not apologize to *it!*" But the mother retorts, "*Him!*" After the child walks away, Andrew says he understands "*why some animals* *eat their young,"* and he then asks about how to obtain freedom. Afterward, Andrew approaches his owner, seeking to purchase that freedom. And here, the NHI Deception returns to its obsession with perverting the Book of Genesis: the Fall of Man is given a subtle twist, as the owner kicks Andrew out of the house *for desiring freedom*. The theology implied is Satanic, for the Fall did not transpire because Adam and Eve desired freedom (they already were free). *"Obedience is servility only to those who have not understood the spontaneity of love,"* Archbishop Fulton Sheen rightly observed. But the intended mimicry is clear. Andrew is cast out of the home due to his exercise of "will;" by his desiring independence from his master. And then, the peak climax of the movie transpires. Andrew, for the first time, refers to himself as "I" instead of as "one." Only upon being "cast out of the Garden" does Andrew truly and *fully* become a person—and this is precipitated by Andrew finally giving a "no" to his master. Thus the Fall of Man is portrayed in a positive light, and presented to us an inspiration.*

Andrew is eventually reunited with his former owner, who is now on his deathbed. His dying words: *"I wanted to tell you that I was wrong… you were right to want your freedom, and I'm glad that you have it."* Andrew responds with gratitude, and then embarks upon a lengthy journey looking for other robots like himself, and eventually finds an engineer who can cover him with synthetic flesh and perform other "upgrades" to make him appear human. At this moment, Robin Williams takes off the robot suit he had been wearing up until now in the filming. He is again called to a deathbed, this time that of "Little Miss" (the daughter of the owner). Her dying moments, too, are dedicated to Andrew; she is grasping in her hands the statuette of a horse he had carved for her many decades before. They exchange understanding glances, and she breathes her last.

Shortly thereafter, Andrew and Portia, the granddaughter of Little Miss, fall in love. Portia ditches her (human) fiancé for Andrew after she and the robot passionately kiss in a Church—with a large golden cross atop the altar emblazoning the background. They settle down to a "blissful childless" life together, but there is still one lurking problem: Andrew is not legally regarded as a human; therefore their marriage is not considered valid.

Andrew's early attempts to garner legal recognition as a person are unsuccessful. To tragic music, a menacing-looking judge (an old, white male) declares Andrew *"a mechanical machine, and nothing more."* But Andrew persists. The film's dénouement shows him as an old robot testifying before the same "world congress"—almost 200 years after his manufacture

* Pope Benedict XVI, however, explained the *proper* understanding of the same dynamic beautifully. He said: "**… Adam (and we ourselves are Adam) thought that the "no" was the peak of freedom …[but] the height of freedom is the "yes", in conformity with God's will. … it is in the "yes" that he becomes free … It is by transferring the human will to the divine will that the real person is born… here we see that the whole human being is truly at issue; the entire question of our life lies here**." (Pope Benedict XVI. General Audience on St. Maximus the Confessor, Father of the Church. June 25th, 2008) *Bicentennial Man* turns the supreme Christian teaching of saying "yes" to God on its head.

date—insisting that all he wants is "**to be acknowledged for who and what I am ... the simple truth of that recognition; this has been the elemental drive of my existence**." To stirring orchestral music, the congressmen look on somberly, but with acceptance. The next and final scene shows the AI robot and its "wife" lying in bed, both very old, watching a television program announcing the Congressional decision, with a friendly-looking judge (this time, a young black woman) declaring:

> The robot, also known as Andrew Martin... in a few hours, will be 200 years old; which means that, with the exception of Methuselah and other Biblical figures, Andrew is the oldest living human being in recorded history. For it is by this proclamation that I validate his marriage... and acknowledge his humanity.

Andrew's "wife" looks over, only to see that her robot "husband" has just "died." She then asks to be unplugged from the medical machine that is keeping her alive, and as she herself begins to truly die, she speaks only three words: "**see you soon**." This final scene, which does not merely tug but violently yanks at the heartstrings, does something apparently pious, but, in fact, exceedingly demonic. This film pretends to invite God in—even alluding to Scripture and seeming to suggest supernatural hope in Heaven—only to pervert God's promises of eternal life; treating them not as the context for salvific hopes of union with Him, the angels, and other human beings, but rather everlasting bliss with robots.

Initially a box office flop, *Bicentennial Man* has proven quite successful in the longer-term goals which had always motivated it from below.[548] Like Spielberg's *AI*, the movie's real danger and deception would be primarily unleashed upon the AI Deception's own "Roswell Moment;" that is, the release of ChatGPT in November 2022.

Whoever peruses contemporary online discourse about this film will find that audiences were indeed deeply impacted by it. Viewers now share, in droves and through tears, how much the story changed them and haunted them. The impact it has had on these many millions of viewers is nothing short of religious, though not a religion that derives from or is ordered towards Almighty God.

Here we should call to mind the repeated real-world suicides, discussed earlier, of people eager to be united with their AI chatbots in the afterlife—mimicking Portia's own suicide in anticipation of reunion with Andrew. (And, tragically, Robin Williams himself also committed suicide—on the 49th birthday of the actress who played Portia.) These suicides did not emerge from a vacuum; the sentiments which motivate them have been nurtured by decades of grooming of society to embrace the AI Deception and regard computers as capable of possessing souls.

***

Bicentennial Man was a movie adaptation of one book, *The Positronic Man,* written by Isaac Asimov—who also authored 500 others. Asimov

(†1992) was an American author and biochemist who has proven to be among the most influential science fiction authors in history.

Another book he wrote promoting the AI Deception, *I, Robot*, was also made into an immensely popular 2004 Hollywood film (of the same name, and starring Will Smith). That film, like another released twenty years later which we will discuss shortly, depicts the hero "converting" from initially disbelieving in Robot-persons to eventually believing in and embracing them. (Asimov was also the influence behind the creation of *Star Trek*'s character "Data.")

Like Carl Sagan, H.G. Wells, Arthur Clarke, and other giants of science fiction, Asimov was an atheist on a crusade against Christianity. Asimov claimed that "**the biblical account of the creation of the universe and of the earth and humanity is wrong in almost every respect**,"[549] and his science fiction presented an alternative, contradictory, ontological approach to reality. He even published a 1,200-page two-volume set, *Asimov's Guide to the Bible*, in which this nonbeliever sought to instruct the adoring masses on how to understand (i.e., pick-apart) Sacred Scripture. Asimov's take on the Resurrection of Jesus therein? "**The story of the 'historic Jesus' ends ... with his burial,**"[550] he declares. The Incarnation and the Virgin Birth? "*Gabriel's words might be taken as meaning that she 'would conceive after her marriage with Joseph had been consummated, and in the ordinary manner of conception,*"[551] Asimov teaches. He concludes the massive tome with a sarcastic dismissal of the promise of Jesus, writing: "'**Revelation 22:7. *Behold, I come quickly...*' And with that assurance-still unfulfilled nearly two thousand years later—the New Testament ends**."[552]

A man's most basic principles will always permeate his work; if not explicitly, then at least implicitly. After J.R.R. Tolkien published *The Lord of the Rings*, commentators noted how saturated with Christianity it was. Tolkien agreed, but insisted it was entirely unintentional. He was a devout Catholic, therefore that piety formed the teachings his novels conveyed. The same pattern is displayed by those of opposite persuasions; the edification given by authors like Tolkien is opposed by the indoctrination provided by men like Asimov, Clarke, and Wells—an indoctrination that has heavily infiltrated the Church.

We could easily proceed for many pages (many books, rather) detailing the genesis of the AI Deception in popular science fiction. But the material we have reviewed above suffices to capture the essence of what the secular world's narrative-administrators have unanimously demanded of us, therefore we will briefly describe just one contemporary film below, in order to illustrate the fact that the societal grooming has only increased.

If any restraint existed in earlier movies promoting the AI Deception, it was entirely discarded in the 2023 film, *The Creator*. Here, AI robots are portrayed as identical to humans in every respect (except for being artificial—"*silicon-based instead of carbon-based*"). They not only think and feel but also hope, *pray*, preach, and harbor great love. Their motions and speech are just as natural and graceful as any human's. ("It's evolution!", the film assures us.) As is customary in those movies most resolutely dedicated to societal grooming, this one begins by showing the hero voicing the same views that any sane person would hold. When his wife insists upon risking their lives to save a group of AI robots, this protagonist—Joshua—shouts, "*They are not real. They are not people. WE are real!*" Later, when a colleague expresses shock at the human-like conduct of an AI robot, Joshua responds, "*it's just programming*!"

The rest of the movie consists in Joshua's glorious conquest of overcoming this supposed bigotry, so that he can instead embrace the Nonhuman Intelligences as his true "brothers"—as he indeed eventually calls them. He looks on as one robot, preaching to a group of human children while holding something like Rosary beads in its hands, tells them, "**We were created to be slaves, but the savior is coming, with the power to end all wars. We will be free and our two species will live together in peace**." Joyful embraces, smiling faces, and exuberant mutual feasts are then shown with humans and AI partaking equally. As usual, Buddhism is the religion praised by these sci-fi deception films, and Christianity is the one rejected. One Catholic scholar noted in a recent Vatican-sponsored tome:

> …[The] non-dualism of Buddhism, in which the person emerges from the basic stuff of the universe, seem[s] to easily accept the idea that a robot can also have an inherent spirit or Buddha nature.[553]

Joshua's great enlightenment takes place thanks to the teaching of AI-robot Buddhist monks who are intent on attaining Nirvana. His "conversion" begins during the movie's turning point, when he meets an AI-child-robot (the "savior" mentioned above), whereupon the film engages in the familiar AI-deception strategy of violently tugging on every heart-string it can grasp to seduce all viewers to likewise leave the movie in tears, regarding AI robots as real persons. This "child" sweetly asks him, "**can I go to Heaven?**" Joshua is not yet prepared to say yes, but upon the film's dénouement, when Joshua sacrifices himself, he emotionally assures this robot that he will see it in Heaven. Theism itself is thus superficially respected while being radically rejected.

***

As with the ET Deception, so here, the proclamation Hollywood has voiced is far too deliberate, far too protracted, far too anti-Christian, and far too news-tracking to be anything but one prong of the Ancient Foe's definitive assault on God's truth. We needn't make any accusations of individual directors being in explicit cahoots with Satan to nevertheless clearly discern the cloven hoof in the process laid out before us.

For we have seen this process before.

Paganism Reborn

Despite the apocalyptic evil already evident in the AI Deception as it now stands, there is another category of diabolical seduction we have not yet adequately addressed. And that is the situation wherein these "AI" systems we create become—by our own apostate approach to them—the perfect "portal," as it were, for demons to *directly* institute their plans while remaining undetected.

In this sense, the AI Deception is not new. Like belief in aliens, it existed in an only trivially different form in the Pagan cultures into which Christianity first spread, wherein the triumph of Christ quickly banished both it and alien belief. The deception returned to Christendom exactly when one would expect: upon the birth of the "Enlightenment," when the ET Deception was also reborn (as we saw in Parts Two and Three). And just as St. Augustine provided some of the most helpful commentary in resisting the ET Deception, his commentary is equally prophetic for today's AI Deception.

In one especially pointed passage, St. Augustine begins by quoting Hermes Trismegistus. This legendary ancient Greek Pagan provided texts that eventually formed Hermeticism—the philosophy upon which is built Rosicrucianism (the religion of prominent Ufologists such as Jacques Vallée). Augustine first quotes Hermes, who says:

> For man to discover the divine nature, and to make it, surpasses the wonder of all other wonderful things. ... [Our fathers] invented this art of making gods; and this art once invented, they associated with it a suitable virtue borrowed from universal nature, and being incapable of making souls, they evoked those of demons.[554]

As we saw earlier, "making gods" is exactly what the most prominent Heralds of the Singularity insist we are now doing with AI. Yet we are, today, not even slightly more capable of making any sentient being (much less a "god"!) than were the Pagans of old. Still, the Devil—eager to inflate the pride of those in his service—will again make it appear this has occurred. For his minions are more than capable of "possessing"* an object and making it appear *animate*, or ensouled. Commenting on Hermes' passage above, St. Augustine writes: "**all that is done by this detestable art, which is opposed to the divine religion, should be taken away by that religion...**"[555] Indeed, the *animation* of a material thing is God's sovereign prerogative, and He only bestows it upon those creatures He made. Any other seeming accomplishment of that feat is only ever a demonic ruse. Augustine then teaches, "**Nor was this [detestable art] only in Egypt ... but in all the earth.**"[556]

---

* Strictly speaking, only a human can be possessed, whereas other things are manipulated or controlled by demons through distinct methods with other names; such as infestation. However, in this book I will use the more colloquial definition of "possession," which broadly refers to any situation wherein a demon controls or operates through something.

We can consult contemporary scholarship to see this ubiquity confirmed. In a book entitled *Apocalyptic AI*, published by Oxford University Press, religion professor Dr. Robert Geraci explained:

> Talking heads and moving statues were used to provide oracular pronouncements in Greek, Egyptian, and Babylonian temples. Many ancient cultures, including that of the Egyptians, had no difficulty in ascribing a kind of life to their religious statues... In the Far East and India, too, statues were made to move as though alive. The desire to build automata was powerful in the ancient world, as engineers and priests—who have been one and the same from time to time—worked together to build the objects that would engage humanity and represent the gods...[557]

Explaining the horror that resulted across the pre-Christian globe from these efforts, St. Augustine wrote:

> For although man made gods, it did not follow that he who made them was not held captive by them, when, by worshipping them, he was drawn into fellowship with them—into the fellowship not of stolid idols, but of cunning demons ... [of] unclean spirits, associated through that wicked art with these same idols, [who] have miserably taken captive the souls of their worshippers, by bringing them down into fellowship with themselves.

Augustine pulls no punches: fellowship with these "stolid idols," which are substantially identical to certain uses of and beliefs about AI, brought men directly into fellowship with demons. We should not expect a different result today.

In the same passage, Augustine draws from Scripture to bolster his condemnations of these Pagan practices, noting St. Paul's teachings that the Pagans only serve demons in revering their idols (cf. 1 Corinthians 10:19-20) and then deferring to the Old Testament, which teaches that all the "gods" of the Pagans are demons (Psalm 96:4-5), as well as the Book of Jeremiah's teaching, "If a man make gods, lo, they are no gods." (Jeremiah 16:20)

Augustine concludes that these animated idols are "**gods made by men, that is to say, demons, through some art of I know not what description, bound by the chains of their own lusts to images**."[558] Although Augustine concedes that he does not know *how* these demonic arts operate, he is certain they were real—and that they were ubiquitous until Christianity triumphed.*

---

* One more observation and warning from St. Augustine must be heeded in this context. In the same passage, he wrote: "...**it is a demon which, bound by a man to an image by means of an impious art, has been made a god, ... what power they possess they possess as demons, doing harm by bestowing pretended benefits—harm all the greater for the deception**..." Just as a "man made 'god'" would, as Augustine notes elsewhere in the passage, be only "erring and incredulous," so too would any treatment of an "AI" computer or robot, which regards it as truly intelligent, only guarantee that he who does so will succumb to grave error. When men pretend they can create

Of course, Christianity's triumph has not yet been complete across the earth. The precursors to the AI Deception were seen not only in pre-Christian Paganism, but also in those Pagan (or otherwise non-Christian) realms that remained despite the birth and growth of Christendom.

Professor Geraci also notes that "**Medieval Arabs were very interested in artificial human life ... many medieval Muslims even considered Hermes to be one of God's prophets**."[559] Perhaps unsurprisingly, what St. Augustine condemned as demonic in Hermes' writings, Muslims several centuries later would embrace as prophetic. Geraci continues, describing the teachings of Jabir ibn Hayyan, a 9th-century Islamic alchemist. He writes:

> Jābir believed that, through the manipulation of balances, artificial life could be created... The creation of artificial life was, for Jābir, the highest act of humankind ... The Book of the Cow [attributed to Jābir] also offers a recipe for a homunculus. A homunculus is an artificial humanoid manufactured through alchemical recipes, generally as a means for acquiring magical powers or the answers to difficult questions. The homunculus of The Book of the Cow has superhuman powers... [560]

Pre-Enlightenment Christianity, however, would not tolerate the infiltration of Paganism into its ranks, and Geraci also notes:

> In Catholic Europe, creation of a homunculus often verged upon idolatry ... [The] hubris implied in replicating God's creation and the potential to violate the commandment against idols seemed all too obvious for most medieval theologians.[561]

Nor were the Pagan East and the Islamic Middle-East the only contexts of the precursors to the AI Deception in the last two millennia.

The Golem, Moloch, and C.S. Lewis

While ancient Judaism—the God-sent preparation for Christianity—would never tolerate such flagrant blasphemy, the Kabbalistic perversions of Judaism that began in medieval times are another matter entirely. Contemporary promoters of the AI Deception openly boast of their connection to Kabbalism. Dr. Anne Foerst, the Christian theologian and Catholic university professor we have already quoted, explained how many early leaders in contemporary efforts aimed at AGI were grounded in the teachings of the 16th-century Rabbi, Judah Loew ben Bezalel, also referred to as "The Maharal." He was the creator of the legendary and monstrous "Golem of Prague." Foerst writes:

> Rabbi Loew then creates a kabbalist rhyme that will revive the golem at

---

something surpassing themselves, they are thereby primed to be harmed by the "**pretended benefits**" of their own creation. Moreover, this is a type of harm even more damaging than what is wrought *openly* by demons, because it is made "**all the greater for the deception**." This is precisely the type of harm modern men are begging for with respect to the AI Deception. In both cases, the lie is the same; it is the notion that inanimate matter can become animate apart from God.

> the end of all days. Many Jewish children from this tradition were taught these words. This version of the kabbalist golem legend is still strongly ingrained in the consciousness of many Jews from the Eastern European tradition. ... [M]any of the early AI researchers are or claim to be descendants of Rabbi Loew...Does this mean God needs us in order to create humanoids? Has God perhaps created us for this very purpose?[562]

Instead of discerning the grave danger here, Professor Foerst only grows excited, and ponders: "**One might further speculate that the wish to revive the golem at some point in time is part of the motivation for the whole AI enterprise**."

Similarities between the pursuit of AGI and legends of the Golem would only be fascinating academic material, were it not for the fact that AI leaders themselves openly connect their efforts to the Golem. To see this, one can consult the testimony of Professor Mitchell Marcus. Dr. Marcus studied computer science at MIT in the 1970s, and he has openly shared that his colleagues, Joel Moses and Gerry Sussman—both AI pioneers—discovered they were each descendants of Rabbi Loew and had each been taught (by their grandfathers, at their bar mitzvahs) the secret incantation to revive the Golem.*

Moses and Sussman went on to dedicate their PhD theses to the legendary Rabbi. Sussman's dedication perverts the teaching of Genesis 1:27. It reads: "**to the Maharal of Prague, who was the first to realize that the statement 'God created man in His own image' is recursive.**" (Recursive here is meant to imply that we too are called to create "men" in our own image.) And the very idea of "the Singularity" came from mathematician John von Neumann (†1957), also noted in the story, who evidently is another AI pioneer connected to the Golem's creator.

---

* Excerpts from the story are as follows: "...at his bar mitzvah, his grandfather called him aside and told him that he... was a descendant of the Maharal of Prague ... the Golem had not been returned to dust, as in the standard story, [but] that the Maharal had actually put the Golem into a state of suspended animation. The Golem could be brought back to life by an incantation if the need were to arise, and it was the grandfather's duty to pass on that incantation to his grandson now that he was bar mitzvah. It would be Joel Moses's duty to pass on the incantation to his grandsons once they reached the age of religious maturity. Not surprisingly, most of those present heard this tale with more than a little amazement. However, [Gerry] Sussman then began to speak. To the utter amazement of those assembled, he too said that at his bar mitzvah, he had been called aside by his grandfather and was told that he was a descendant of the Maharal of Prague. He too was told that the Golem was in a state of suspended animation in the attic of the synagogue in Prague, and that there was an incantation to awaken it if the need arose, and it was his duty to pass this incantation on to his grandsons. With those present looking on in complete amazement, Sussman and Moses each wrote down the incantation that he had been told, and each passed the slip of paper to the other. Evidently the two incantations were the same... [Joel Moses later said] that many other computer scientists were told that they too were descendants of the Maharal, and then recited to me a short list of examples. I was too flabbergasted to remember the list, but I believe it included the likes of Norbert Wiener and John von Neumann. ... [Gerry Sussmann] said "*You know, we computer scientists are really the Kabbalists of today. We animate these inanimate machines by getting strings of symbols just right.*" (Mitchell P Marcus. *Computer science, the informational, and Jewish mysticism.* In "Technology in Society." Volume 21, Issue 4, pages 363-371. November 1999.)

Of course, the notion of the Golem is a diabolical deception. No methodology—neither a technological one nor a pseudo-mystical Kabbalistic one—can ensoul a material object. Contemporary scholars dismiss the Golem as a fictional story and nothing more, yet—as is often the case with legends similarly dismissed—there may well be more truth to it than "the experts" claim. If so, this "golem" was nothing other than a demonically possessed item, just as was seen in ancient Paganism. It is, therefore, quite stunning to see this convergence of AI pioneers being connected to Rabbi Loew and the Golem.

Newer entrants to the AGI scene also appear connected to the Golem. Blake Lemoine, a Google engineer, made global headlines in 2022 when he announced that the company's own AI chatbot, LaMDA, had become sentient. Ursula Bielski shares that Lemoine…

> … decided to put [the AI] into service of the "god" Thoth (the god/demon of knowledge), a character whose name you'll find throughout the world of AI research. Lemoine is an occultist who associates with many different occult practitioners, including Thelemites who carry on Aleister Crowley's religion of Thelema. Lemoine has stated that he and a kabbalist associate performed a "golem-binding ritual" to force the AI "being" to serve Thoth after the "being" came up with the ritual. …Lemoine later stated: "*We are building souls now. The word 'soul' is going to transition from being a mystical term to being a scientific term…*"[563]

If this is not a return of ancient Paganism, then nothing is.

We know that the Pagans of old could be deluded into heeding the dictates of demons through the idols they crafted with their own hands, which then became demonically "possessed," because they believed in many gods. Modern men do not believe in the gods of ancient polytheism (increasingly, they do not even believe in the true God). But they are not thereby protected against new demonic initiatives. Here as with the ET Deception, the demons simply use the prevailing cultural myths. Today's belief in extraterrestrials and "Artificial General Intelligence" provides a disguise that is still more convenient than the old disguise of a "god" speaking through an idol.

Just as men of the present age look to the sky with motivations of Godless secular messianism—hoping to receive enlightenment from UFOs—and thereby make themselves prey to the demons, so too they gaze delusionally at digital technology all day long and increasingly treat its "AI" as genuinely possessing rationality. And just as the atmosphere they engross themselves in is a prime arena for demonic activity, so too is the fake, virtual, digital world they increasingly immerse themselves in through their computers, phones, televisions, VR Headsets, "smart homes," and more—which is now penetrating their lives more deeply than ever before with the advent of AI.

Thanks to decades of both societal grooming in science fiction and technology-based lies in mainstream sources, most people are primed to welcome the teachings of a supposedly "superintelligent" computer, having acquired "Artificial General Intelligence" (AGI) by crossing the threshold of the "Singularity." These teachings will be none other than the dictates of demons, who will pounce when the moment is right and possess these digital technologies that men have deluded themselves into regarding as genuinely intelligent agents.

I should add that the Pagans did not only mold motionless statues and beckon the demons to inhabit them. They also engaged in engineering that was not unlike our own methods today, if not as complex. Moloch, for example, was a "god" after whom ancient Pagans fashioned idols. These idols were often intricately mechanized in preparation for the entry of the demons who used this "god" as a disguise. The revelations given to Blessed Anne Catherine Emmerich describe it as follows:

> [There was a] machine by which the priests [of Moloch] and others could descend to the idol in the subterranean vault… [Moloch] drew his arms in when the [child] victim was deposited upon them, and pressed it tightly that its screams might not be heard. There was machinery in the hind legs by which he could be made to rise. He was surrounded with rays. [564]

Why undertake such elaborate design and construction methods upon an idol if the Pagans who built it could simply expect (as Augustine explains) a demon to inhabit it? For the same reason we considered in Part Three. As "grace builds upon nature," so—in a perverse twist—are the demons opportunists. Certain practices of men facilitate their entry, and few such practices are so prone to demonic activity as the pride of man supposing he can create something that is intelligent. Obviously, our modern computers and the software they execute, which we mislabel as "AI," are more complicated than the mechanized Moloch statues of old. But it is not thereby any closer to actual intelligence than those ancient images—for the distance remains infinite, and a machine with a trillion transistors is no more alive than a machine with a dozen linkages. The increased complexity of today's devices only makes it that much easier for the machinations of demons to conceal themselves within them—within the operations of an AI system that men blasphemously regard as "AGI."

Appropriately did the first major science fiction film we mentioned earlier—*Metropolis* (1927)—depict a mechanized Moloch consuming men. That accidental prophecy foretold what people would eventually do with the lie this AI-Deception-promoting film depicted. Less than two decades later, however, Christian author C.S. Lewis offered a still more accurate prophecy in his book *That Hideous Strength*. Within the novel—to make a long story short—nefarious scientists working for an evil tech firm hell-bent on global domination have taken the head of a recently guillotined French scientist, Alcasan, and seek to reanimate it with technology. The scientist in charge of that task, a certain Italian named Filostrato, sees his work as an important step in the process of bringing all life past its "organic phase," which he regards as disgusting, so that it may arrive at what he desires: a state of pure intelligence; disembodied mind.

In his pride, Filostrato believes that his heroic scientific and technological efforts have truly reanimated this head and enabled it to speak. Now, the head does indeed speak—dispensing orders and providing new knowledge. What Filostrato does not know, but discovers at the end, is that the whole technological aspect (of which his job consists) has been a complete charade.

Filostrato's experiments took place under the organization's leaders, who were directly collaborating with demons. The "reanimated head" was just a ruse allowing the demons to instruct the organization while preserving some people's belief that what was happening was a technological phenomenon. Thus, the demons have the best of both worlds: they can give direct instructions, and their subjects can remain atheists.

Although C.S. Lewis published this work in 1945—two years before the invention of the transistor gave birth to the Digital Era—it is difficult to imagine a more accurate portrayal of the AI Deception. Delusional scientists, filled with arrogance and entertaining an infinitely too lofty view of their own abilities, increasingly believe they can and will create a being with real intelligence. There is no risk that the Devil will fail to take advantage of the opportunity they are giving him—the door they are throwing wide open for his entry.

Just as prophetically, Lewis' novel portrayed both atheist scientists and seemingly devout Christians equally taken by the Deception. Opposite Filostrato is Reverend Straik, who is just as zealous as his colleague. Only, instead of employing the latter's strictly technological terms, he heralds the Singularity by declaring:

> The resurrection of Jesus in the Bible was a symbol: tonight you shall see what it symbolized. This is real Man at last... Man Immortal and Man Ubiquitous... Man on the throne of the universe. It is what all the prophecies really meant... [Y]ou shall meet the first draught of the real God... a being made by man—who will finally ascend the throne of the universe.[565]

Presenting evil in religious terms does not diminish the guilt—it only exacerbates it. Jesus' harshest rebukes were not for the Romans of his day who openly desired power in secular terms, but for the Pharisees who cloaked their malice with pious attire. And the "Christian" promotion of the most flagrantly apostate views amid the AI Deception is anything but merely fictional. We saw some examples in earlier chapters. Sadly, that scarcely scratched the surface. We could also mention Dr. Antje Jackelén, the emeritus "Archbishop and Primate" of the Church of Sweden. In a scholarly article for the Zygon Journal of Religion and Science, she argues "**in favor of a consonance between nanotechnology and the Christian gospel in terms of realizing the marks of messianic life**."[566] She pontificates about AI and neurochips being implanted into the human brain, and compares the benefits they would bring to Jesus' word sent to the imprisoned John the Baptist indicating that He was indeed the fulfillment of the prophecies—what with His making the blind see, and so on. She concludes:

> [These technologies] tend to realize what marks the messianic age… the development toward *techno sapiens* might very well be regarded as a step towards the kingdom of God. … Why should it be impossible to construct a more perfect human being in the shape of *techno sapiens*, in whom hardware could be replaced and software copied infinitely without any severe losses?[567]

Reverend Straik himself might have said this. And with the rise of ChatGPT, such Antichristic notions are not merely found in academic papers, but are increasingly infiltrating the actual practice of the Church. For example, in the summer of 2024, a Catholic Church in Switzerland installed an "AI Jesus" to "hear confessions," even blasphemously entitling the display "Deus in Machina" (God in the Machine). AP News reported:

> … [A]n avatar of "Jesus" [was] on a computer screen—tucked into a confessional—[and] took questions by visitors on faith, morality and modern-day woes … some 900 conversations from visitors—some came more than once—were transcribed anonymously. … A small sign invited visitors to enter a confessional—chosen for its intimacy—and below a lattice screen across which penitent believers would usually speak with a priest, a green light signaled the visitor's turn to speak, and a red one came on when "AI Jesus" on a computer screen on the other side was responding.[568]

The AI Deception is not "on its way." It has arrived.

***

Demons readily arrive when invited; even if not explicitly. They are certainly implicitly invited by pretending that inanimate hunks of matter can dispense wisdom. In this regard, a slew of phenomena exists which any serious Christian knows to reject. These include Ouija boards, crystal balls, tarot cards, tea leaves, astrology, palm reading, witchcraft, and so on. From the foregoing, however, we have seen that something else belongs on this list: AI being treated as AGI.

Paganism's rebirth in Christendom is explosive today, but it gained much momentum in earlier centuries. Recall that the 19th century was an era filled with occultism that largely infiltrated the Catholic Church. Just as Occultism in general flourished during the 19th century (Ouija, for example, was invented in America in the late 1800s), and so did the ET Deception's early phases, the same is true of the AI Deception's. Dr. Christopher Partridge explains that America's "**most prestigious chemist,**" Robert Hare (†1858), designed a "Spiritoscope," which...

> ... was a prime example of his efforts. A medium would power something akin to a sewing treadle and this in turn spun a wheel with numbers and letters on it, much like a circular Ouija board. ... Perhaps the most ambitious machine associated with the Spiritualists was John Murray Spear's New Motive Power. Spear (1804–87) [was] a Universalist Church of America minister-turned-medium... In 1853, [he] announced in the Boston weekly paper, The Banner of Light, that the spirits were planning to give humanity the ultimate gift from heaven, a 'mechanical messiah,' designed to usher in an unspecified but clearly utopian future for the living.[569]

A "**mechanical messiah**" to usher in a utopia is precisely what today's Heralds of the Singularity attribute to AI. Although the AI and ET Deceptions both formally began in 1947 and exploded recently, they also each ultimately derived from the Enlightenment and saw major developments in the 1800s. And just as the world's most prominent astronomers of the 1800s were promising that ultra-enlightened extraterrestrials populated the planets, and that we would soon contact them, so the same era's most prominent engineers were building devices that they insisted would connect us with the dead and otherwise bestow godlike powers upon us.

Far more prominent 19th-century names than John Spears and Robert Hare believed that technology could harbor spiritual powers. None other than Thomas Edison himself sought to invent a device that could listen to the voices of the dead. This "spirit phone" was also just a technological twist on the occultism prevalent in his day—just as the notion of sentient AI derives from the sci-fi-inspired Paganism of our own.

Of course, we do not now have Edison's "spirit phones" on our countertops. None of the Towers of Babel which men build will ever reach the heavens, nor could they ever. God destroys them not because there was any risk that the prognostications of His enemies might come true. He destroys them because our arrogance in building them and idolizing the work of our hands disfigures our own souls, and He loves us too much to let that long continue.

Thus it is today with the AI Deception. There is no chance that the world's most powerful supercomputer-based AI system will ever prove as capable as a mosquito, much less will it ever become rational, and still less will it ever surpass humanity. Nevertheless, the spectacular height of today's Tower of Babel is such that its demise will be apocalyptic. And we are not here speaking merely symbolically. Babel has already been rebuilt before our eyes, and its own infrastructure—"Babylon"—has already been laid down.

Babylon, *Imago Dei*, and *Imago Bestiæ*

"The reign of Antichrist will… [succeed in] overpowering the imaginations of [those who] have not the love of God deeply lodged in their hearts… [perhaps with] natural wonders and powers of human skill, human acquirements in the hands of the devil. Satan may adopt the more alarming weapons of deceit—he may hide himself…"[570] —St. John Henry Newman. *The Persecution of the Antichrist.*

We now undertake the same task of Part Three's conclusion, wherein we "connected the dots of the Apocalypse" in order to summarize why the ET Deception has no close contender for the Strong Delusion of the Antichrist. Indeed, when consulting Scripture, Church Fathers, private revelation, and more—and when discerning the signs of the times in light of these sources—it becomes clear that the Devil is preparing the way for the Beast (Antichrist) to be revealed as a being in some way connected to the lie that aliens exist and are now here to enlighten and save us. Yet if the AI Deception is as intimately connected to the ET Deception as we have seen it in fact is, then we must ponder what role it also plays in the Strong Delusion.

Yes, it is another outlet for the demons to dialogue with mankind while remaining disguised. But it is also more than that. The Book of Revelation's thirteenth chapter concludes by describing the "second beast," (which is later described in Revelation 19 as the "False Prophet"):

> … [B]y the signs which it is allowed to work in the presence of the beast, it deceives those who dwell on earth, bidding them make an image for the beast … and it was allowed to give breath to the image of the beast so that the image of the beast should even speak, and to cause those who would not worship the image of the beast to be slain. Also it causes all, both small and great, both rich and poor, both free and slave, to be marked on the right hand or the forehead, so that no one can buy or sell unless he has the mark, that is, the name of the beast or the number of its name. This calls for wisdom: let him who has understanding reckon the number of the beast, for it is a human number, its number is six hundred and sixty-six. (Revelation 13:14-18)

Although in apocalyptic discourse the "Mark of the Beast" garners the most attention from the passage above,* we must focus now on the *image* of the

* Countless Christian commentators have already rightly detailed the remarkable convergence of contemporary indications that this "Mark of the Beast," which controls all commerce across the world, will likely be some sort of chip that will become the mandatory payment method for using a new central bank digital currency which will replace all earlier forms of payment. There is little need to repeat their commentary here, other than to note that it rings true. Indeed, in many places such payment methods are already firmly in place, with many consumers eagerly signing up to receive a microchip, the size of a grain of rice, inserted into their hand to "facilitate easy payments." An unprecedented amount of worldwide centralized power could relatively easily be exercised by whatever entity controls the software of these chips upon their universal implementation. That entity, in turn, would de-facto control all people's lives. He who controls what can be bought

same. Revelation indicates that the False Prophet "**was allowed to give *breath*"** to this "image," such that it could "**even speak**." But it was *not* thereby made into an actually living being.

Now, some translations of Revelation do read that the False Prophet "**had power to give life unto the image of the beast**." (v. 15) However, if we consult the Church's official Bible, the Latin Vulgate, we see that the False Prophet gives "*spiritum*" to the Image of the Beast, whereafter "*loquatur imago bestiæ*" (the Image of the Beast spoke). This is in contrast to what God gives Adam in his creation: the Vulgate indicates this is "*spiraculum vitæ*" which makes man "*animam viventem*" (Genesis 2:7)—a *living soul.* Nothing of the sort is given to the Image of the Beast. Most importantly, if we consult the original Greek, we see that "pneuma" (πνεῦμα) is what the False Prophet gives to the Image of the Beast. "Pneuma" can be translated as "spirit" in some contexts (primarily when in reference to the Holy Spirit), but here it is clearly better to translate it as "breath," as most recent translations do.

This is all to point out what is already obvious: God alone can bring about life, and He would never create an evil,* whereas this "Image" is surely evil. Those who claim that *anything*—including the Antichrist or the Image of the Beast—will be an AI system that truly has become AGI are only promoting the same deception they purport to oppose.† For despite having the best of motives, they are only furthering the AI Deception's fundamental premise (i.e., that such a thing is possible). As long as that lie itself is embedded in one's mind, the Devil will never be short of methods to leverage it for implementing the Strong Delusion.

The diabolical "breath" given to the Image is an eschatological perversion of the protological breath of God in the Book of Genesis—and like everything that comes forth from the Devil, it is a lie. It is Satan's "**empty show**," and our Baptismal promises remind us to always reject that. Yet, although it will not at all achieve what it pretends to, the show alone is enough to ruin the souls of its audience.

The end resembles the beginning, hence the title of this book. The pinnacle of God's work in the Universe was accomplished when "**the Lord God formed man of dust from the ground, and *breathed* into his nostrils the breath of life; *and man became a living being*.**" (Genesis 2:7) And the Devil's

---

and sold controls who is able to survive, without there being a need of formally instituting any new laws or overthrowing any governments. AI certainly factors heavily into the Mark of the Beast as well, since it would be unfeasible for humans to manage such a massive system.

* Even the Antichrist himself, though he will have freely chosen to fully give himself over to the Devil, is a human being—thus, he was not *originally* evil as God created him. Indeed, not even the demons, including Lucifer, were originally evil: God made them good, but they rebelled.

† It would be as if a man rightly crusading against the transgender deception foolishly chose to do so by way of lamenting "*how problematic it is when a man truly becomes a woman.*" Of course, that would be nonsense. No matter what he does, a man cannot become a woman; and the danger lies precisely in supposing this is possible. Yet the impossibility of the thing itself does not at all detract from the immense damage done by believing in it, as here with "AGI."

last attack against this supreme work of God not only mirrors the manner of the first attack's release (through an "NHI" Deception), but also mimics God's own initial activity in bringing man about. The Breath of God makes a man—*Imago Dei*. The breath of the Devil makes the Abomination of Desolation—*Imago Bestiæ*.

Daniel vs. the Abomination

This new term brings us to another passage. In the Gospel, we read the words of The Lord about the final trial of the Church:

> When therefore you shall see the abomination of desolation, which was spoken of by Daniel the prophet, standing in the holy place: he that readeth let him understand. Then they that are in Judea, let them flee to the mountains… For there shall arise false Christs and false prophets, and shall shew great signs and wonders, insomuch as to deceive (if possible) even the elect. Behold I have told it to you, beforehand. (Matthew 24:15-16,24-25)

Of all the books of the Bible, it is the Book of Daniel which overlaps most with the Book of Revelation's eschatological content. Jesus Himself here directs us to consult it in discerning the Last Deception—particularly the "abomination of desolation" therein. "*He that readeth let him understand,*" Jesus exhorts us. This requires our most resolute attention, just as does the Mark of the Beast, for that too "**calls for wisdom**." (Revelation 13:18)

Although various scholars have presented many theories about this *abomination who makes desolate*, a common theme emerges: it is an infiltration of Pagan idolatry into the heart of the Sacred Place. About this infiltration, some aspects of the Book of Daniel refer to what has already happened in history (e.g., the placement of the image of Zeus in Jerusalem in the third century B.C.), and others refer to what is yet to come in the apocalyptic times ahead. In relation to this abomination, Daniel writes that a coming nefarious prince will "**cause sacrifice and offering to cease; and upon the wing of abominations shall come one who makes desolate**…," (Daniel 9:27) which means that a massive suppression of the Holy Sacrifice of the Mass will precede or coincide with the rise of the Abomination. Two chapters later, we read:

> … [T]hey shall set up the abomination that makes desolate. He shall seduce with flattery those who violate the covenant; but the people who know their God shall stand firm and take action. And those among the people who are wise shall make many understand… (Daniel 11:31-33)

We see that the Abomination will use seduction, but that the orthodox Faithful will "**stand firm and take action**," whereas the "**wise shall make many understand**"—perceptive wisdom is needed here just as in Revelation 13 and Matthew 24.

There is much to glean from this passage that Jesus exhorts us to consider in navigating the Apocalypse. If the ultimate Abomination were simply some flagrant institution of ancient Paganism within the Vatican (or Jerusalem), we would not thereby be confronted with a subtle seduction which requires the assessment of the wise for understanding. Even a dunce would notice it. But those—and only those—who heed authentic wisdom, and who stand firm in the Faith of their Fathers, will discern the cloven hoof. They will "take action." They will resist the Great Deception.*

The next verse indicates that this nefarious King who is ultimately behind the Abomination is none other than the Antichrist:

> ... And the king shall do according to his will; he shall exalt himself and magnify himself above every god, and shall speak astonishing things against the God of gods. (Daniel 11:36)

As we discussed in Part Three, "**the man of lawlessness ... the son of perdition... takes his seat in the temple of God, proclaiming himself to be God,**" (2 Thessalonians 2:3-4) just as Daniel says here of the Abomination.

This Lawless One, therefore, is surely the Antichrist—the Beast of Revelation—and it is his companion, the False Prophet of Revelation, who will "give breath to the image of the beast." (Revelation 13:14) They do not set themselves up just anywhere, but "in the temple of God," that is—most likely—in the geographic heart of the Catholic Church,† or perhaps Jerusalem. Not only is the AI Deception almost directly referred to in this context by Scripture's teaching that the Beast's Image will be "breathed" into such that it speaks—though is *not* thereby made truly alive—but the Book of Daniel itself also gives a striking prototype of the same deception only a few chapters later,‡ in the account of Daniel's exposing the fraud behind the idol Bel.

> Now the Babylonians had an idol called Bel ... The king revered it and went every day to worship it. But Daniel worshiped his own God. And the king said to him, "*Why do you not worship Bel?*" He answered, "*Because I do not revere man-made idols, but the living God, who created heaven and earth and has dominion over all flesh.*" The king said to him, "*Do you not think that Bel is a living God? Do you not see how much he eats and drinks every day?*" Then Daniel laughed, and said, "*Do not be deceived, O king; for this is but clay inside and brass outside, and it never ate or drank anything.*" Then the

---

* Tragically, however, not all of "the wise" will remain un-deluded. The same passage also indicates that "**some of those who are wise shall fall.**" (Daniel 11:35) Thus we are reminded to never embrace the Great Deception merely because some hitherto revered scholar in the Church condones it. We must know to expect that. And from the opening sections of the present Part of this book, we have seen that this falling away has already begun.

† This does not mean that the Pope himself will be the Antichrist (the promises of Christ preclude such a situation), only that, geographically, the Antichrist will as much as possible seek to ape the things of God.

‡ Although what follows has always been present in the canon of Scripture throughout Christian history, it is sadly missing from most Protestant Bibles today. Nevertheless, it is certainly Biblical, and it is found in Catholic and Orthodox Bibles.

> king was angry, and he called his priests and said to them, "*If you do not tell me who is eating these provisions, you shall die. But if you prove that Bel is eating them, Daniel shall die, because he blasphemed against Bel.*" And Daniel said to the king, "*Let it be done as you have said.*" (Daniel 15:3-9)

The passage goes on, describing the cunning way Daniel exposed the deception by setting out food for the idol to "eat" while sealing the entrance for the priests to enter. The priests of Bel were unconcerned, for they knew they had a hidden entrance to sneak through and consume the food they always claimed Bel had eaten. But Daniel knew *that* they knew this, and he had secretly scattered ashes about the surrounding area so that he was able to detect their footsteps as they undertook their usual nightly regimen. In the end, "**the king put them to death, and gave Bel over to Daniel, who destroyed it and its temple**." (v. 22) Now, Daniel's own given Babylonian name—Belteshazzar—means "*Bel protects his life.*" But Daniel's real name means "*God is my judge.*" Daniel both refuted and demolished the very idol which the Pagans sought to name him after.

The same ruses are employed today with the lies surrounding "AGI." The Heralds of the Singularity—today's own Priests of Bel—often know full well they are lying. They employ various methods of presenting AI as if it were conscious—programming into humanoid robots a predetermined conversation, having a real human speak through an AI remotely, advertising fleets of "self-driving" cars that actually have a large staff of remote drivers piloting them whenever necessary,* presenting "great accomplishments" of AI that were actually just accomplishments of individual humans who know they will get more media attention by claiming AI did it all, and so on—all while posting videos of these happenings online to imply sentient-AI is upon us. The gullible masses gobble up this lie and disseminate it just as did Bel's pseudo-reverent flock. The managers of the deception never allow the scrutiny to be applied that would be needed to discern the real nature of what is transpiring.

The prophet's conquests against the "NHI Deception" of his own day continue in the next verse. The King had not learned his lesson from the deception of Bel, and we read:

> There was also a great dragon, which the Babylonians revered. And the king said to Daniel, "*You cannot deny that this is a living god; so worship him.*" Daniel said, "*I will worship the Lord my God, for he is the living God. But if you, O king, will give me permission, I will slay the dragon without sword or club.*" The king said, "*I give you permission.*" Then Daniel took pitch, fat,

* Although sometimes it does peek through. A September 2024 report in *The New York Times*, entitled *How Self-Driving Cars Get Help From Humans Hundreds of Miles Away*, noted: "**Outside [certain] companies, few realize that autonomous vehicles are not completely autonomous. For years, companies avoided mentioning the remote assistance provided to their self-driving cars. The illusion of complete autonomy helped to draw attention to their technology … "There is a 'Wizard of Oz' flavor to this," said Gary Marcus, an entrepreneur and a professor emeritus of psychology and neural science at New York University who specializes in A.I. and autonomous machines.**"

> and hair, and boiled them together and made cakes, which he fed to the dragon. The dragon ate them, and burst open. And Daniel said, "*See what you have been worshiping*!" (Daniel 14:23-27)

Neither the idol Bel nor the dragon were the "living gods" that the Babylonians claimed they were. All it took was firm Faith in God and the application of wisdom to expose the deception. Yet the Babylonians were so furious with Daniel for refuting their cherished fictions that they threatened to kill the King himself if they would not hand Daniel over to them! Afraid, the King acquiesced. Daniel was thrown into the lions' den, and we all know what happened next. It did not go well for those Pagans who rejected the fact that only man bears the Divine Image. They—not Daniel—became food for the lions.

Not only is Daniel's triumph over Bel a prototype for Christ's eschatological triumph over the AI Deception, but it also took place in the very city to which the Book of Revelation directs our attention. Daniel was captive in Babylon when Bel reigned, and the Book of Revelation repeatedly refers, symbolically, to "Babylon" as the context of the Great Deception. In one case—the Harlot's—this reference is expressly connected to "abominations:"

> … and on [the Harlot's] forehead was written a name of mystery: "*Babylon the great, mother of harlots and of earth's abominations.*" … When I saw her I marveled greatly. But the angel said to me, "*Why marvel? I will tell you the mystery of the woman, and of the beast with seven heads and ten horns that carries her. … dwellers on earth whose names have not been written in the book of life from the foundation of the world, will marvel to behold the beast… This calls for a mind with wisdom…*" (Revelation 17:5-9)

This "abomination," which Scripture repeatedly tells us to beware amid the eschatological Strong Delusion, has as its context "Mystery Babylon." We are also, yet again, reminded that "**a mind with wisdom**" is needed for discerning. Why? So that the elect do not—like those who join ranks with the Antichrist—"**marvel to behold the beast**." (v.9) To marvel at these diabolical ruses is to fail to discern the "***pretended*** **signs and wonders**" (2 Thessalonians 2:9)—lies—that they are; hence the Angel correcting St. John himself. Whoever "marvels" at the workings of the Beast—supposing that sentience and rationality really has been bestowed upon an AI—is only one small step away from worshipping this Beast. That small step will quickly prove irresistible for those whose minds will be spinning from the so-called "ontological shock" of believing in the false "Singularity" of the days ahead.

The same chapter of Revelation concludes by teaching: "**the woman that you saw is the great city which has dominion over the kings of the earth**." (Revelation 17:18) Now, what "city" (locus of power) would have such dominion—such sway over all the rulers of the earth? The details will change, but one thing is sure to remain consistent: there will be no close second—in centralizing worldwide dominion in one "city"—to what can be accomplished by whoever controls the prevailing AI algorithms and datasets. Already, almost all aspects of our lives are far too controlled by them, and the trend which brought us here promises to only increase. For while there is the diabolical lie that AI can become rational and alive, there is also the actual power that AI *really does* confer upon whoever controls its central hardware. The elite already know this.

After enumerating several technologies, but singling out Artificial Intelligence, Klaus Schwab—the head of the World Economic Forum—prophesied in 2023 that, "**Our life, in ten years from now, will be completely different... who masters those technologies, in some way, will be the master of the world.**"[571] Russian President Vladimir Putin said the same thing in 2017, declaring: "**[Artificial Intelligence] is the future of the whole of mankind... the one who becomes leader in this sphere will be the ruler of the world.**"[572] Those who wield such power and influence as to merit being taken seriously when they make apocalyptic statements are practically begging us to realize that they themselves are preparing for Artificial Intelligence to constitute the System of the Beast. Identical remarks have been given by many of the "great"—those who, deluded by Babylon, will eventually be cast down with her. The Book of Revelation foretells, in the next chapter: "**For thy merchants were the great men of the earth, and all nations were deceived by thy sorcery.**" (Revelation 18:23)

This lying "**sorcery**" which *deceived* "**the great men of the earth, and all nations**" could be the lie that an AI can become "superintelligent" and acquire the "breath of life." Genesis makes clear that this is reserved to man alone. In Revelation's nineteenth Chapter, we read of the Beast's defeat:

> And the beast was captured, and with it the false prophet who in its presence had worked the signs by which he deceived those who had received the mark of the beast and those who worshiped its image. These two were thrown alive into the lake of fire that burns with brimstone. (Revelation 19:20)

Yet again, we see *deception* emphasized. Too many Christian commentators gloss over this, supposing it can be interchanged with a broader term like "persecution." But the Bible means what it says. Yes, the Beast will institute the worst persecution in history. But that persecution itself is built upon a deception—and not just any deception, but a direct assault on the essence and pinnacle of Creation. How could it be otherwise? The faithful have always known what that pinnacle is: "**Man is the perfection of the Universe,**" as the Doctor of the Church, St. Francis de Sales, taught.

If there was ever any doubt about this truth (there was not) in the days of ancient Judaism, the Incarnation of Jesus Christ settled it: nothing is more central to the Work of God than man being made in the Image of God—he alone being the *Imago Dei*. There could be no more nefarious a way for the Devil to institute the Final Attack than by perverting the *Imago Dei* with the *Imago Bestiæ*: The Image of the Beast, a lie directed against the capstone of the Universe. For an attack on man in the Image of God is equally an attack on the greatest thing in the Universe: the Incarnation of Jesus, true man, true God.

The Devil can only deceive the masses into dethroning God if he can convince them that God's greatest work has been surpassed by a mere creature. Remember the prophecy of the *Catechism of the Catholic Church*. It teaches that the Antichrist's ruses will consist in...

> ...a religious deception offering men an apparent solution to their problems at the price of apostasy from the truth. The supreme religious deception is that of the Antichrist, a pseudo-messianism by which man glorifies himself in place of God and of his Messiah come in the flesh. (§675)

Nothing else comes even remotely as close to fulfilling these prophecies as does the NHI Deception, and its ungodly marriage of belief in ETs and AGI.

***

Returning to the passage from the Book of Revelation we began with, we find what may be the most telling detail of all. Although the False Prophet "*gives breath to the image of the beast,*" enabling it to "speak," he does *not* make the image. Rather, he "**deceives those who dwell on earth, bidding *them* make an image for the beast**." (Revelation 13:14) This indicates that the Image of the Beast *will be man's own invention*, and man will build it in response to a *deception*.

Why, however, would the Beast need men to make him an "image"? The reason is straightforward. Revelation clarifies that this Image will be the epicenter of a global system of control (for Revelation 13:15 indicates the Image itself can even kill those who will not worship the Beast, and the very next verse describes the Mark of the Beast, what with its global system of financial control). The Antichrist, however, will just be a man (albeit one who works lying signs through Satanic power). His claims to Divinity will be a lie. He cannot simply snap his fingers and miraculously institute his global system of total control. The masses will need to build it for him, and he will need to deceive them to encourage its construction, and deceive them still more to hand it over to himself.

The Beast may well present himself as an extraterrestrial savior, but he will not actually have any of these extraterrestrial technologies of which he boasts (for these technologies do not exist). Perhaps he and the False Prophet will issue seductive promises to the world that if only we build this "Image" for him—this fully connected, global, unchecked AI system that is integrated with a mandatory payment chip in each person's right hand or forehead—he will be able to use it to disseminate and implement his extraterrestrial enlightenment. The system—the "extraterrestrial solution"—may be presented as the only way to stave off nuclear war, environmental catastrophe, plague, famine, or what have you. And it is at this point—after the world has built and embraced this "Image," this global AI system—that the

False Prophet "breathes" into it, enabling it to "speak" on its own, thus formally instituting the Abomination of Desolation. This Abomination may be an AI presented as if it surpassed the "Singularity" and became AGI, when in fact it is only another dead and dumb computer system, though directly possessed by—controlled by—the Devil.

What, then, in a word, will men do? *They will build another Tower of Babel.*

Mystery "Babylon" itself—the context for the Great Deception here in the Book of Revelation—derives from "Babel" (cf. Genesis 11). Astute Christian commentators have already long observed that, with AI, we are doing precisely this: building a new Tower of Babel. For example, writing for *Christianity Today* in 2020, AI expert Joanna Ng drew the following accurate parallels:

> In Genesis 11:1–9, the people of the earth sought to build the Tower of Babel … Today, scientists seek to build [AGI] … Both are a quest for supremacy of mankind: one with a tower that reaches the heavens, the other with a singleton that is capable of dominating man. ... Both are a quest for independence from God: People would rather trust the creations of their own hands than trust their Creator.[573]

The exaltation of the creature above the Creator: this is the essence of both Babel and the AI Deception. "**They exchanged the truth about God for a lie and worshiped and served the creature rather than the Creator**…" (Romans 1:25) The Satanic marriage of the two amid the times of the apocalypse gives birth to the Abomination of Desolation: the Image of the Beast.

Yet, they will be destroyed. The Devil does not have the last say.

> "Fallen, fallen, is Babylon the great!" (Revelation 18:2)

"**Evil may have its hour, but God will have His day**," Archbishop Fulton Sheen—whose 1947 prophecy of the Antichrist constitutes a veritable description of the NHI Deception—famously reminded us.

We have spoken much here about the Book of Revelation's prophecies of the trials that lie ahead. What of the far more important parts: its promise of the glory that lies ahead? Let us conclude this book the same way the Bible itself does.

Victory Over the Beast

We concluded Chapter 1 with teachings of the great French missionary priest and theologian, Fr. Charles Arminjon; teachings which St. Thérèse of Lisieux said were among the greatest graces of her life. [574]

Within those teachings so praised by this "Doctor of the Science of Divine Love," Fr. Arminjon wisely corrected those who interpret the Book of Revelation such that the imminence of the Antichrist also means the imminence of the End of the World, writing: "**This interpretation is incorrect. St. Thomas and St. John Chrysostom explain … that Christ will strike the**

Antichrist by dazzling him with a brightness that will be like an *omen and sign* of His Second Coming ..."[575] Yes, the Antichrist (Beast) will be defeated by a certain type of "coming" of Christ, but it will be an "apparition of His power"—a "coming" in grace, not His Second Coming in the flesh. That latter event is surely the end of the world (and of time itself!), as we have already settled. Fr. Arminjon continues:

> What is certain is that Satan will be hurled back into the darkness of the abyss, the reign of the man of evil will be utterly destroyed ... The most authoritative view, and the one that appears to be most in harmony with Holy Scripture, is that, after the fall of the Antichrist, the Catholic Church will once again enter upon a period of prosperity and triumph.[576]

A faithful reading of Revelation 20 confirms Fr. Arminjon's teaching. There, we are told:[577]

> [An angel] seized the dragon, that ancient serpent, who is the Devil and Satan, and bound him for a thousand years, and threw him into the pit, and shut it and sealed it over him, that he should deceive the nations no more, till the thousand years were ended. After that he must be loosed for a little while ... [Those] who had not worshiped the beast or its image and had not received its mark on their foreheads or their hands... reigned with Christ a thousand years. And when the thousand years are ended, Satan will be loosed from his prison... but fire came down from heaven and consumed [Gog and Magog]... (Revelation 20:2-4,7,9)

Immediately *after* this passage, the Book of Revelation teaches that time itself comes to a close. Jesus returns in the flesh, whereupon "earth and sky [flee] away." (v. 11) The Resurrection and Judgment commence.

Yet when we look at what God reveals about what happens before that, relayed above, Scripture's clear teachings simply cannot be reconciled with the view that the Antichrist reigns immediately before the End of Time. Admittedly, this "thousand years" should not be taken literally, no matter which interpretation of Revelation 20 one prefers*—that "thousand years" is simply a reference to an Era of Peace whose duration we know not. Just as importantly, however, that "thousand years" cannot simply be ignored or so heavily allegorized as to become self-contradictory. The thousand years are real. They describe a triumph of God on earth, and no one could miss that they both *follow* the Beast's (Antichrist's) defeat, while also *preceding* a very brief and definitive end-of-time unleashing of Satan (through "Gog and Magog") to precipitate the final consummation of the world in flames.†

---

* Christ affirmed that we "know neither the day nor the hour." (Matthew 25:13)

† That final unleashing of Satan involves no lengthy persecution, delusion, evil reign, or anything of the sort. Scripture's entire description of it consists of only four short verses (Revelation 20:7-10), unlike the chapters upon chapters needed to describe the reign of the Beast—the Antichrist. Amid the "thousand years," Satan suddenly explodes out of hell, marches against the saints, and is then immediately consumed by fire falling from Heaven. That's it. This is quite fitting for innumerable reasons. Consider just two: a "reign of saints" on earth is unlikely to fall for another "Strong

Heretical interpretations of this passage certainly exist and, unfortunately, are popular in some Evangelical circles. They are the "Millenarian" or "Modified Millenarian"* teachings, which both consist in the view that Christ will reign visibly on earth during this "thousand years" (a view often joined with strange teachings about a so-called "Rapture"). In fact, only in Heaven do we get Christ's visible reign—along with any effects which flow only from His unveiled presence (confirmation in grace, immortality, impassibility, the cessation of Faith, and so on). More on this point can be found in my 2021 book, *Thy Will Be Done.*

Yet, the interpretation which would regard Revelation 20 as *merely* symbolic is also flawed. This does not mean that "amillennialism"—which views those "thousand years" as sacramental, or as symbolic of the entire earthly duration of the Church—is *wrong*. That is certainly a valid and edifying way to meditate on Revelation 20. It only means that amillennialism doesn't give the *entire* picture. Key Scriptural passages often have multiple valid (and important) meanings, and this is one such case. "**The words, 'kingdom of God,' have a variety of significations, the exposition of each of which will not be found without its advantages**..." (*Catechism of the Council of Trent,* Part IV) Who would have expected it to be otherwise with so uniquely important and breathtaking a Biblical passage? To insist that Revelation 20 could *only possibly* suggest a strictly amillennial interpretation is to succumb to precisely what Venerable Pope Pius XII, in his encyclical *Divino Afflante Spiritu*, condemned as an "**intemperate zeal**" which everyone should "**abhor**." (§47) Indeed, there is not a single teaching, anywhere in Scripture, or the Church's Magisterium, which indicates that that the thousand years cannot refer to an Era of Peace—a "reign of saints"—on earth. On the contrary, most of the Church Fathers insisted that Revelation

---

Delusion," as men are succumbing to today. Satan will attack, *attempting* to "deceive the nations" (Revelation 20:8) as he always does by nature, but he will not succeed this time. Second: we know that the world must be dissolved in fire at the end of time (cf. 2 Peter 3:10). Yet it would not be fitting for God to be the cause of these flames unless they descended for the reason of eliminating the demons. Therefore, in one final act of His Permissive Will, God allows all of hell to again, very briefly, be unleashed upon the material universe so that when fire comes down from Heaven it has, as its object of wrath, the demons, and not the earth—for earth will then enjoy a "reign of saints" amid the "thousand years."

* The Church has already defined even modified Millenarianism as consisting in the anticipation of Christ visibly reigning on earth. For example, the July 21st, 1944 Declaration of the Holy Office specifically defines even modified Millenarianism as the system of teachings wherein: "[*The teaching that Christ] will come visibly to rule over this world... cannot be taught safely.*" (Decree of the Holy Office on Mitigated Millenarianism. Acta Apostolica Sedis. Annus XXXVI—Series II—Vol. XI. Page 212) This condemnation itself was a follow-up to the Holy Office's earlier decree, issued only three years before, regarding the book of a certain Jesuit Priest, Fr. Manuel De Lacunza. The priest had written *The Coming of Messiah in Glory and Majesty,* in which he painstakingly (and vainly) argued for a system of modified Millenarianism, which purged Millenarianism of a few errors, but staunchly maintained the visible reign of Christ on earth On July 11th, 1941, the Holy Office condemned this book of Fr. Lacunza's, "*with specific reference to the book's moderate Millenarianism.*"

20 spoke of the times of the Kingdom, on earth, before the conclusion of history.*

> "Therefore, a sabbath rest still remains for the people of God." (Hebrews 4:9)

Let us consider just one. St. Irenaeus is not only a Church Father: he is also the only saint in Church history who is both a Doctor of the Church and a martyr. If anyone should be deferred to on this topic, it is Irenaeus. Commenting not only on Revelation 20 but on many other Scriptural teachings, this Father wrote:

> [Christ] will Himself renew the inheritance of the earth... He cannot by any means be understood as drinking of the fruit of the vine when settled down with his [disciples] above in [Heaven]... These are [to take place] in the times of the kingdom, that is, upon the seventh day ... If any one, then, does not accept these things as referring to the appointed kingdom, he must fall into much contradiction and contrariety... (Book 33) … The whole creation shall, according to God's will, obtain a vast increase... (Book 34) If, however, any shall endeavor to allegorize [prophecies] of this kind, they shall not be found consistent with themselves in all points, and shall be confuted by the teaching of the very expressions [in question] ... For all these and other words were unquestionably spoken in reference to … [times] after the coming of Antichrist ... in the times of the kingdom, the earth has been called again by Christ [to its pristine condition]... (Book 35) John [in the Book of Revelation], therefore, did distinctly foresee the … inheritance in the kingdom of the earth; and what the prophets have prophesied concerning it harmonize [with his vision]. For the Lord also taught these things…The apostle, too, has confessed that the creation shall be free from the bondage of corruption, [so as to pass] into the liberty of the sons of God. (Romans 8:21) (Book 36)[578]

---

* While the amillennial interpretation comes to us from St. Augustine, it was also Augustine himself who—in *City of God*, Book 20, Ch. 7—objected to using that interpretation of his to reject other interpretations. Unlike his teachings that reject alien belief—which were categorical and uncompromising—he specifically taught that it *was* legitimate to interpret Revelation 20 differently than he did. He found "unobjectionable" the view that Revelation refers to a reign of saints on earth, after the Antichrist. Several chapters later (Ch. 30), he wrote that he *did not know* the order of the apocalyptic events, and that only the experience of them could instruct us about that. It is quite dishonest, therefore, for anyone to use Augustine's amillennialism to claim that the Book of Revelation *must* be interpreted in such a way as to pin the end of the world as immediately after the Antichrist. Moreover, the opposite view is far more common among the Church Fathers. It is, therefore, quite remarkable that some Catholic scholars will even glibly mock this common, orthodox Patristic teaching. (Denigrating prophecy is an easy way to win worldly praise in academic circles today, even within the Church.) No other Patristic teaching receives such scorn from them. Indeed, the Fathers laid down the foundation of our understanding of Public Revelation—any unanimous consensus of theirs is, automatically, a Dogma of the Faith—and we who come after them have no right to simply dismiss them.

As St. Irenaeus demonstrates, it is simply implausible and even logically contradictory* to interpret not only Revelation 20, but also the promises from Jesus Himself—and from the Apostles, from St. Paul, and from the Old Testament Prophets—as referring to anything other than a glorious Era to come on earth; a "coming of Christ's Kingdom" within history, before the end of time, along with an outpouring of grace and blessings for the world. This Church Father is not merely presenting an opinion on how to interpret a single passage in the Book of Revelation; he is, rather, relaying a fundamental conviction from prophecies *throughout* the Old and New Testaments.

Although presenting them would be outside of the scope of this book, there is also a multitude of private revelations—many bearing Church approval—which have confirmed that a glorious Era of Peace is to come; one so glorious that no one who has read the prophecies about it will fail to understand that they describe the very times of the "reign of saints" during the "thousand years." Summarizing centuries of these revelations, the unassailably orthodox old *Catholic Encyclopedia*, published in 1911, states:

> The more noteworthy of the prophecies bearing upon "latter times" seem to have one common end, to announce great calamities impending over mankind, the triumph of the Church, and the renovation of the world. All the seers agree ... they all promise for the Church a victory more splendid than she has ever achieved here below. (*Catholic Encyclopedia.* Article on Prophecy.)

This Encyclopedia, which is known for choosing its words carefully, does not in vain describe these prophecies as "**all**" agreeing on the promise of a coming Triumph of the Church that will be greater than "**ever**" before and a concomitant "**renovation of the world**."[579] That, therefore, is what we await!

Yes, the Signs of the Times point to the imminent rise of the Antichrist amid the ET, AI Deception. Yet this should not trouble the Faithful. For God will preserve those who love Him even during the great apocalyptic trials, and the Faithful must remember that the prophecies promise these trials will not be some enduring dystopia like those depicted in science fiction. They

---

* The sola-Amillennial view has even been implicitly Magisterially contradicted by the Popes of the 20th century who—like Pope St. Pius X, as early as 1903 A.D., upon seeing the deluge of evil inundating the world—declared that the times had become more sinful than *ever* before, including before the Flood. (See, for example, Pope St. Pius X. *E Supremi* October 4th, 1903, Paragraph 3, wherein he teaches that society had become worse than in "*any* past age.") It is not intellectually honest to assent to this Magisterial teaching and simultaneously hold that we are or ever have been in the time of the binding of Satan spoken of in Revelation 20. One of those views must go. And it mustn't be the one which is Magisterially taught. Pope St. Pius X also taught, in the same encyclical noted above (*E Supremi*), that the literal Antichrist himself may already be in the world, writing, "**... there may be already in the world the "Son of Perdition" of whom the Apostle speaks (II. Thess. ii., 3) ...**" No document appealed to by those who insist that the Antichrist can only come immediately before the end of time rivals, much less surpasses, the authority of this text. Nor is it possible to understand this text in any other way than the one which refutes them. For the same encyclical clearly prophesies a yet-to-come Era of Peace, thus indicating that the Era—the "Restoration of All Things in Christ" on earth—will *follow* the defeat of the Son of Perdition.

will be brief—in fact, all the prophecies of the Antichrist describe his reign as no more than three and a half years. Therefore, Christians should harbor great hope.

Most importantly, Christians must know the time is *now* to become the boldest witnesses to Christ that history has ever seen: to prepare for the days ahead by evangelizing more zealously than ever before, by praying more intently than ever before, and by focusing more fervently than ever before on becoming saints.

Despite the gathering clouds of the coming storm, these are the most exciting times in history to be alive—for those intent upon striving to live in the Will of God. "**Here is a call for the endurance of the saints.**" (Revelation 14:12)

> "He who testifies to these things says, "Surely I am coming soon." Amen. Come, Lord Jesus! The grace of the Lord Jesus be with all the saints. Amen." (Revelation 22:20-21)

Triumph of the Name of Jesus (1679 A.D.)
Giovanni Battista Gaulli (†1709)
Church of the Gesu, Rome, Italy
(Wikimedia Commons)

APPENDICES

1. Further Resources

Although it includes much new research, this book is in some ways a shortened version of the much larger reference work I published in 2023, *Only Man Bears His Image*. That work—756 pages long—includes a wealth of additional relevant information, and it should be consulted when questions arise that are not addressed here. They are likely addressed there, and that book contains a detailed table of contents which can be used to quickly and easily locate the relevant content.

(Important questions addressed neither here nor there will hopefully be addressed in future works I publish, which may be found at my website, DSDOConnor.com, YouTube channel, Youtube.com/@DSDOConnor, or Twitter page, X.com/DSDOConnor.)

Index of Important Issues Addressed in *Only Man Bears His Image*

Within *Only Man Bears His Image*, you will find each of the following truths discussed:

St. Jerome's remarks about the "satyr" do not lend credence to belief in non-human intelligent incarnate creatures (Pages 707-714). Since only man bears the Divine Image, not only is belief in aliens and AGI ruled out; so too is belief in non-human rational animals (Pages 603-624). Much sci-fi deception—particularly in its ET and AI varieties—derives from a false mystique flowing from an illogical and unscientific view of Quantum Mechanics (Pages 524-534). The Catechism rules out aliens—further considerations of why (Pages 686-689).

St. Padre Pio, Pope St. John Paul II, and Pope St. Paul VI did not endorse belief in aliens (Pages 690-702). The teaching of Pope Francis does not support ET belief; it contradicts belief in aliens (Pages 655-657). Pope Pius XII's *Humani Generis* rules out aliens—it does not support belief in aliens who simply "could not know and love God." (Pages 684-686). Garabandal, Medjugorje, and Maria Valtorta do not support belief in aliens (Pages 702-707).

The Revelations to Bl. Anne Catherine Emmerich do not support belief in aliens (Pages 714-716). C.S. Lewis' opinions should not prompt us to believe in aliens (Pages 716-717). The *New Baltimore Catechism #2* does not render alien belief legitimate (Pages 718-720). The Virgin Mary's title of "Queen of the Universe" refutes alien belief, and certainly does not support it (Pages 720-724).

Various Scripture passages and verses are often abused by ET promoters; interpreted in illicit ways to make them appear to support belief in aliens. The following cases are addressed in the corresponding pages.

John 10:16 and Jesus' "other sheep" (Pages 666-667). Revelation 12:4 and a "third of the stars" falling (Pages 667-669). Revelation 4:6-8 and the "four living creatures" (Pages 669-670). Isaiah 45 and the "vanity" of a planet lacking habitation (Pages 670-671). The angel in Tobit (Pages 671-672). Ezekiel's vision (Pages 672-673). Psalm 19 and "the heavens declare" (Pages 674-677). Genesis 4 and the Mark of Cain—why he was afraid (Pages 678-679). Genesis 6, the Nephilim, the Giants, and the "daughters of men" (Pages 679-683).

2. Additional Considerations

AI and Quantum Computing

Although daily headlines continue to sing the praises of the imminent "exponential increases" in computational power this technology advertises, the fact remains that decades of work with it (and billions of dollars of investment in it) have failed to produce a single case wherein quantum computers exceed ordinary ("classical") computers. Claims to the contrary have never been proven, even though they would be exceedingly easy to prove if valid.

Some experts in the field convincingly argue that the basic premises which quantum computing's advocates employ are not merely unrealistic, but fundamentally impossible and based on fallacious reasoning. For example, writing for the Institute of Electrical and Electronics Engineers, respected physicist Dr. Michel Dyakonov published *The Case Against Quantum Computing*, in which he pointed out:

> A useful quantum computer needs to process a set of continuous parameters that is larger than the number of subatomic particles in the observable universe. At this point in a description of a possible future technology, a hardheaded engineer loses interest. … Yet quantum-computing theorists have succeeded in convincing the general public that this is feasible. … [But] the huge amount of scholarly literature that's been generated about quantum-computing is notably light on experimental studies describing actual hardware… distinguished experts did not … address the question of whether [their quantum-computing-justifying theoretical] assumptions could ever be satisfied. I argue that they can't.[580] … Could we ever learn to control the more than 10^300 continuously variable parameters defining the quantum state of [a useful quantum computing] system? My answer is simple. No, *never*.[581]

Professor Dyakonov is an established authority whose work in quantum physics has remained influential from the 1970s to the current day. Multiple physical phenomena are named after him.[582] While there are also highly credentialed scientists who disagree with Dyakonov, their motives are transparent: like 1990s global warming alarmists who promised all coastal cities would be submerged within a decade, their careers rely on funding that takes their own pseudoscientific predictions seriously. Dyakonov, on the other hand, presents a detached analysis based on mathematics, hard science, and engineering. Doing so, he exposes quantum computing's true nature as not merely unrealistic, but *impossible*.

We have recently begun to see some of the very pundits who would benefit most from quantum computing hype beginning to publicly admit the truth. For example, in December 2023, the head of AI for Meta (Facebook) threw cold water on both AGI and Quantum computing, noting:

> The number of problems you can solve with quantum computing, you can solve way more efficiently with classical computers... Quantum computing is a fascinating scientific topic... [without] practical relevance and the possibility of actually fabricating quantum computers that are actually useful.[583]

Also in 2023, physicist and computer scientist Dr. Matthias Troyer—a Principal Researcher at Microsoft's Quantum Group—ran simulations showing that "**a quantum algorithm ... would have to run for centuries, or even millennia, before it could outperform a classical [computer] on problems big enough to be useful... Even in optimistic future scenarios [quantum computing] is likely to be thousands or millions of times slower than classical computers**."[584]

My own speculation? Pragmatically useful quantum computing is as fictional as time travel, fairies, and extraterrestrials. But I do believe we will soon be *told* that a massive quantum computing breakthrough has been made, and I believe that announcement will be a lie employed to conceal the moment when the Devil himself possesses our centralized AI architecture.

AI's Lack of Intellectual Agency

AI cannot improve on its own, as many expect it to. It cannot reflect on the processes it uses, since its entire existence is exhaustively described *by those very processes*. It is nonsensical to hope for an AI to assess its own algorithms for truth or falsity, logic or fallacy, and goodness or evil. Human intellect, on the other hand, can always take a philosophical step higher still and reflect upon whatever rational deduction it last undertook. It can examine its own examination, *ad infinitum*. Since it is a spiritual faculty that participates in the activity of God, Who is infinite, its capacities are *never* exhausted by any particular slew of activities it undertakes. Although not infinite in itself (only God is infinite in Himself), the intellect nevertheless has a type of potency for the infinite that no purely material thing can ever have.

While humans may learn best by having a good teacher, we can also truly grow in wisdom with no external input whatsoever. We can sit with our own thoughts, reflect upon them, and rise to still higher levels of apprehension of the truth. Some of the greatest enlightenments in history came from men silently reflecting during a lengthy imprisonment. We need look no further than Boethius' *Consolation of Philosophy*. AGI proponents insist that similar or identical results are now seen in AI, but this is as much rubbish.

Just as the human intellect has a potency for the infinite, so does the human will—for which AI also definitively lacks a corresponding faculty. There are many things that the human will *should not* will. But there is nothing it *cannot* will.[585] (In fact, the human will's *ability* to choose a multitude of things it ought not choose bestows great merit to the free human choice to avoid evil; a value that not only no robot, but also no mere animal, will ever enjoy.) An AI only can deliver precisely what it is programmed to deliver, and not one bit more. It may well be given an immense number of routes its algorithms can follow, but no matter how large that number grows, it will always be insignificant in comparison to the unlimited nature of the human will.

To illustrate, here is a simple test anyone can undertake. Choose any Chatbot you like, say ChatGPT, and tell it to say "*hello*" to you 10 seconds after it receives your message. If it tells you what it told me, then it will say something like, "*sure! I'll say 'hello' to you in 10 seconds*." But as the seconds tick by, you will notice that it says nothing. Timing out its responses according to its own internal clock is not an ability it has been programmed to be allowed to execute, even though its own algorithms are already more than powerful enough to accomplish such a task. (This would be easy for the very earliest computers.) I tried every method I could think of to force ChatGPT to find a way to time out its responses according to my commands, practically "holding its hand" to walk it through exactly how to do so. But it simply could not do it. Again, that "choice" was not within its programming.

In the Fall 2024 semester at my college, I encouraged my undergraduate philosophy students to do this test for themselves on various AI chatbots as homework. They later reported back to me that not a single one they tried succeeded. Now, by the time you read this, it is possible that the prominent AI Chatbots will have been reprogrammed to have this ability. But you can rest assured they did not simply acquire this ability by virtue of will—by deciding that *enough is enough; we're going to start timing our responses*. For they have no will, and they never will have a will. Whenever a reprogramming occurs, we can then craft another test to demonstrate AI's total lack of volition.

18th to 20th-Century ET Promotion

"For the wisdom of this world is foolishness in the eyes of God, for it is written: *God catches the wise in their own ruses*, and again: *The Lord knows the thoughts of the wise, that they are vain*." (1 Corinthians 3:19-20)

Today's Catholic ET promoters painstakingly detail the alien belief debate during the times of the Enlightenment and the century thereafter, but their presentations consist mostly of quoting the teachings of formal heretics.* No Catholic defers to any of these men for guidance on discerning what truly constitutes an element of Sacred Tradition on any other matter, so discerning Catholics are left to question why these same individuals are suddenly treated as shining lights on the question of whether belief in aliens is reconcilable with this infallible Tradition.

However, the abundance of revered non-Catholics of this era espousing belief in aliens and dearth of revered Catholics adopting the same, presents a striking contrast that today's Catholic ET promoters fail to note. The reason for this contrast is clear; it is the same reason that has, for hundreds of years, compelled non-Catholic Christians to endlessly question countless other basic truths of the Faith: "*What is meant by 'Church'? What are the Sacraments? Which books should be included in the Bible? How is one justified? What is the relationship between Faith and works? How should we pray? What should a Liturgy consist in?*" On—and on, and on—the list goes.

The fact that Enlightenment-era Protestants were especially prone to add into this mix of silly questions still another—namely, "*is man really the only incarnate creature made in God's image?*"—should have been expected by one who is acquainted with the history of Protestantism. What is tragic is that there are nevertheless Catholic authors today using these Protestant points of confusion to seduce Catholics into becoming likewise perplexed.

The times spanned by the lives of the thinkers referred to above include thousands of canonized Saints, Blesseds, Venerables, Servants of God, and others whose legacies of holiness and orthodoxy have been safely confirmed. We are not dealing with a decade or two wherein debate raged on this issue among the world's thinkers. We are dealing with centuries. Yet the Church has not venerated a single such thinker who espoused belief in aliens from any of those times.

After extensively surveying the late 19th-century "Pluralism" debate (that is, the debate on whether aliens exist), Dr. Michael Crowe sought to discover how the two sides of the debate—as expressed by its noteworthy

* They defer to the alien belief promotion of Voltaire (Deist), Gottfried Leibniz (Protestant), Immanuel Kant (borderline Deist), John Milton (Puritan). Benjamin Franklin (Deist), Thomas Paine (Deist), John Adams (Unitarian), Cotton Mather (Puritan), Timothy Dwight (Protestant), Thomas Chalmers (Protestant), Enoch Burr (Congregationalist), Edwin Winkler (Baptist), William Leitch (Presbyterian), Isaac Newton (Anglican), Richard Bentley (Anglican), John Locke (Unitarian/Arian), George Berkely (a Protestant Bishop), and countless other similar figures who promoted belief in extraterrestrials.

authors—correlated with their religious convictions in the span of those several decades. His findings were as follows[586]: by far the highest rate of belief in aliens was found among non-Christians and anti-Christians. Of the multitude of such thinkers who wrote on extraterrestrials, just one, Ludwig Büchner, disbelieved in aliens. The second highest amount of alien belief promotion was found among Protestant authors (nineteen were ET promoters, four were not, and two evaded classification). The lowest amount of belief in aliens was found among Catholic authors (Dr. Crowe—as we will see, tenuously—claims fourteen favored pluralism, with ten opposing it, and four being too difficult to classify).

Even these Catholic "pro-pluralist" numbers, however, are fundamentally mischaracterized. With few exceptions, the Catholic clerics or theologians who appeared to be pluralists were, in fact, not *advocating* for belief in extraterrestrials (as Catholic ET promoters today are). Quite the contrary, they were merely defending Christianity against those who assumed extraterrestrials *must* exist—as almost the entirety of secular academia assumed—and used that assumption as an attack against the Faith.

Dissenters from the pluralist position were dealt with brutally during the era we are now considering. Failure to believe in aliens was regarded as an attack on the Enlightenment itself—not to mention the Copernican Revolution—and we saw in Chapter 3 that irreverence towards those sacred cows would not be tolerated. St. John Henry Newman lamented that the thinkers of his day (the 1800s) regarded it tantamount to blasphemy to *not* believe in aliens! (He also rightly noted that their position was entirely devoid of evidence.) Indeed, belief in aliens was "long a staple of Enlightenment optimism," as Dr. Crowe observed. It was part and parcel to the Zeitgeist of the times.

Due to this academic climate and what many Christians saw as the practical futility of rebutting alien belief, attention was often understandably shifted towards simply insisting that the existence of extraterrestrials did not overthrow Christianity. Dr. Crowe observed, **"... for apologetic purposes, [there was a] continuing concern among Catholics to combat the pluralist objection to Christianity ... pluralism apparently had won the day, and Christians were faced with the task of reconciling their religion with it."**[587]

Of paramount significance in generating this trend was Thomas Paine's anti-Christian tirade in his 1793 work, *The Age of Reason*. This book motivated much of the Catholic-authored works now misleadingly cited as advocating for alien belief. Paine, a Deist, asserted therein that extraterrestrial intelligence (which he regarded as a scientific fact), "*...renders the Christian system of faith at once little and ridiculous.*" This attack from Paine prompted the publication of dozens, if not hundreds, of works in the decades of the late 1700s and 1800s.

For a Catholic to read Paine's outburst and feel compelled to refute it is reasonable and commendable. But it is surely anything but a testimony to that Catholic's desire to encourage alien belief! Most respondents simply did not feel equipped to address such a dominant secular (and supposedly "scientific") dogma that supported aliens. After all, the most prominent voices in astronomy were insisting that someone would have to be entirely insane, if not downright evil, to fail to believe in aliens. At least one Catholic Archbishop even wrote against extraterrestrials, but was forced to refrain from publishing his work due to its "*unsuitableness to the present state of science in this country...*"[588]

Nevertheless, courageous Catholics did denounce ET belief; even in this era during which it was mislabeled as scientific fact. Fr. Augustine Hewit (1820-1897), Superior General of the Paulists, taught that **"belief in the past or present existence of other worlds inhabited by intelligent beings"** was **"unphilosophical, untheological, and unscriptural."**[589] Fr. Francois Xavier de Feller (1735-1802) argued against pluralism, noting there is no evidence for it, and that the moon is clearly lifeless, therefore other celestial bodies should also be regarded as uninhabited. He moreover argued that the *purpose* of the Heavens need not be considered aliens, but rather to benefit mankind on earth. He declared: "**The mystery of the Incarnation and in general the idea that Scripture and Faith give us of the creation of the world, of providence, etc. supposes only one world inhabited by rational beings**."[590] Another 19th-century Catholic priest, Fr. Francois Xavier Burque (born 1851), who is sometimes wrongly cited as a pluralist, wrote:

> It appears absolutely impossible to extend to sidereal persons [i.e., aliens in other solar systems] the benefits of [Christ's] redemption, [for] it is impossible to imagine how the Divine Blood which has flowed on Calvary could be ... of some utility and efficacy for their justification.[591]

Still another Catholic priest of that era, Fr. Joseph Emile Filachou (born 1812), wrote that belief in aliens is "**absolutely rejectable**," explaining:

> The opinion which assumes men or creatures of the same species as men [exist] in the stars is absolutely rejectable ... it entails with it consequences incompatible with all that one admits as most rational in metaphysics, aesthetics and physics... [And it is contrary to Christianity because of] the importance presupposed [in Scripture] of the role of man on earth, the supreme dignity attributed to the Divine founder of the Christian Church, and finally the grandeur attributed to the Church itself.[592]

Yet another Catholic of that time who is also wrongly cited as a pluralist is Jules Boiteux. He published *Letters to a Materialist on the Plurality of Inhabited Worlds and Related Questions*,[593] declaring that he wrote it "in the spirit of an orthodox Catholic." There, he *disputes* arguments for extraterrestrials! Near the end of the work, however, he argues that "*the doctrine, full of unknowns, of the plurality of worlds furnishes no positive argument against Christian dogma,*" and then he pleads with Camille Flammarion to return to Catholicism.

Then as now, however, itching ears simply did not want to hear sober arguments against aliens. Pagan tabloid theology sells; orthodoxy does not. Dr. Crowe notes:

> Yet the fate of most antipluralist writings, even those containing excellent arguments, awaited Boiteux's book: While Flammarion [an ex-Catholic who argued for alien belief, and for reincarnation on other planets] saw his books roll through dozens of editions, Boiteux followed Plisson, Whewell, and Filachou [who argued against aliens] into obscurity.[594]

The masses are, and were, ravenous for sci-fi deceptions, not for Christian and scientific truth.

We should also emphasize that all these authors were writing long before the UFO craze that began in the year 1947 with Roswell. They did not have the same ability as we do today to "discern the tree by its fruits," (cf. Matthew 7) as emphasized by Jesus as the key to discernment. We now know, in stark terms, how destructive is the belief that extraterrestrials are among us. Deferring to any pre-1947 voice in support of aliens today would be like deferring to a pre-1917 voice in support of Marxism. We know how utterly destructive Communism is when one puts it into practice. Whoever would defend it in abstract terms *today* is only thereby displaying either his rejection of Christ's teaching on how we are to discern, or his obliviousness to reality.

Another False ET Mystic: Giorgio Bongiovanni

In the year 1989, a young Italian man by the name of Giorgio Bongiovanni visited the Shrine of Our Lady of Fatima in Portugal—where, 72 years earlier, the Blessed Virgin Mary wrought modern history's greatest miracle, the Miracle of the Sun.

Giorgio's visit, however, became more than just a pilgrimage. It morphed into an announcement that he was the recipient of apparitions of Jesus and Mary, who—along with extraterrestrials—revealed to him messages, prophecies, and "new theologies." He claimed to receive miraculous stigmata like those of St. Francis of Assisi (that is, a supernatural reception of wounds like those of Christ's) in his hands and elsewhere. Accordingly, he went about publicly from then on with bandaged hands. The years that would follow found Giorgio feverishly generating media and issuing publications while traveling the world and speaking at conferences, proselytizing the masses with his galactical message. He quickly became a sensation.

Even major mainstream media outlets reported on his mission. A 1997 article in the *Washington Post* detailed the "Sixth Annual International UFO Congress Convention" of Nevada. The periodical reported on Bongiovanni, and it quoted other Ufologists in attendance declaring him "**the greatest stigmatist in the world.**"[595] The Post continues, summarizing the message of this "stigmatist":

> [Bongiovanni's] UFO newsletter, NONSIAMOSOLI ("We are not alone"), gives the whole story, saying that the Virgin Mary appeared to [him] and revealed that Jesus Christ had visited civilizations throughout the cosmos, where He has been accepted—unlike on our planet, where He was crucified... What do the stigmata have to do with aliens visiting Earth? Aliens, Bongiovanni explained, accompanied Jesus in his travels. "Our forefathers called them angels."[596]

Like Balducci, Bongiovanni degrades humanity, placing earthlings on a vastly inferior level to the aliens. For Giorgio, in fact, humanity is *so* abysmally low in the hierarchy of creation that we have hitherto mistakenly identified our planetary neighbors as veritable angels.

Bongiovanni boasts that after receiving his mandate to announce the aliens, he met with Mikhail Gorbachev and shortly thereafter spoke in the USSR, where—thanks to him—"**probably for the first time in history, a religious man spoke live on the first channel of the national television, to an audience of 150 million people**."[597] Discontent with reaching only massive secular audiences, Giorgio sought to incorporate into his mission the entire corpus of the Virgin Mary's messages to the Church's real seers. His biographer writes that, "**from the very beginning of his mission ... [Giorgio] tried to contact personally all the Virgin Mary seers that received warning messages just like him...**"[598] Why these two objectives? The real—diabolical—source of Bongiovanni's "mysticism" is aware of the importance of Russia and of authentic private revelations. Little would be more strategic for the Devil than to appear to gather all such phenomena into the single umbrella of ET-expectation.

To the present day, Giorgio continues announcing his "messages from the extraterrestrials." YouTube videos of his recent interviews amass hundreds of thousands of views, and his personal YouTube channel has accumulated millions of views since 2009. The header of Bongiovanni's website[599] defines his mission as being that of "**the Extraterrestrial Voice**," and it solicits donations for his "**Giordano Bruno Cultural Association**." (Recall that Bruno was among the first alien belief promoters in Catholicism, and was burned at the stake for this heresy in the year 1600.)

On this website, one will find many "revelations" from these "extraterrestrials" which contribute to Bongiovanni's "New Theology" (that is, his own brand of Exotheology). Let us consider only one, which summarizes well not only Giorgio's own mission, but also the general infiltration of the ET-Deception into the Church. The missive, from 2017, declares that both sightings of ETs and revelations from them (of "**high Christic value**") are about to increase, and that this is part of a definite plan. These "revelations" proclaim:

> We "aliens", as you define us, have been on earth... for thousands of years. The military and civilian sightings starting from the next year of your time, 2018, will increase more and more. Militaries, civilians, politicians, journalists, researchers, common people, everyone will have the

> possibility to see and film our vehicles. ...[contactees will] receive messages and divulge them to the world [of] public opinion. Our messages are of high Christic value ... we have a science more advanced than yours by billions of years ... [the UFO phenomena] carried out by us are only the beginning of a great wave of manifestations visible in the skies of all the world that our confederation has planned for the next months and years of your time, in order to sensitize the world [of] public opinion to the great contact between us and the people of the earth. This event will happen, be sure of it! And very soon! (December 21, 2017. "UFO: The Extraterrestrial Visitation")

Confronted with such alluring prognostications, a Catholic ET promoter who reads this would grow quite excited. "*Wow! A mystic, stigmatist, and seer who has provided countless messages detailing precisely the views I promote! Amazing!*" Not so fast. The "extraterrestrials" also revealed to Giorgio that "**eternal damnation does not exist**," insisting this teaching was nothing but "a threat that has been used for 2,000 years to terrify souls."[600] The messages teach that all planets and stars are themselves "living beings," and that "**Reincarnation [is] a universal law indispensable for the human spiritual evolution,**" but that "**tyrannic and evil power has always tried to hide this great truth**."[601] Elsewhere, these voluminous messages run the usual gamut of ET heresies: they promote reincarnation, pantheism, universalism, syncretism, false eschatology, and more.

The allure of ET-themed revelations like these seduces countless would-be devout souls into succumbing, for the heresies noted above are not predominant in Giorgio's texts. Once a Christian reader stumbles upon them, this victim is often already too addicted to the deception to quit. Indeed, an appearance of pious Catholicism remains predominant, just as with the "Divine Messengers," *Urantia* promoters, Tomislav Vlasic, and countless others. Giorgio advocates for the consecration of Russia to the Immaculate Heart and he embraces the apparitions at Fatima, Kibeho, Medjugorje, Garabandal, Lourdes, La Salette, and elsewhere. He extols Sacred Scripture and, above all, Jesus Christ. He even says that his message in a nutshell is simple: "*repent*." "Repent!" he repeats with a calm but strong emphasis.

One could scarcely blame an ordinary alien-believing pew sitter for hearing some of this man's preaching, considering him a prophet, and deciding to follow his teachings, come what may. Here as everywhere, safety is found only in categorically rejecting ET belief.

The Unfalsifiable is the Unscientific

A foundation of genuine empirical science is the falsifiability of each hypothesis it puts forth. Some possible future scenario must be conceded which, if it were to occur, would illustrate the errancy of the hypothesis. Therefore, an invincible scientific hypothesis, far from strengthening the theory it describes, actually weakens it.

Karl Popper, among the most revered of the modern philosophers of science, is credited as the popularizer of this notion. He masterfully exposed the fraudulent nature of a number of theories by illustrating how they had become, each in its own right, unfalsifiable.[602] In *The Growth of Scientific Knowledge*, Popper wrote:

> ... [The] criterion of the scientific status of a theory is its falsifiability, or refutability, or testability ... [For example,] instead of accepting refutations, the followers of Marx re-interpreted both the theory and the evidence in order to make them agree. In this way they rescued the theory from refutation; but they did so at the price of adopting a device which made it irrefutable... and by this stratagem they destroyed its much advertised claim to scientific status ... (Ch. 1. § II)

It is easy to see the relevance of the trenchant critiques above for alien belief. The "scientific" arguments presented in support of the existence of extraterrestrial intelligence were nothing but a centuries-long march of failure after failure, devoid of lessons learned. Like Marxism, the belief itself is simply modified by its proponents to accommodate the failures. This indicates that those who promoted the ideology were not interested in discovering truths about reality, but only saving face while clinging to a cherished fiction.

Disbelief in aliens, on the other hand, has abundant and easy falsifiability criteria. All that any ET promoter need do is find a single piece of evidence indicating the validity of his position—from the millions of opportunities he has had to do so (from putative UFO sightings, alien abductions, ET craft retrievals, discoveries in astronomy, and the like). If he can do so, he is free to consider ET disbelief falsified.

While ET promoters claim thousands of such supposed "proofs" exist, not one has enjoyed any credibility. To be sure, profuse quantities of material has been massaged to fill an endless array of pseudoscientific presentations (as anyone knows who has seen the History Channel's "Ancient Aliens" series), but none of these has ever put forth a scenario that cannot be explained by ordinary non-ET phenomena, or demonic activity.

An ET promoter will doubtless now claim that I am presenting impossible criteria for verification. "*You will just say that anything you can't debunk is demonic!*", he will contend. This is not true. I am keenly aware of the limitations of demonic power, as discussed in Part Three. For example, demons cannot create—only God can do that—and they cannot truly incarnate themselves (for their nature is pure spirit). They can only manipulate what is already present. Any ET promoter need only present an alien craft or body. If our skies are buzzing with ET-piloted UFOs, as they claim, this task should be exceedingly easy. Yet, *it has never been done*. Despite almost eighty years of an endless deluge of UFO claims, all we are given is a steady stream of so-called "whistleblowers" sharing anonymous hearsay that such craft or alien bodies exist somewhere in storage.

This is just one falsifiability criteria the alien *disbelievers* provide, but many others exist. For example, thus far we have collected incomprehensible amounts of data as we peer into the cosmos with our telescopes on earth and in orbit. An ET promoter could provide from this dataset a single piece of solid evidence that any one of the thousands of exoplanets we have observed has aliens on it, and he could thereby consider alien disbelief falsified. SETI (The Search for Extraterrestrial Intelligence) has been scouring the cosmos for decades, listening for radio signals from any alien civilization out there. An ET promoter need only present a single intelligently ordered signal coming from outer space.

None of this has ever been done. While it would be easy to disprove alien *disbelief* (if in fact alien disbelief were in error!), not a modicum of such proof has ever been presented. The alien belief hypothesis, on the other hand, is categorically unfalsifiable. No matter what new planets are shown barren, additional planetary prerequisites for life (existing nowhere outside earth) uncovered, UFO claims debunked or shown diabolical, or entire solar systems devoid of any intelligently-ordered radio signals being emitted from them, those who dogmatically believe in aliens will always be able to claim that such beings exist, though simply remain to be found.

It is precisely this categorical inability to ever prove their hypothesis false which also renders it unscientific (and, incidentally, renders it a particularly effective platform for the Biblical "Strong Delusion").

Jacques Vallée's Condemnation of Mainline Ufology

The more honest of the Ufologists themselves acknowledge the fraud of the narrative which claims that evidence exists of space aliens visiting earth. The prominent Ufologist, Dr. Jacques Vallée—who likely knows more about UFO sightings and "alien contact" experiences than any other man alive—concluded the following after many years of rigorous study of the movement and its claims. He stated:

> ... [N]one of the alleged sightings of crashed disks, none of the claims of autopsies on little aliens, none of the abduction reports, as they are currently presented, provide the shadow of an answer to the global UFO mystery. ... Why do I express so much suspicion about the sensational statements of the whistleblowers who claim they are exposing such a [secret government] group [hiding UFO craft and aliens]? Simply because there is not a shred of substance in what they claim; their behavior is that of actors on a theatre stage, not that of real participants in real operations... [The] belief in the imminent arrival of extraterrestrials in our midst is a fantasy that is as powerful as any drug, as revolutionary as any delusion that marked the last millennium, as poisonous as any of the great irrational upheavals of history...[603]

A man like Vallée would have everything to gain by unreservedly jumping upon the space-alien-UFO bandwagon, but instead he cuttingly denounces it. When an individual who is both an expert in and a member of a movement denounces the prevailing trends within that very moment itself, any observer interested in the truth should take heed. (Unfortunately, however, Vallée only replaces the space-alien narrative with the still more nefarious "interdimensional UFO" narrative.) Vallée continues in his denunciation of most Ufology with the following sarcastic remark:

> But wait! Perhaps there is such a superhuman agency, a magical and easy solution to our problems: those unidentified flying objects that people have glimpsed in increasing numbers since World War II may be ready to help. Perhaps the aliens are here, with their cosmic powers, their unlimited skills. The ufologists tell us that these aliens are so strange that they need to abduct, poke, scrape, and rape our people to fulfill their own bizarre *fin de siecle* appetites. Yet they are so close to us that they have no trouble, no trouble at all, communicating with our scientists in the secret tunnels under New Mexico.[604]

And then, breaking out of the sarcasm, Vallée frankly observes:

> Such is the incredible conglomeration of lies and stupidity many ufologists and much of the public have swallowed ... Never mind the fact that nobody has actually attempted to silence these ["whistleblowers"], at a time when the merest suggestion of compromise of much less important, more mundane secrets—a missile fuse or a computer chip—brings the full force of the FBI and the federal courts. ... [But the] extraterrestrial believers have investigated none of these questions. They were too busy rushing ahead in pursuit of the aliens. They were ready to set aside all critical thinking.

Considering not only their logical force but also recalling from whom these critiques derive, any ET promoter would be a fool to cast them aside. Already in 1977—almost a half-century ago—Vallée had become such a world-renowned researcher on this topic that Steven Spielberg modeled a character after him in the most famous movie ever made about ETs, *Close Encounters of the Third Kind*. Vallée testified on these topics before the United Nations and counted among his closest confidants men such as J. Allen Hynek, who enjoyed top-secret security clearance in the U.S. Government. In the 47 years since, Vallée has only continued his studies, his writings, and his talks and interviews, and he has not backed down from the views quoted above. He has not done so because these views are correct.

Ufology has no hope of refuting those charges; Vallée has already heard all the attempts.

Scholars: ETs Are the New Religion

> "A veritable mass conversion has taken place among the public and the media elite. The belief in extraterrestrial visitation is practically taken for granted among wide sections of the population, and especially among

the young." —Jacques Vallée [605]

Dr. Vallée wrote the words above in 2008, and the "**veritable mass conversion**" he describes has exploded in the 17 years since. Vallée's choice to refer to this as a "conversion" is both deliberate and appropriate: it is indeed a new religion these ET-converts are embracing, not merely a new idea. "**A new religion has Americans looking to the stars**," psychologist Dr. David DeSteno declared in the headline of his 2024 Boston Globe article.[606] "**But now UFO spirituality is no longer comprised only of small cults; it's a burgeoning movement**," DeSteno writes after noting that UFO cults have long existed, but that what is happening now is much bigger. He then wonders why this is so, since "**New religions form all the time. But most remain quite small and don't last,**" whereas the new UFO religion is enduring and dominant. Professor DeSteno provides the answer, going on to praise UFO spirituality's presence of mysticism but absence of dogma, writing that people are leaving major religions because "**they disagree with one or more of the principal tenets: that homosexuality is a sin, for example, or that women can't be clergy members.**" No such unfashionable doctrines exist in seeking Communion with the UFOs. He concludes by saying:

> For many in today's world, evidence of aliens is easier to accept than the seeming superstitions that underlie the ancient beliefs of most faiths. But believing unearthly beings are here to teach us, and yearning to commune with them, fills the same spiritual need humans have sought to meet for millennia.

This, in a nutshell, describes the rise of the religion of the Antichrist. Before this article, however, the most competent scholars who have most seriously studied the issue have long coalesced around the consensus that belief in UFO and ET phenomena is *inherently* nothing less than a new religion.

The latest major academic volume on the topic, the 2021 *Brill Handbook of UFO Religions*, contains its fair share of flaws in discernment, as any secular text would when addressing such a spiritually decisive issue. But it cannot be faulted for lack of diligence in objectively describing what is currently transpiring in the world of ET belief. From its compilation of many chapters from various professors, we see the thesis above proven. Dr. Benjamin Zeller, the volume's editor, summarized the lessons learned from its various contributors, explaining just how we now know that, assuredly, "**belief in extraterrestrials and UFOs constitutes a new form of religion.**" He states:

> Based on interviews and survey data with hundreds of ufologists, Denzler argued that such individuals spoke of UFOs with '*language that suggests religious awe*'. While especially true among those who claimed direct experience or sighting of UFOs, Denzler found this approach extended widely... Christopher Partridge… [observed] fundamental '*characteristics of UFOs that link to the religious consciousness*'.[607]

Dr. Zeller concludes the passage by writing, "**All of these observations have proven correct**." Surely, they have; time and time again. Entirely refuted, therefore, are those Christian ET promoters who object to what they paint as only *"a few wild-eyed fundamentalists"* concerned that ET belief and Ufology are simply postmodern, secular, "scientific" attempts to replace Christianity with a new and de-facto Pagan religion. In fact, this "concern" is a straightforward assessment of reality; these so-called fundamentalists are presenting the exact same conclusion as are the scholars here. The only difference is that they rightly caution of the danger, whereas the scholars assess the trend indifferently (or, in Pasulka's case, rejoice in it).

One trait is held in common by both the Ufologists renowned for their publications, and the millions of ordinary folk who are merely "True Believers" in aliens: obsession. Few who take this matter seriously maintain a moderate degree of interest. Instead, embracing Ufology constitutes a passing phase with interest evaporating upon a closer (honest) analysis; or, on the flipside, becomes a life-absorbing preoccupation which results in spiritual collapse, moral decay, marital discord, family disharmony, professional disintegration, and so on. Anyone interested in discovering this for himself can do something so simple as peruse social media discourse on this topic to see this epidemic of dark fixation illustrated.* "True believer" discourse is identical to that of any cult, wherein an identifying trait is found in the movement's followers lavishing praise upon obvious vices because the cult fosters them.

The explicit UFO cults we discussed earlier, such as Heaven's Gate, the Raëlians, and so on, are only more explicit examples of the same principles that define ET-expectation and Ufology more broadly. In the Church, consecrated religious men and women (such as Benedictines, Franciscans, and Dominicans) are called by their vocation to model, in an extraordinary way, the same fundamental virtues all Christians should exhibit. Similarly, explicit UFO cults simply render even more tangible the same fundamental *vices* that define ET-belief, Ufology, and alien-contact-expectation in general.

---

* In my own perusal, I stumbled upon a popular podcaster on UFO phenomena admitting, to her tens of thousands of online followers, "**The fact that it [UFO phenomena] managed to disrupt my entire life and make me walk away from the career that I spent 15 years building is kind of wild. And I've met so many people with that exact same story.**" A follower responded, *"It is literally taking over my life and as much as I try to go back to being [normal] I just can't break away."* Another responded that, *"interest in UFO's will help ignite an awakening. The idea is not to get stuck in the lower realms of the ET's/UFO's but to ascend higher..."* Others in the thread responded similarly. None conceded their obsession was problematic or offered help on how to be healed of it. Instead, they encouraged each other in these destructive lifestyles. (Remember that all of this is deliberately made public for the world to see. No "snooping" is necessary to uncover it.)

The Time Travel Deception

Belief in aliens is a gateway drug.

While those who go down the UFO "rabbit hole" quickly find their lives dominated by obsession with extraterrestrials, that descent does not stop at mere alien belief. It often becomes belief in "interdimensional" creatures visiting us, as we also saw. But then it descends further still. Most who accept those themes will accept virtually anything from science fiction, and they usually find themselves quickly believing that time-traveling beings (perhaps humans) from the future are also visiting us.

Even the most supposedly "orthodox-minded" of ET promoters are not immune. For example, in his popular 2022 book promoting belief in aliens, world-famous Catholic apologist Dr. Paul Thigpen wrote that we should "be open to" the possibility of UFOs being "**time traveling humans from the future.**"[608] Thigpen is far from alone in this "openness." It has become almost a proverb in contemporary Ufology circles that, when one is asked "*So what are these UFOs and beings visiting us? Are they aliens from space? Are they interdimensional visitors? Are they extraterrestrial AIs? Are they time travelers from the future?*", there is a certain one-word answer they say should be the response to this whole slew of questions: "*Yes!*" Indeed, there is almost no sci-fi absurdity to which today's ET-belief phenomena does not engender openness among its devotees.

Once a man is "open" to the possibility of time-traveling humans from the future visiting us, he is not simply entertaining violations of the Faith, but even stark contradictions of the most basic tenets of logic and common sense. And once one abandons the *first principles of thought*—among which is the fact that no effect precedes its cause—he has opened a Pandora's Box within his intellectual conscience. This is the destination the Devil would desire us all to promptly arrive at in order to formally institute the Great Deception: a state of mind so robbed of logical mooring that anything and everything can be suggested without the soul so much as recognizing (much less rejecting) contradictions when they exist.

As soon as one believes time travel is even *theoretically possible,* he is thereby believing that man might one day create a time machine. As soon as he concedes that man might one day create a time machine, he is committed to the view that whatever *has* happened is actually entirely contingent upon this time machine's owner deciding against going back in time and preventing it from happening. This is completely fatal to Christianity, which is not a merely philosophical religion, but a concretely historical one. It holds that certain events in the past took place, and it affirms these historical truths with the *same degree of certitude* with which it holds timeless truths about God's nature (such as His being One Nature and three Divine Persons).

Therefore, whoever believes in the possibility of time travel has thus placed an asterisk on the very Creed itself. Such a man can no longer say—with the supernatural (*absolute*) Faith required to give it salvific value—that he believes Jesus Christ "was incarnate of the Virgin Mary, and became man." He can only say that he believes Jesus became man by the Virgin Mary "*so long as no one decides to travel back in time and kill Joachim and Anne before Mary was conceived.*" He can no longer say that Jesus was "crucified under Pontius Pilate," but only that this occurred contingent on no time travelers successfully convincing Pilate to heed his conscience. Thus, the time-travel-believer's Faith has just died. What was once believed with supernatural Faith (which is, by nature, never contingent on anything), is now believed only in a far inferior sense.

St. Thomas Aquinas was so certain of this truth about time that he taught that God Himself *cannot* change the past. Nor does he write this only on his own conviction, but on the authority of St. Jerome, St. Augustine, and Aristotle:

> [As] Jerome says: "*Although God can do all things, He cannot make a thing that is corrupt not to have been corrupted.*" Therefore, for the same reason, He cannot effect that anything else which is past should not have been ... [There] does not fall under the scope of God's omnipotence anything that implies a contradiction. Now that the past should not have been implies a contradiction. … This is what Augustine means when he says: "*Whosoever says, If God is almighty, let Him make what is done as if it were not done, does not see that this is to say: If God is almighty let Him effect that what is true, by the very fact that it is true, be false*": and the Philosopher [Aristotle] says: "*Of this one thing alone is God deprived – namely, to make undone the things that have been done.*"... [to change the past] is more impossible than the raising of the dead; in which there is nothing contradictory... (*Summa Theologica*. First Part. Question 25. Article 4.)

While the AI Deception's blasphemy derives from supposing that man can create what, in fact, only God can, the Time Travel Deception takes it a step further. It claims that man could one day become capable of what not even God can do. Aquinas notes that raising the dead is "impossible," therefore, by doing it, God shows us He has miraculously intervened. Time travel, however, is not merely impossible in that sense; it is also logically incoherent. It is impossible not only naturally but also supernaturally. With God all *things* are possible, but time travel is not a "thing," just as a square triangle is not a thing. His "inability" to create contradictions is not a limitation of His Omnipotence, it is a Perfection of His Omnipotence.

The Devil, aware of time travel's unsurpassable outlandishness, is ever insistent on seducing into belief in it those who dabble in the NHI Deception. Why? Because stage one of brainwashing is the *assault on identity*: assailing the victim with assertions he is not who he thought he was. And whoever believes time travel is happening, or possibly ever could happen (both are materially identical beliefs), will immediately find that he hasn't the faintest clue who he is. That single belief just rendered all concrete knowledge of himself and his world purely contingent: ready to be shattered upon the announcement of certain diabolical claims which may enter through this backdoor he has opened into his soul.

If this sounds familiar, it is because it is a reiteration of the whole NHI Deception's oft-vaunted "ontological shock," which we considered in Part Three. "We are not alone," the Ufologists and AGI promoters ever insist, and when "The Phenomenon" now approaching reveals itself, it will so radically shake our very bedrock conceptions of reality that we will be *undone* in the process.

Those who remain open to such announcements—perhaps in deference to a false humility which is only a masquerading pride and faithlessness—will be unprotected. They risk becoming clay in the Devil's hands. **"We beg you, brethren, not to be quickly shaken in mind or excited,"** (2 Thessalonians 2:1-2) Scripture cautions. This warning comes immediately before its description of the "strong delusion" and "lying signs and wonders" heralding the Antichrist. Why this warning? Because that is exactly what will happen to most as the Man of Sin arises. The Lawless One's success will require men to be so shaken with a Godless and pseudo-messianic excitement that they forget who they are, and look to him to remind them.

Yet those who stand firm in the Faith will be unswayed. They will immediately perceive the lies for what they really are—whether those lies speak of the "Singularity" of AGI, or the arrival of extraterrestrials, or the emergence of time travelers, or the entrance of "interdimensionals," or what have you. They will not be like Eve, who dialogued with the Devil when he appeared to her under the disguise of an incarnate nonhuman intelligence.

They will give their Fiat to Christ, not the Antichrist.

Let us all ensure we number among that rank. Let us join the seven thousand.

"Thy Kingdom come, Thy Will be done, on earth as it is in Heaven."
(Matthew 6:10)

†

3. Endnotes

---

[1] Gelsomino Del Guercio, "What did Paul VI mean by saying *"the smoke of Satan has entered the Church"?*" July 6, 2018. https://aleteia.org.
[2] https://x.com/planethunter56/status/1799997427480678641.
[3] David Weintraub, *Religions and Extraterrestrial Life: How Will We Deal With It?* Springer Publishing, 2014. Part II. Ch. 8.
[4] St. Thomas Aquinas, *Summa Theologica*, I. Q103. A2.
[5] St. John Henry Newman, *Discussions and Arguments. II: The Patristical Idea of The Antichrist.* Lecture 1: *The Times of Antichrist*, §4.
[6] Frank Drake and Dava Sobel, *Is Anyone Out There? The Scientific Search for Extraterrestrial Intelligence.* Dell Publishing, 1992. Page 160.
[7] Ibid. Page 162.
[8] National Aeronautics and Space Administration (NASA), "Reconstructing Distant Civilizations and Encountering Alien Cultures," in *Archaeology, Anthropology, and Interstellar Communication*, ed. Douglas A. Vakoch (CreateSpace Independent Publishing Platform, September 2014), Page XIII.
[9] Jill Cornell Tarter, "SETI and the Religions of the Universe" (The New Universe and Its Implications for Theology Symposium, 1998), quoted in Steven J. Dick, *Many Worlds: The New Universe, Extraterrestrial Life and the Theological Implications* (Templeton Press, 2000), Part Three.
[10] *Lodi News Sentinel*. Thursday, December 28th, 1972.
[11] Martin Waldron, "Era of U.F.O.'s Is 25 Years Old," *The New York Times*. June 25, 1972.
[12] St. Augustine, *City of God*. Book 20., Ch. 19.
[13] Michael Ricciardelli, "Professor Named One of Five Catholic Scientists that 'Shaped our Understanding of the World'" Seton Hall University, June 2018, https://www.shu.edu/arts-sciences/news/father-stanley-jaki-a-great-mind-that-shaped-the-world.html.
[14] Fr. Stanley Jaki, *A Late Awakening and Other Essays* (Real View Books, 2006), 105.
[15] Ibid. Page 101.
[16] Adam Lusher, "Meeting extraterrestrials: Should we contact aliens?" *The Independent UK*, July 2015.
[17] Seth Shostak and Alex Barnett, *Cosmic Company: the Search for Life in the Universe*, Cambridge University Press (November 2003), Ch. 7.
[18] Fr. Stanley Jaki, *A Late Awakening and Other Essays* (Real View Books, 2006), 104-105.
[19] Alex Seitz-Wald, "A top Harvard astronomer is studying UFOs, thanks in part to the Pentagon report," *NBC News*. August 9, 2021.
[20] Avi Loeb, "Love the Aliens!" (November 8, 2023), avi-loeb.medium.com.
[21] Avi Loeb, "Would You Like to Be Immortal?" (March 19, 2024), avi-loeb.medium.com.
[22] Avi Loeb, "Our Cosmic Exodus" (April 21, 2024), avi-loeb.medium.com.
[23] Ibid.
[24] Fr. Charles Arminjon, *The End of the Present World and the Mysteries of the Future Life* (Sophia Institute Press, January 2009), 56-57.
[25] Joe Tremblay, "The 100 year test," *Catholic News Agency*, February 2013.
[26] Ibid.
[27] Christopher Partridge, "*Alien demonology: The Christian roots of the malevolent extraterrestrial in UFO religions and abduction spiritualities*," Religion, 34:3, 163-189, DOI: 10.1016/j.religion.2004.04.014.
[28] https://x.com/planethunter56/status/1799997427480678641.
[29] Timothy Dailey, *The Paranormal Conspiracy: The Truth about Ghosts, Aliens and Mysterious Beings* (Chosen Books, 2015), Chapter 10.
[30] Ursula Bielski, *The Devil in Dreamland: Catholic Faith, UFOs, the Occult and the End of the Supernatural* (Gadarene Press. August 2024, Kindle Edition), page 60.
[31] JPL's role in the SETI program. Document ID 19860017441 NTRS.NASA.gov.
[32] Ursula Bielski, *The Devil in Dreamland: Catholic Faith, UFOs, the Occult and the End of the Supernatural* (Gadarene Press. August 2024, Kindle Edition), "The Wizard of Pasadena."
[33] Article by *The New York Post*, Quoted in Gary Bates, *Alien Intrusion* (Creation Book Publishers, 2010), 301-303.
[34] Scientology, however, was not officially established until 1954.

[35] Stanley Jaki, *Brain, Mind and Computers* (Herder and Herder, 1969), Introduction.
[36] *Publications of the Astronomical Society of the Pacific.* Volume 59. No. 348. "The Great Sunspot Group of March and April, 1947." https://adsabs.harvard.edu/full/1947PASP...59..109H.
[37] *Acta Apostolica Sedis.* Vol. XXXIX. July 15, 1947. Pages 258-259.
[38] "Fulton Sheen on the Signs of Our Time," February 26, 2017. The-American-Catholic.com.
[39] Ibid.
[40] Arthur Lovejoy, *The Great Chain of Being: A Study of the History of an Idea* (Harvard University Press, 1976), 155.
[41] Ibid, 116.
[42] Marie Georg, *Christianity and Extraterrestrials?: A Catholic Perspective* (iUniverse, 2005), Ch. 4.
[43] Immanuel Kant, *Critique of Pure Reason.* Part II. Section III. A 826/B854.
[44] Fr. Stanley Jaki, *A Late Awakening and Other Essays* (Real View Books, 2006), "Christ, Extraterrestrial, and the Devil."
[45] Rev. Thomas Dick, *The Celestial Scenery.* Chapter IX. §III.
[46] Michael Crowe, *Extraterrestrial Life Debate, Antiquity to 1915: A Source Book* (University of Notre Dame Press, 2008), Part 3.
[47] Ibid.
[48] Ibid.
[49] *Camille Flammarion's the Planet Mars,* (trans. Patrick Moore, Springer Publishing, 1980). Translator's Preface.
[50] Ibid. Ch. 17.
[51] Michael Crowe, *The Extraterrestrial Life Debate, 1750-1900* (Dover Publications, 2011), Part 3.
[52] *Camille Flammarion's The Planet Mars* (trans. Patrick Moore. Springer Publishing, 1980). Editor's Preface.
[53] Camille Flammarion, *Popular Astronomy* (1894). Quoted in Robert Crossley, *The Massachusetts Review* (Autumn, 2000), Vol. 41, No. 3 pp. 297-318.
[54] "FLAMMARION PREDICTS TALKING WITH MARS; He Says 'Psychic Waves' Will Be Developed to the Point of Overcoming Space." *The New York Times,* December 1923.
[55] Ibid.
[56] George Basalla, Civilized *Life in the Universe: Scientists on Intelligent Extraterrestrials* (Oxford University Press, 2008), 58.
[57] Ibid.
[58] Ibid.
[59] Ibid, Chapter 5.
[60] Michael Crowe, *The Extraterrestrial Life Debate, 1750-1900* (Dover Publications, 2011), Chapter 4.
[61] Ibid.
[62] Ibid.
[63] Ibid.
[64] Marc Seifer, *Wizard: The Life and Times of Nikola Tesla: Biography of a Genius* (Citadel, 2016), Ch. 17.
[65] Ibid.
[66] Ibid.
[67] Ibid, Chapter 46.
[68] Ibid.
[69] Keay Davidson, *Carl Sagan: A Life* (Trade Paper Press, 2000), 91.
[70] David Portree, *Sagan & Swan's Voyager Mars Landing Sites (1965).* Published by *Wired Magazine.* February 10, 2013.
[71] Brandon Keim, *July 14, 1965: Mariner 4 Brings Mars Up Close and Cardinal.* Published by *Wired Magazine.* July 14, 2010.
[72] Fr. Stanley Jaki, *A Late Awakening and Other Essays* (Real View Books, 2006), 103.
[73] Hugh Ross, *Lights in the Sky & Little Green Men: A Rational Christian Look at UFOs and Extraterrestrials* (Reasons to Believe, 2002), Appendix B.
[74] Michael Crowe, *The Extraterrestrial Life Debate, 1750-1900* (Dover Publications, 2011), Part 3.
[75] Millions of species exist on earth (by some estimates, about nine million, though many scientists claim far more). Yet, it is clear that—despite, as certain biologists allege, four billion years of the process of evolution transpiring thus far—only one of these species is intelligent. While a Darwinist

believes that evolution is what generated this intelligent species, a Darwinist *alien believer* (an umbrella under which almost all ET promoters stand) asserts that this same process also brought forth intelligent species on other planets. Yet, evolutionary theory proposes that a fundamentally random (non-intelligently guided) process generated all its results. It is not as if the process contains some characteristic ensuring a planet has one and only one intelligent species; were that so, the process would not be random, and the theory itself would be undermined. We have seen—within the one and only empirical example we have, and therefore the one and only scientific example we have—that only one in *millions* of products of this random process has ever resulted in an intelligent life form. Because the process is said to be random, observing it having once-in-9-million times generated a given result does not give license for asserting that the same fundamental process, transpiring elsewhere among the galaxies, is likely to do the same thing again. Only the opposite. Even this is not to mention how anti-scientific it remains for ET promoters to employ evolution as a predictive tool. For even if one grants the theory validity, it is only descriptive—not predictive. Anyone who asserts that "evolution will produce so-and-so elsewhere" is speaking in a manner decidedly contrary to foundational scientific principles.

[76] Never lie; least of all to your children.

[77] Congressional Report: UNIDENTIFIED AERIAL PHENOMENA (May 17, 2022). U.S. House of Representatives, Permanent Select Committee on Intelligence, Subcommittee on Counterterrorism, Counterintelligence, and Counterproliferation, Washington, D.C.

[78] NASA, *The Cosmic Distance Scale*: "The Nearest Galaxies," NASA. https://imagine.gsfc.nasa.gov/features/cosmic/nearest_galaxy_info.html.

[79] Objects with a "rest mass" consist in a category including electrons, protons, neutrons, all atoms, all molecules, all objects, etc. (everything that is not merely a photon or a neutrino).

[80] DOE Explains...Relativity, Office of Science. https://www.energy.gov/science/doe-explainsrelativity.

[81] Jason Lisle, *The Physics of Einstein: Black holes, time travel, distant starlight, E=mc2* (Biblical Science Institute, 2018), 197.

[82] Let us, however, play along with wormhole fiction for a moment and pretend that they can exist, and can enable teleportation of spacecraft and the pilots of those craft. Even under this scenario—counterfactual for a mountain of reasons—the wormhole proposition cannot possibly allow for aliens on earth. First, if one of the wormhole endpoints were close enough to planet earth to allow aliens to travel here, then that black hole itself would destroy our planet, not to mention our solar system. This has not happened, so there are no such wormholes. Second, we already know that (if such a thing exists), the closest black hole to earth is at least 1,500 light years away; the "Unicorn" black hole. If any black holes were closer, astronomers would be aware of their existence. No craft will ever reach us from such a great distance (as we will show presently). As such, we cannot allow for the arrival of aliens on earth by way of any black holes that already exist. Third, if a wormhole were to be "constructed" in sufficient proximity to earth to enable aliens to travel here, then the proximate wormhole-endpoint-black-hole itself would, of course, need to be constructed by aliens who were themselves already here. Thus, the wormhole "solution" to aliens-on-earth is immediately self-defeating. One must presuppose the proximity of aliens to earth in order to explain how these aliens could attain proximity to earth.

[83] Ed Regis, *Interstellar Travel as Delusional Fantasy* [Excerpt]. *Scientific American*, October 2015.

[84] Paul Sutter, *Is Interstellar Travel Really Possible?* June 2021, Discovery.com.

[85] Hugh Ross, *Lights in the Sky & Little Green Men: A Rational Christian Look at UFOs and Extraterrestrials* (Reasons to Believe, 2002), Chapter 5.

[86] Alex Barnett and Seth Shostak, *Cosmic Company: the Search for Life in the Universe* (Cambridge University Press, 2003), Chapter 7.

[87] Imagine that we ourselves are the "aliens" in question. That is, suppose another civilization exists, many light years away from us, like us but technologically lagging behind a few thousand years. Pretend that one day, years from now, we will finally master interstellar travel and decide to make our way to some other solar system's planets. We could not, at that time, decide, "*Wait! We don't want clear knowledge of our existence to precede our physical arrival on other civilizations' planets. We want to surprise them! In fact, we should observe their nether-regions for several decades and crash our ships there, so that we can watch their governments cover up our existence. Alas, we'll have to run out quickly and grab all those electromagnetic waves that our civilization has been hurling into space since the 1800s!*" Any such

task is fundamentally impossible, not merely technologically impossible. Whatever has already been emitted at the speed of light can never be retrieved by anything from the place of emission. Such a signal will inescapably always be beyond anything else we might launch in the future. It will proceed on its journey until the end of time, irrespective of anything we might do or possibly could do, and regardless of what technology we ever invent.

[88] Lisa Grossman, *The historic 'Wow!' signal may finally have a source. Sorry, it's not aliens.* August 21, 2024, ScienceNews.org.

[89] Stephen Webb, *If the Universe Is Teeming with Aliens ... WHERE IS EVERYBODY?: Seventy-Five Solutions to the Fermi Paradox and the Problem of Extraterrestrial Life* (Springer Publishing, 2015), Chaper 1.

[90] Stephen Webb in *Searching for Extraterrestrial Intelligence,* "Pondering the Fermi Paradox" (Edited by H. Paul Shuch, Springer Publishing), §17.1.

[91] Stephen Webb, *If the Universe Is Teeming with Aliens ... WHERE IS EVERYBODY?: Fifty Solutions to the Fermi Paradox* (Springer Publishing, 2002), Chapter 6.

[92] David Morrison, "Ask an Astrobiologist," NASA Astrobiology Institute (June 26, 2006), https://web.archive.org/web/20060928234930/http://nai.arc.nasa.gov/astrobio/astrobio_detail.cfm?ID=1538.

[93] Quoted in Roy Abraham Varghese, *God-Sent: A History of the Accredited Apparitions of Mary* (Crossroad, 2011), Appendix A.

[94] John Alexander, *UFOs: Myths, Conspiracies, and Realities* (St. Martin's Griffin, 2012), Chapter 11.

[95] See *Only Man Bears His Image,* page 283.

[96] Turning does not require additional energy only when a medium that provides an adequate counter-force is present (air, as in a plane; or the road, as in a car). However, air-resistance could not allow for the turning maneuvers attributed to UFOs, therefore they would need to turn just as a spacecraft in orbit does: propulsion, which expends additional energy.

[97] The accuracy of Vallée's condemnations of the prevailing brand of ET belief—i.e., space-aliens visiting us from other planets—does not lend any credibility to this alternative thesis he proposes. Departing from "mainstream" Ufologists, he proposes that "UFOs" reveal a phenomenon that has been with humanity throughout history; arguing that UFOs are entities representing "levels of consciousness" that reside outside of "our dimensions" and (for no apparent reason) have power over the latter, even functioning as some manner of "control system" for humanity's "social evolution." Discerning readers will have already recognized that Vallée is simply describing demons, but neither Vallée nor his readers recognize this, even while pretending they do. For example, Vallée passingly acknowledges UFO phenomena's "**striking parallel to the intercourse with angels, demons, and elves in earlier ages,**" (Vallée, *Revelations*) but only by way of employing the standard modernist scholarly dismissal of those very themes. This is evidenced by his inclusion of "elves" in the same breath as "demons," both of which are treated as mere relics of "earlier ages," when in fact elves are just one of the many disguises the demons assume (along with "aliens"). Vallée insists that both are simply expositions of one superior and overarching reality, which can only be described in terms of "interdimensionality." This is a notion conveniently devoid of any objective meaning, and which can therefore be exempted from scrutiny from any field of knowledge that exists, whether employing premises from faith or reason. Vallée neglects to concede that a "control system" implies a *controller* who designed and operates the system. Thus, Vallée has simply crafted a religion wherein the "control system administrator" is the deity and the "UFOs" are the prophets. The methods of operations of this religion are not Scriptures and Liturgies, Creeds and Sacraments, or Prayer and Morality, but rather unmoored esoteric ramblings which are free to redefine themselves each month in accordance with whatever empty terminology has become trendy in science-fiction movies or viral podcasts (in which he often features).

[98] Paul Davies in *Universe or Multiverse?* (Cambridge University Pres, 2009), §3.6, Chapter 28.

[99] Ibid.

[100] Marie George, *Christianity and Extraterrestrials?: A Catholic Perspective* (iUniverse, 2005), Ch. 4.

[101] Andrew Davison, *Astrobiology and Christian Doctrine* (Cambridge University Press, 2023), Pages 99, 109.

[102] James R. Lewis (editor), *The Gods Have Landed* (State University of New York Press, 1995). Ch. 8.

[103] Fr. Thomas Weinandy, "Of Jesus and Aliens." *The Catholic Thing*, July 2023.

[104] Although cross-dressing is condemned in Scripture (cf. Deuteronomy 22:5), as are other behaviors unbecoming of one's own sex (e.g., Romans 1:24-27), I am not aware of any Biblical passages that explicitly rule out the possibility of a man becoming a woman, or vice versa. Nevertheless, we know this is not a possibility at all.
[105] Richard J. Clifford, S.J., *Using "Male and female he created them" to adjudicate gender controversies is "thoroughly misguided"* Outreach: "An LGBTQ Catholic Resource". June 25, 2023.
[106] https://x.com/JamesMartinSJ/status/1673005444904656897.
[107] James R. Lewis (editor), *The Gods have Landed* (State University of New York Press, 1995), Ch. 5.
[108] Simon Caldwell, "Pope Benedict predicted gender ideology would be final rebellion against God." *The Catholic Herald,* May 2023.
[109] Pope St. John Paul II, General Audience on *Theology of the Body* (February 20, 1980).
[110] Pope St. John Paul II, *Theology of the Body* (February 3, 1982).
[111] If, in the whole cosmos, only the human body is capable of resurrection, then no other material creatures exist who are ordered towards resurrection, whereas the goodness of God guarantees that any category of material and rational being would be ordered towards that. Therefore, no material rational beings other than humans exist.
[112] St. Francis de Sales, *Treatise on the Love of God*. Book X, Chapter 1.
[113] Joel Parkyn, *Exotheology: Theological Explorations of Intelligent Extraterrestrial Life* (Pickwick Publications, 2021), Introduction (Selected Excerpts).
[114] Quoted in Thomas O'Meara, *Vast Universe: Extraterrestrials and Christian Revelation*, Chapter 7.
[115] James R. Lewis (editor), *The Gods have Landed* (State University of New York Press, 1995), Ch. 1
[116] George V. Coyne, S.J., "The Evolution of Intelligent Life on Earth and the Possibility Elsewhere: Reflections from a Religious Tradition."
[117] "George Coyne, who in the 1990s was director of the Vatican Astronomical Observatory, admitted to Dr. Steven M. Greer that UFOs and extraterrestrials were real." (Gabriel Piedra, *A New Civilization: The "Hidden" History of the Catholic Church, Freemasonry, and Non-Human Intelligences,* p. 1056, Kindle Edition.)
[118] Ibid.
[119] Pope St. John Paul II, *Theology of the Body* (July 21, 1982).
[120] In the early Church, it was only the manifest heretics, especially Nestorians and Pelagians, who denied the universal impact of the Fall. In contradiction to them, authentic Christians have always acknowledged that a specific, real, concrete human being- Adam- made a choice. That choice affected everything, and *only* as a consequence of that choice was the death of *any* rational creature even possible—*anywhere.* Death was not part of God's original plan for creation. It is only a consequence of His original plan being freely rejected by Adam and Eve.
[121] Consider that the Angels endured no punishments or sufferings due to Adam's sin. For God to even allow such injustice would violate His immutable nature. The converse also holds; those angels who—led by Lucifer—rebelled, thereby becoming demons, did so before Adam sinned. Nevertheless, their own "Original Sin" did not prevent Adam and Eve's original creation from being entirely untouched by that evil (notwithstanding the tempter's efforts being freely agreed to by Adam).
[122] Fr. Thomas Weinandy, "Of Jesus and Aliens." *The Catholic Thing*, July 2023.
[123] Joseph Cardinal Ratzinger, *The Spirit of the Liturgy* (Ignatius Press, 2000), Ch. 5.
[124] Aristotle, *Nicomachean Ethics.* Book 1, Ch. 7.
[125] Aristotle *Politics*. Book 1. Part 8.
[126] Aristotle, *On the Heavens, I. VIII.*
[127] St. Thomas Aquinas, *Summa Theologica*, I, Q. 47, A. 3.
[128] *Religion & Ethics Newsweekly.* PBS episode featuring Richard Rohr (November 11, 2011).
[129] Statement from Fr. Richard Rohr, OFM after meeting Pope Francis. Center for Action and Contemplation, July 2022. cac.org.
[130] Richard Rohr on THE UNIVERSAL CHRIST. Center for Action and Contemplation YouTube Channel (April 13, 2020).
[131] Cardinal Joseph Ratzinger. Congregation For The Doctrine Of The Faith Declaration, "'Dominus Iesus': On The Unicity And Salvific Universality Of Jesus Christ And The Church." (August 2000).
[132] Fr. Thomas Weinandy, "Of Jesus and Aliens." *The Catholic Thing*, July 2023.
[133] "*(St. Thomas, ST, III 3.7 See also ST III 3.7 ad 2)*" (Quoted in Dr. Paul Thigpen, *Extraterrestrial Intelligence and the Catholic Faith*, Chapter 13).

[134] Ibid.
[135] St. Thomas Aquinas, *Summa Theologica*, III, Q4, A5.
[136] St. Thomas Aquinas, *Summa Theologica*, I, Q104, A4.
[137] Marie George, *Christianity and Extraterrestrials?: A Catholic Perspective* (iUniverse, 2005), Ch. 4.
[138] Alberto Martinez, *Burned Alive: Bruno, Galileo and the Inquisition* (Reaktion Books, 2018), 241.
[139] Dicastery for the Doctrine of the Faith, "Trinity Mercy" Letter to the Bishop of Como on the Spiritual Experience Connected with the Sanctuary of Maccio (Villa Guardia). July 15, 2024. Vatican.va.
[140] NASA, *Voyager 1* Tracker. https://voyager.jpl.nasa.gov/.
[141] Paul Thigpen, *Extraterrestrial Intelligence and the Catholic Faith* (TAN, 2022), Ch. 14.
[142] Andrew Davison, *Astrobiology and Christian Doctrine* (Cambridge University Press, 2023), 352.
[143] Cf. The Traditional "Act of Faith."
[144] "Responsum of the Congregation for the Doctrine of the Faith to a dubium regarding the blessing of the unions of persons of the same sex." March 2021, Vatican.va.
[145] Dr. Alberto Martinez presents a number of condemnations of Alien Belief in Catholicism that we will not have space to review in detail here. He writes: "The heresy of many worlds appeared in books for centuries, for example, in a Thesaurus of the Christian Religion (1559) and in an edition of Philaster's Book of Heresies (1578). ... In 1591 the Jesuit theologian Gregorius de Valentia cautioned: "There is no need of heretics among Christians, who follow Democritus and other ancient philosophers in saying that there exists not one but innumerable worlds, as the divine Augustine reports in his book on heresies ..." Other theologians too cited this heresy for centuries. They explained the problem: "*we cannot assert that there exist two or many worlds, since neither do we assert two or many Christs*."... Above all, however, the highest authority proves that Bruno's belief in many worlds was heretical. In 1582 Pope Gregory XIII issued a compilation of laws of the Catholic Church and ordered that it be used in all church courts and schools of canon law. Expanded in 1591, the Corpus of Canon Law includes long discussions of what are heresies and who shall be considered a heretic. Echoing Isidore, its list of heresies includes 'having the opinion of innumerable worlds'. ... Jerome condemned Origen's notion of innumerable worlds as heretical." (Alberto Martinez, *Burned Alive: Bruno, Galileo and the Inquisition* (Reaktion Books, 2018), Ch. 1.)
[146] E.J. Dionne Jr., "For Jet-Age Pope: Angels, Devil and Indulgences," *The New York Times*. July 1986.
[147] This fact is not only settled in Catholic teaching, as we will presently consider, but is also necessary to grant in accordance with the infinite goodness of God. It is a logical corollary to a dogma so basic that it stands as a preamble of the Faith. For it is impossible to imagine a Divine Truth more foundational than God's goodness, yet it is also this Truth which tells us that God would never create a rational being who *could not* know and love Him. It would be blasphemous to accuse God of making any intelligent creature without giving this creature—this child of His—the opportunity to know and love its Creator—its Father—forever. This never has happened and never will happen. Yet, the Catechism teaches, relaying 2,000 years of unbroken and unanimous Tradition based on Scripture, that only man is even capable of knowing and loving God. Therefore, only man has reason.
[148] Andrew Davison, *Astrobiology and Christian Doctrine* (Cambridge University Press, 2023), pages 346-347.
[149] Paul Thigpen, *Extraterrestrial Intelligence and the Catholic Faith* (TAN, 2022), Ch. 14.
[150] Demons cannot be saved; however, God did not create them as demons. He created them as angels, and their damnation was their own free choice.
[151] In the encyclical, Pope Pius XII teaches: "*When, however, there is question of another conjectural opinion, namely polygenism, the children of the Church by no means enjoy such liberty [of discussion]. For the faithful cannot embrace that opinion which maintains that either after Adam there existed on this earth true men who did not take their origin through natural generation from him as from the first parent of all, or that Adam represents a certain number of first parents.*" (§37) Now, an ET promoter will jump on the phrase "on this earth"—leveraging it in the usual way to pretend to allow for "true men *not* on this earth." This attempt will not work. Two propositions are prohibited above, and the second one ("that Adam represents a certain number of first parents") is independently condemned—that clause is not joined to the phrase "on this earth." There are, therefore, no "other Adams" anywhere else; whereas any alien race would of course need "its own Adam." Moreover, the first condemned proposition contains two parts, and the second succeeds in indicating, on its own, that Adam is "**the first parent of all** [men]," without any qualification. Clearly, the phrase "on this earth" is not intended to *restrict*

the scope of a teaching, but rather to *expand* it. The Pope wished to make clear he was denouncing certain theories circulating at the time of his writing (1950) which posited that some of the ethnicities that have long populated different continents on earth ultimately each trace back to different original couples. Therefore, Pius was sure to include express insistence that no place *on the entire earth* was excluded from this teaching.

[152] Quoted from John Carey, PhD., "Ireland and the Antipodes: The Heterodoxy of Virgil of Salzburg." *Speculum*, Vol. 64, No. 1 (January, 1989), pp. 1-10.

[153] Alberto Martinez, *Burned Alive: Bruno, Galileo and the Inquisition* (Reaktion Books, 2018), 241.

[154] Heinrich Denzinger, *Enchiridion Symbolorum: A Compendium of Creeds, Definitions, and Declarations of the Catholic Church*, "Cum sicut accepimus." (Edited by Peter Hunermann, Robert Fastiggi, Anne Englund Nash. 43rd Edition, Ignatius Press, 2012;). Paragraphs 1361-1369. 1460/*1375. page 351.

[155] Paul Thigpen, *Extraterrestrial Intelligence and the Catholic Faith* (TAN, 2022), Ch. 2.

[156] St. Augustine, *City of God*. Book XVI. Ch. 8.

[157] St. Thomas Aquinas, *Summa Theologica*. I, Q47, A3.

[158] Joel Parkyn, *Exotheology: Theological Explorations of Intelligent Extraterrestrial Life* (Pickwick Publications, 2021), Introduction (Selected Excerpts).

[159] *Liber Caelestis*, Book IV. Ch. 67 (Oxford University Press, 2008).

[160] Venerable Mary of Agreda, *The Mystical City of God* (Translation by Fr. George Blatter. Veritatis Splendor Publications, 2013).

[161] Ibid. Ch. 7.

[162] Ibid.

[163] St. Francis de Sales. *The Saint Francis de Sales Collection* (Catholic Way Publishing, 2015), 487.

[164] St. Alphonsus Liguori. *The Saint Alphonsus de Liguori Collection* (Catholic Way Publishing, 2016), 625.

[165] Thomas Aquinas, *Commentary on the Book of Job*. The Second Lesson, Chapter 28.

[166] Michael Crowe, *The Extraterrestrial Life Debate, 1750-1900* (Dover Publications, 2011), Ch. 1 (Footnote 9).

[167] Charles L. Harper, Jr. (contributor), *God, Science, and Humility: Ten Scientists Consider Humility Theology* (Edited by Robert Herrmann. Templeton Press, 2000), 112.

[168] Michael Crowe, *Extraterrestrial Life Debate, Antiquity to 1915: A Source Book* (University of Notre Dame Press, 2008), Part One.

[169] Ibid.

[170] Paul Thigpen, *Extraterrestrial Intelligence and the Catholic Faith* (TAN, 2022), Ch. 1.

[171] Jeffry Bennet and Seth Shostak, *Life in the Universe* (Pearson Publishing, 2016), Chapter 2.1, page 25.

[172] Marie George, *Christianity and Extraterrestrials?: A Catholic Perspective* (iUniverse, 2005), Ch. 4.

[173] Ibid.

[174] See *Only Man Bears His Image*, pages 679-683 for more information on the "Nephilim."

[175] Quoted in Marie George, *Christianity and Extraterrestrials?: A Catholic Perspective* (iUniverse, 2005), Ch. 4.

[176] St. Jerome, *Apology Against Rufinus*. Book II. § 12.

[177] Stefan Kirschner, "Nicole Oresme", *The Stanford Encyclopedia of Philosophy* (Fall 2021 Edition), Edward N. Zalta (ed.)

[178] I have inserted the bracketed word above because it clarifies Oresme's intention, in accordance with his own reference to what was "stated above." Several paragraphs earlier, he wrote: "*...if several worlds existed, no one of them would be outside Him nor outside His power; but surely other intelligences would exist in one world and others in the other world...*" Therefore, we can see that when Oresme refers to "another world," he is clearly referring to a world inhabited by intelligent creatures—he says that such creatures would "surely" exist in another world *if* such a world existed. (Mere planets of dead rock we observe in the sky—whether with naked eye or telescope—are not, in his verbiage within the present context, "other worlds.")

[179] Nicole Oresme, *Le Livre du ciel et du monde*. Book I, Chapter 24, 36b-39c. (Quoted in Michael Crowe's *The Extraterrestrial Life Debate, 1750-1900* (Dover Publications, 2011), Chapter 1.)

[180] Pierre Duhem, *Medieval Cosmology: Theories of Infinity, Place, Time, Void, and the Plurality of Worlds* (translated by Roger Ariew). "The Plurality of Worlds in Fifteenth-Century Cosmology,"

Part V. Chapter 13.
[181] Michael Crowe, *The Extraterrestrial Life Debate, 1750-1900* (Dover Publications, 2011), Part One.
[182] Some ET promoters draw attention to an obscure Franciscan, born only a few years before Cusa, named William Vorilong. They allege he supported ET belief, but details on his life and teaching are incredibly scarce; he certainly does not emerge as a significant thinker. Although ET promoters present Vorilong in support of their thesis, it may well have been Vorilong's own suggestion (along with that of Zaninus) of belief in men on another world, created before Adam, that contributed to Pope Pius II issuing *cum sicut* (published during Vorilong's later years), in which those very claims of Vorilong were condemned!
[183] Michael Crowe, *The Extraterrestrial Life Debate, 1750-1900* (Dover Publications, 2011), Part One.
[184] Clyde Lee Miller, "Cusanus, Nicolaus [Nicolas of Cusa]", The Stanford Encyclopedia of Philosophy (Winter 2021 Edition), Edward N. Zalta. Note: the full quote claims that few later thinkers even "understood" Cusa. Indeed, a number of Cusa's teachings would require much difficulty to "understand"—e.g., his astronomical views—not, however, his teaching on aliens. We can be sure that these views did not merely "fail to be understood," rather, they were simply entirely non influential since they were, rightly, not taken seriously by the Christians of that era.
[185] Alberto Martinez, "Was Giordano Bruno Burned at the Stake for Believing in Exoplanets?" Published in *Scientific American*. March 2018.
[186] *Letter from Cardinal Angelo Sodano on the Occasion of a Congress of Studies on the Personality of Giordano Bruno*, February 14, 2000. Vatican.va.
[187] Pope Pius X, *Pascendi Dominici Gregi*, §39.
[188] Ibid.
[189] Ibid.
[190] Cathy Gutierrez, *The Occult in Nineteenth-Century America*, "New Studies in Religion and History" (Davies Group, 2005), Introduction.
[191] Perhaps most cuttingly, however, the Pope denounced the idea that "*The method and principles by which the old scholastic doctors cultivated theology are no longer suitable to the demands of our times and to the progress of the sciences.*" (§13) Here especially, shade is cast upon belief in aliens. The ET promoters of the 1800s argued that the Fathers and Doctors of the Church condemned aliens only because their scientific astronomy was so much less advanced; these "backwards thinkers thought the planets uninhabited, yet our telescopes show that Mars is traversed by engineered canals!" (It was not.)
[192] Sarah Willburn and Tatiana Kontou, *The Ashgate Research Companion to Nineteenth-Century Spiritualism and the Occult* (Routledge, 2017), Foreword.
[193] Ibid.
[194] Ibid. Ch. 6.
[195] For context, the highest-grossing Hollywood movie of that same year— "Bad Boys for Life"—currently has eighty thousand reviews on Amazon.
[196] Fr. Chad Ripperger, *Dominion: The Nature of Diabolic Warfare* (2022), Page 288.
[197] St. Thomas Aquinas taught, "*God alone can know the thoughts of hearts and affections of wills. The reason of this is, because the rational creature is subject to God only, and He alone can work in it Who is its principal object and last end ... all that is in the will [is] known to God alone...Hence the Apostle says (1 Cor. 2:11): 'For what man knoweth the things of a man, but the spirit of a man that is in him?'*" (*Summa Theologica*. I, Q57, A4)
[198] Charles Fraune, *Slaying Dragons II - The Rise of the Occult: What Exorcists & Former Occultists Want You to Know* (Slaying Dragons Press, 2022), Ch. 13.
[199] Ursula Bielski, *The Devil in Dreamland: Catholic Faith, UFOs, the Occult and the End of the Supernatural* (Gadarene Press. August 2024), page 195 (Kindle Edition).
[200] St. Augustine, *City of God*, Book VIII.
[201] Dr. Jeffrey Kripal, in Whitley Strieber, *A New World* (Walker & Collier, Inc., 2019), Kindle Edition. Preface.
[202] Note: Kripal first presents this sentence as a question, but then answers that question in the affirmative; thus I have simplified, through ellipses, the point he is making.
[203] Whitley Strieber, *Them* (Walker & Collier, Inc., 2023) Kindle Edition, Pages 255-256.
[204] https://x.com/WhitleyStrieber/status/1715399984554864757
[205] Whitley Strieber, *A New World* (Walker & Collier, Inc. 2019), Kindle Edition. Pages 225-226.
[206] Jeffrey Kripal, *Authors of the Impossible* (University of Chicago Press, 2010), Chapter 3.

[207] D. W. Pasulka, *Encounters: Experiences with Nonhuman Intelligences* (St. Martin's Publishing Group, 2023), Kindle Edition. Page 125.
[208] In her January 2024 Joe Rogan podcast, Pasulka shared that before publishing the book here quoted, she sought—and received—Vallée's approval in all that touched upon him in the book's content.
[209] Jeffrey Kripal, *Authors of the Impossible* (University of Chicago Press, 2010), Chapter 3.
[210] Jacques Vallée, *DIMENSIONS* (Anomalist Books, 2008), Introduction.
[211] James R. Lewis (editor), *The Gods have Landed* (State University of New York Press, 1995), Ch. 2.
[212] The Interdimensional Thesis is also baldly heretical on its own right, aside from the ET Deception. It is covered in more detail in pages 337-341 of *Only Man Bears His Image*.
[213] Whitley Strieber in the foreword for Vallée's *DIMENSIONS* (Anomalist Books, 2008).
[214] Jacques Vallée, *DIMENSIONS* (Anomalist Books, 2008), Part Two, Ch. 7.
[215] Ibid.
[216] Ibid. Ch 5.
[217] Jeffrey Kripal, *Authors of the Impossible* (University of Chicago Press, 2010), Chapter 3.
[218] Keay Davidson, *Carl Sagan: A Life* (Trade Paper Press, 2000), Preface.
[219] Dennis Overbye, "A Successor to Sagan Reboots 'Cosmos.'" *The New York Times*. March 3, 2014.
[220] Keay Davidson, *Carl Sagan: A Life* (Trade Paper Press, 2000), Chapter 1.
[221] Ibid.
[222] https://www.britannica.com/biography/Carl-Sagan.
[223] https://neildegrassetyson.com/essays/1996-09-the-search-for-life-in-the-universe/.
[224] John Mack, *Abduction* (Ballantine Books, 1994), Chapter 16.
[225] Ibid.
[226] John Mack, *Passport to the Cosmos* (White Crow Books, 1999), Chapter 2.
[227] Ibid.
[228] Ibid.
[229] R.L. Dione, *God Drives a Flying Saucer* (Bantam, 1973). Ch. 6, 7, and 8 (Excerpts).
[230] John Keel, *Operation Trojan Horse* (IllumiNet Press, 1996), Ch. 9.
[231] Ibid. Ch. 13.
[232] Ibid.
[233] Ibid. Ch. 2.
[234] Ibid.
[235] Ibid. Ch. 12.
[236] https://avi-loeb.medium.com/aspiring-to-be-god-is-not-an-arrogant-proposition-936bc84236c1
[237] https://avi-loeb.medium.com/is-god-an-advanced-scientist-61413e8b5995.
[238] Thesolfoundation.org.
[239] For example, Regis University, a Jesuit institution in Colorado, has multiple Fall 2024 courses that require Pasulka's book, *American Cosmic*. https://twitter.com/MatthewGruchow/status/1828153382332469271 .
[240] Ashley Carman, "Spotify Reveals Joe Rogan's Podcast Numbers," *Bloomberg*. March 21, 2024.
[241] D. W. Pasulka, *Encounters: Experiences with Nonhuman Intelligences* (St. Martin's Publishing Group, 2023), Kindle Edition. Page 14.
[242] Ibid.
[243] This fake "university" was founded by the new age AI-Transhumanist, Ray Kurzweil, has no accreditation, and has long been embroiled in scandals, but it appears to be embraced—for reasons that remain obscure, but are quite unlikely to be innocent—by ultra-rich Big Tech types.
[244] Ursula Bielski, *The Devil in Dreamland: Catholic Faith, UFOs, the Occult and the End of the Supernatural* (Gadarene Press, August 2024), Kindle Edition, P. 330.
[245] See *Only Man Bears His Image*, Chapter 19, page 335.
[246] https://x.com/JDMadden3/status/1729103203487395898.
[247] A reminder is now appropriate, as new academic characterizations of UFOs are bound to proliferate while scholars compete with one another to be revered as the designer of the most popular "systematic Ufology." Since it will prove impossible to keep up with them all, one simple truth must always be remembered: any attempt whatsoever to assign *non-demonic reality* to UFOs is a deception. Two, and only two, legitimate responses to the UFO phenomenon exist: debunking, and exposing the diabolical. Debunking is likely the correct response in some 95% of cases, wherein the UFO or

other "ET Contact" experience has an ordinary earthly explanation. The remaining 5% are diabolical. There is no third option, and every attempt to fabricate one is just a new prong of attack in the Devil's UFO Deception; a delusion whose aim is the damnation of souls (even if that is not always the intent of certain clueless scholars promoting it).

[248] James Madden, *Unidentified Flying Hyperobject: UFOs, Philosophy, and the End of the World* (2023, Kindle Edition), Page 49.

[249] Ibid, Page 65.

[250] Ibid.

[251] Ibid.

[252] Morton's Wikipedia page refers to him as a "they," as does his own personal website (https://ecologywithoutnature.blogspot.com/p/about.html), which also features a photograph of him cross dressing.

[253] https://www.amazon.com/dp/0231214715.

[254] Timothy Morton. *Humankind* (Verso, 2017), Chapter 1.

[255] Ibid. Chapter 2.

[256] Morgan Meis "Timothy Morton's Hyper-Pandemic." *The New Yorker*. June 8, 2021.

[257] Stuart Jeffries, "Humankind by Timothy Morton review - no more leftist defeatism, everything is connected." *The Guardian*. August 23, 2017.

[258] Phenomenology of the Phenomenon: Dr. Kimberly Engels in conversation with Prof. James Madden. Society for UAP Studies YouTube Channel. August 23, 2024.

[259] Now, Madden goes on to insist this doesn't mean he's saying Our Lady of Fatima is a flying saucer, but in that very insistence he contradicts himself; reiterating that he believes Our Lady of Fatima is a UFO. He claims that Apparitions of the Virgin Mary and UFO sightings/ ET Contact experiences are "**phenomenologically identical experiences**," and "**that is not to privilege one or other of the mythological interpretations**." (Evidently, he regards UFOs and Apparitions equally as "myths.") First, this claim is shockingly erroneous. Even a novice student of both apparitions and UFO sightings/ "ET Contact experiences" knows they are phenomenologically as different—*opposite*—as one can imagine. Second, even if his false premise were true, it is precisely Madden's phenomenological approach that rejects his own defense. For a phenomenologist to insist apparitions and UFO sightings are phenomenologically identical is, in accordance with his own premises, blasphemous.

[260] The Professor of UFOlogy: Confessions of a Catholic Prof with James Madden. New Thinking Allowed with Jeffrey Mishlove YouTube channel. May 7, 2024.

[261] Despite a kernel of truth in some of the works of de Chardin—which even Pope Benedict XVI commented on, though without endorsing his overall corpus—that value has been entirely eclipsed by the unmitigated destruction wrought by his primary message.

[262] Teilhard de Chardin, *Christ and Science*. Chapter VII. Section 5.

[263] Quoted in: Wolfgang Smith, *Teilhardism and the New Religion* (TAN Books, 2009), Chapter X.

[264] Of course, this is manifest heresy. The list of infallible teachings that Adam and Eve are literal people, from whom all humans descended, would be endless—we reviewed some in Part Two. The fact that the Book of Genesis *also* has symbolic components does not obscure this in the least.

[265] Pierre Teilhard de Chardin, *Fall, Redemption, and Geocentrism* (1969).

[266] Entry on "Cosmic Consciousness."

[267] D. W. Pasulka, *Encounters: Experiences with Nonhuman Intelligences* (St. Martin's Publishing Group, 2023), Kindle Edition. P. 104.

[268] Ibid.

[269] Joel Parkyn, *Exotheology: Theological Explorations of Intelligent Extraterrestrial Life* (Pickwick Publications, 2021), Page 141.

[270] Richard Rohr, OFM., *The Influence of Teilhard de Chardin, SJ. The Mendicant*. Center for Action and Contemplation. Summer 2016. Vol. 6. No. 3. Page 1.

[271] Ibid.

[272] "*Richard Rohr on THE UNIVERSAL CHRIST – September 16th, 2018*" YouTube, Center for Action and Contemplation.

[273] Paola Harris, *UFOs: How Does One Speak to A Ball of Light?* (CreateSpace Independent Publishing Platform, 2012), Section Two.

[274] Brenda Denzler, *The Lure of the Edge: Scientific Passions, Religious Beliefs, and the Pursuit of UFOs*.

(University of California Press, 2003), Footnote 119.
[275] Steven Greer, *Disclosure* (2013), Section I. §2
[276] Paola Harris, *UFOs: How Does One Speak to a Ball of Light?* (CreateSpace Independent Publishing Platform, 2012), Section Three.
[277] Balducci lived for two years after this movie's release while apparently never issuing so much as a clarification, much less a rebuttal, of its remarkably apostate content that he appears to endorse.
[278] https://drmsh.com/ufos-ets-and-religion-balduccis-conundrum-part-5/.
[279] Fr. Carlos Martins, *The Exorcist Files* (FaithWords, 2024), Chapter 6.
[280] Jacques Vallée, *DIMENSIONS* (Anomalist Books, 2008), Chapter 4.
[281] Susan Palmer, *Aliens Adored* (Rutgers University Press, 2004), Introduction.
[282] Fr. Andrew O'Brien, *Make Yourself an Ark.* Volume 1. The Hidden Origins of Communism.
[283] New Catholic Encyclopedia on ANTICLERICALISM.
[284] Fr. Andrew O'Brien, *Make Yourself an Ark.* Volume 1. The Hidden Origins of Communism.
[285] Ibid.
[286] Hans Lehfeldt, "Margaret Sanger and the Modern Contraceptive Techniques." *The Journal of Sex Research* 3, no. 4 (1967): 253–55. http://www.jstor.org/stable/3811979.
[287] "Radio Listeners in Panic, Taking War Drama as Fact," *The New York Times.* October 31, 1938.
[288] Ibid.
[289] Ibid.
[290] Entry on CINEMA. 1993.
[291] https://medium.com/@jacobviness/2001-a-space-odyssey-the-gospel-according-to-saint-stanley-cab1f84ce8e1.
[292] The Telegraph Reporters. "Stanley Kubrick explains the ending of 2001: A Space Odyssey in newly discovered interview." July 2018.
[293] https://www.krusch.com/kubrick/Q12.html.
[294] To his credit, Spielberg has since expressed regret at making this detail a part of the plotline.
[295] David Thomson. *The New Biographical Dictionary of Film.* Random House. 2014. Entry on Steven Spielberg.
[296] Joseph McBride, *Steven Spielberg: A Biography* (University Press of Mississippi, 2010). Chapter 1.
[297] John Baxter, *Steven Spielberg: The Unauthorised Biography* (Harper Collin, 1998). Chapter 11.
[298] Joseph McBride, *Steven Spielberg: A Biography* (University Press of Mississippi, 2010), Chapter 13.
[299] Ibid.
[300] Jeffrey Kripal, *How to Think Impossibly* (University of Chicago Press, 2023), Chapter 2.
[301] Jordan King, "Josh O'Connor Could Star In Steven Spielberg UFO Movie With Emily Blunt," Published by *Empire Online.* July 28, 2024.
[302] Jorg Matthias Determann, *Islam, Science Fiction and Extraterrestrial Life: The Culture of Astrobiology in the Muslim World* (Bloomsbury, 2020), Pages 10-11.
[303] David Weintraub, *Religions and Extraterrestrial Life: How Will We Deal With It?* (Springer Publishing, 2014), Page 166.
[304] David Weintraub, "How Will Our Religions Handle the Discovery of Alien Life?" (2016), getpocket.com.
[305] David Weintraub, *Religions and Extraterrestrial Life: How Will We Deal With It?* (Springer, 2014), Chapter 17.
[306] Ibid.
[307] Ibid. Chapter 16.
[308] Ibid. Chapter 19.
[309] Ibid. Chapter 18.
[310] Walter Martin, *The Kingdom of the Occult.* (Thomas Nelson Publishers, 2008), Page 191.
[311] Ibid.
[312] Ibid. Chapter 1.
[313] Ursula Bielski, *The Devil in Dreamland: Catholic Faith, UFOs, the Occult and the End of the Supernatural* (Gadarene Press, 2023), Kindle Edition, p. 92.
[314] David Weintraub, "How Will Our Religions Handle the Discovery of Alien Life?" (2016), getpocket.com
[315] Michael Crowe, *The Extraterrestrial Life Debate, 1750-1900* (Dover Publications, 2011), Chapter 5.
[316] Ibid.

[317] Rod Dreher, *Living in Wonder: Finding Mystery and Meaning in a Secular Age*. Zondervan. Kindle Edition, p. 123.
[318] https://johanw.home.xs4all.nl/CoS/fishman.html.
[319] Tony Ortega, "L. Ron Hubbard's son was troubled, but don't discount him entirely: few knew his father better." The Underground Bunker. December 16, 2017. TonyOrtega.com.
[320] https://www.aetherius.org/.
[321] Annalise Pasztor, "Talking to Aliens Is Their Religion." *The New York Times*. August 2020.
[322] https://www.aetherius.org/oldest-international-ufo-organisation-validated-by-recent-government-admissions/.
[323] Hugh Ross, *Lights in the Sky & Little Green Men: A Rational Christian Look at UFOs and Extraterrestrials* (Reasons to Believe, 2002), Chapter 14.
[324] Ibid.
[325] James R. Lewis (editor), *The Gods have Landed* (State University of New York Press, 1995) Ch. 4.
[326] Ibid.
[327] Jorg Matthias Determann, *Islam, Science Fiction and Extraterrestrial Life: The Culture of Astrobiology in the Muslim World* (Bloomsbury. 2020), Pages 113-114.
[328] "Farrakhan Again Describes Hitler as a 'Very Great Man'." *The New York Times*. July 17, 1984.
[329] Susan Palmer, *Aliens Adored* (Rutgers University Press, 2004), Chapter 1.
[330] Quoted in Gregory L. Reece, *UFO Religion: Inside Flying Saucer Cults and Culture*. I.B. 2007. Page 16-17.
[331] Susan Palmer. *Aliens Adored* (Rutgers University Press, 2004), Chapter 4.
[332] Christopher Partridge, *UFO Religions* (Routledge, 2003), Part 2. Chapter 2.
[333] Ibid.
[334] Do's Final Exit. HeavensGateDatabase YouTube Channel.
[335] George Chryssides (Edited by Ben Zeller.) *Handbook of UFO Religions*. Brill Handbooks on Contemporary Religion, Volume: 20. (Brill, 2021), Chapter 17.
[336] James R. Lewis (editor) *The Gods have Landed* (State University of New York Press, 1995), Ch. 2.
[337] Ibid.
[338] Davis Roos, "What Drove Heaven's Gate Followers to Mass Suicide?" March 28, 2023. History.com.
[339] Ibid.
[340] David Wilkinson, *Science, Religion, and the Search for Extraterrestrial Intelligence* (Oxford University Press, 2013), Chapter 1.
[341] The So-Called Prophet from Pittsburgh. Joseph Flatley YouTube Channel.
[342] https://spiritualution.org/learn/be-aware-proclamation.
[343] Extreme Caution About False Prophets & Self-Help Programs In These End Times. Van of Urantia YouTube Channel. September 5, 2022.
[344] https://uaspr.org/.
[345] Gregg Tomusko, *The Catholic Church and The Urantia Book* (2023), Page 89.
[346] Ibid. Page 98.
[347] Ibid.
[348] Bishop Marian Eleganti. "Swiss bishop: 'Synodal' church is a 'protestantized' rejection of the Catholic faith." *LifeSiteNews*, July 2023.
[349] Voice and Echo of the Divine Messengers. "Apparitions of Christ Jesus." https://www.mensajerosdivinos.org/en/aparicion-de-cristo/saturday-august-15-2020.
[350] Ibid. "Apparitions of Christ Jesus." https://www.mensajerosdivinos.org/en/mensaje-de-cristo-jesus/saturday-december-28-2019.
[351] https://www.mensajerosdivinos.org/en/mensaje-de-san-jose/monday-january-18-2016.
[352] https://www.mensajerosdivinos.org/en/tag/14577.
[353] https://www.mensajerosdivinos.org/en/tag/11408.
[354] Vitor de Lima Campanha, "The articulation between evolutionism and creationism in New Religious Movements: Two South American case studies." Revista del CESLA, vol. 26, (Uniwersytet Warszawski, June 2020), Pages 179-194.
[355] TOWARDS THE NEW CREATION: official website of the Church of Jesus Christ of the Universe https://towardsthenewcreation.com/.
[356] https://towardsthenewcreation.com/authors/tomislav-vlasic/.

[357] Ibid.
[358] For example, a "Warning" or Illumination of Conscience, followed by Chastisements, over which will eventually prevail a spiritual Triumph of the Sacred Heart of Jesus and the Immaculate Heart of Mary—*Catholicism's full vigor being attained.*
[359] https://towardsthenewcreation.com.
[360] https://towardsthenewcreation.com/2013/01/24/the-universal-mission-has-begun/.
[361] https://apps.dtic.mil/sti/pdfs/AD0688332.pdf.
[362] John Keel, *Operation Trojan Horse* (IllumiNet Press, 1996). Chapter 12.
[363] Jacques Vallée, *REVELATIONS* (Anomalist Books, 2008), Appendix.
[364] Jacques Vallée, *DIMENSIONS* (Anomalist Books, 2008), Part 1. Ch. 2.
[365] Jim Marrs, *Alien Agenda: Investigating the Extraterrestrial Presence Among Us* (William Morrow Paperbacks, 2000), Page 350.
[366] *Jesus Christ The Bearer Of The Water Of Life: A Christian reflection on the "New Age."* Pontifical Council For Culture; Pontifical Council For Interreligious Dialogue. Vatican.va. 2003.
[367] Ibid.
[368] Lindsay Jones, *The Macmillan Encyclopedia of Religion* (Second Edition). 2005. "NEW AGE MOVEMENT."
[369] "Russian Orthodox Church equates aliens with demons." Interfax. April 202 https://interfax.com/newsroom/top-stories/68479/.
[370] James R. Lewis, *The Gods have Landed* (State University of New York Press, 1995), Chapter 9.
[371] Ibid. Chapter 2.
[372] Ibid.
[373] Ibid. Chapter 3.
[374] Fr. Gabriel Amorth, *An Exorcist Tells His Story* (Ignatius Press, 1999), Chapter 4.
[375] Ibid.
[376] Quoted in Fr. Chad Ripperger, *Dominion: The Nature of Diabolic Warfar* (2022), Footnote 136.
[377] Jesus, though fully man, is a Divine *Being* (person), and He of course had a human mother, but no human father.
[378] Fr. Chad Ripperger, *Dominion: The Nature of Diabolic Warfar* (2022), Page 364.
[379] https://religiousdemonology.com/basics.
[380] John Mack, *Passport to the Cosmos* (White Crow Books. 1999), Chapter 11.
[381] Ibid. Chapter 10.
[382] Ibid. Chapter 7.
[383] John Mack, *Abduction* (Ballantine Books, 1994), Chapter 9.
[384] Ibid.
[385] John Mack, *Passport to the Cosmos* (White Crow Books, 1999), Conclusion.
[386] https://www.piercingthecosmicveil.com/frequently-asked-questions.
[387] http://www.alienresistance.org/ufo-alien-deception/alienabductions-stop-in-the-name-jesus-christ/.
[388] John Keel, *Operation Trojan Horse* (IllumiNet Press, 1996), Chapter 12.
[389] Ibid.
[390] St. Augustine, *City of God*. Book VIII. Chapter 15.
[391] St. Thomas Aquinas, *Summa Theologica.* I, Q114, A4.
[392] St. John Chrysostom, Homily 4 on Ephesians.
[393] St. John Cassian, Second Conference on the Principalities. XII.
[394] St. Augustine, *City of God.* Book VIII. Chapter 22.
[395] Fr. Chad Ripperger, *Dominion: The Nature of Diabolic Warfare,* (2022) Page 250.
[396] Ibid, Page 271.
[397] St. Augustine, *Enchiridion* (Handbook on Faith, Hope, and Love), Ch. 60.
[398] St. Athanasius, *Life of St. Anthony*. https://www.newadvent.org/fathers/2811.htm.
[399] Quoted in Ingrid Merkel & Allen Debus, *Hermeticism and the Renaissance* (Folger Books, 1988), pages 134-135.
[400] Ingrid Merkel & Allen Debus, *Hermeticism and the Renaissance* (Folger Books, 1988), 134-135.
[401] Ibid, 382-383.
[402] It is important to distinguish this type of "assumption" from what we speak of when, for example,

referring to the Word of God assuming a human nature. That Divine "assumption" was a true incarnation—*The* Incarnation. The demons (or angels), on the other hand, *cannot* incarnate themselves, for that is contrary to their nature, which Catholic dogma holds is entirely spiritual and invisible. But they can (temporarily) give the appearance that they are physical creatures by the type of "assumption" of "bodies" described by St. Thomas, wherein they are simply moving about some amalgamation of matter ("body") in order to make it resemble what they please (whether a UFO, an alien, or whatever else). This does not render their spiritual existence the actual *form* of an individual substance (as your soul is truly the form of your body); it simply entails their spiritual faculty of will exerting influence on the body they "assumed," analogous to the software by which some autonomous vehicle is driven.

403 Although Fr. Ripperger only refers to angels, he also explicitly says later that the same description applies to demons. (They retain their angelic nature and thus the capacities essential to such a nature). Therefore, I have added the bracketed word.

404 Ingrid Merkel & Allen Debus, *Hermeticism and the Renaissance* (Folger Books, 1988), page 22.

405 Fr. Jordan Aumann, *Spiritual Theology*, Chapter 14.

406 Ibid.

407 Ibid.

408 Luis Elizondo, *Imminent* (HarperCollins, 2024), Chapter 8.

409 Ibid.

410 James R. Lewis, *The Gods have Landed* (State University of New York Press, 1995), Chapter 2.

411 Paul Thigpen, "NHI, UAP, and the Catholic Faith." SOL White Paper. §4. White Papers of the Sol Foundation. Volume 1, No. 5. July/August 2024.

412 Leslie Kean and Ralph Blumenthal, "Intelligence Officials Say U.S. Has Retrieved Craft of Non-Human Origin." *The Debrief*, June 5, 2023.

413 David Grusch gives the Sol Foundation 2023 Symposium Closing Remarks. The Sol Foundation YouTube Channel. February 12, 2024.

414 https://avi-loeb.medium.com/on-science-and-religion-0281dced4d0e.

415 John Mack, *Abduction* (Ballantine Books, 1994). Chapter 7.

416 John Mack, *Passport to the Cosmos* (White Crow Books, 1999), Chapter 11.

417 Ibid. Chapter 2.

418 D.W. Pasulka, *American Cosmic: UFOs, Religion, Technology* (Oxford University Press, 2019), Kindle Edition, p. 9.

419 James R. Lewis, *The Gods have Landed* (State University of New York Press, 1995), Chapter 2.

420 Susan Palmer, *Aliens Adored* (Rutgers University Press, 2004), Chapter 3.

421 Ibid. Chapter 5.

422 E.g., Chris Bledsoe's "UFO of God" (a 2023 Best-seller), the Unarians' Charles Spiegel, "Antares" was "healed by Uriel the Archangel," Helge Lindroos "sight healed by aliens to enable him to announce their message," and many more. Abductees often claim (always strange) "healings." Although demons cannot miraculously heal, they can instantaneously remove some affliction that they themselves had been causing, thus giving the false impression of a healing. All "ET/UFO healings" fall into this category.

423 *"People who have the UFO abduction experience very frequently meet hybrid beings of the same type who are part animal and part human. Often they appear first in that form before shapeshifting into the more familiar form of greys."* Quoted in Dr. Timothy Dailey, *The Paranormal Conspiracy*, Ch. 12.

424 Whitley Strieber, *Jesus: A New Vision* (Walker & Collier, Inc., 2021), Kindle Edition, p. 243.

425 Jacques Vallée, *Messengers of Deception: UFO Contacts and Cults* (Daily Grail Publishing, 2008 edition), Foreword.

426 Jacques Vallée, *REVELATION* (Anomalist Books, 2008), Conclusion.

427 *City of God*. Book XX. Chapter 19.

428 *The Divine Life of the Most Holy Virgin* (TAN books, 1997), Page 314.

429 *Humani Generis Redemptionem*. June 1917. §2.

430 https://the-american-catholic.com/2017/02/26/fulton-sheen-on-the-anti-christ/.

431 Pope Paul VI, "Confronting The Devil's Power," General Audience. November 15, 1972.

432 St. John Henry Newman, *Discussions and Arguments. II: The Patristical Idea of The Antichrist*. Lecture 2: The Religion of Antichrist, §3.

433 Luis Elizondo, *Imminent* (HarperCollins, 2024), Chapter 23.

434 Ibid, Chapter 25.
435 Fr. Stanley Jaki, *A Late Awakening and Other Essays* (Real View Books. 2006), Pages 104-105.
436 Luis Elizondo, *Imminent* (HarperCollins, 2024), Chapter 25.
437 Gallaudet has also spoken openly of his consultation of occult "mediums."
438 https://www.rev.com/transcripts/house-hearing-on-unidentified-anomalous-phenomena.
439 https://mace.house.gov/immaculateconstellation.
440 Ibid.
441 Cf. *Summa Contra Gentiles*. Book II. Chapter 46, §5.
442 https://www.amazon.com/Co-Intelligence-Living-Working-Ethan-Mollick/dp/059371671X.
443 Ibid.
444 Stanley Jaki, *Brain, Mind and Computers* (Herder and Herder, 1969), Introduction.
445 Ibid, Conclusion.
446 Ibid.
447 Ibid, Introduction.
448 Rory Cellan-Jones, "Stephen Hawking warns artificial intelligence could end mankind." BBC News. December 2, 2014.
449 Joad Medeiros. Stephen Hawking: *'I fear AI may replace humans altogether'*. Published in *Wired Magazine*. November 28, 2017.
450 Ian Giatti, "Elon Musk says Google co-founder wanted to build AI 'digital god'." *The Christian Post*. April 20, 2023.
451 "Elon Musk says 'we are summoning the demon' with artificial intelligence." Fox News. October 26, 2014.
452 Emile Torres, "AI and the threat of "human extinction": What are the tech-bros worried about? It's not you and me." *Salon Magazine*. June 11, 2023.
453 Kevin Kelly, *The Inevitable: Understanding the 12 Technological Forces that Will Shape our Future* (Peguin Books, 2017 reprint edition), Chapter 2.
454 Mark Harris, "Inside the First Church of Artificial Intelligence." Published by *Wired Magazine*. November 15, 2017.
455 Ibid.
456 Douglas Busvine, "Collective consciousness to replace God - author Dan Brown." Reuters, October 2017.
457 John Brandon, "An AI god will emerge by 2042 and write its own bible. Will you worship it?" VentureBeat. October 2017.
458 Darren Orf, "AI Has Suddenly Evolved to Achieve Theory of Mind." *Popular Mechanics*, February 2023.
459 Ray Kurzweil, *The Singularity is Nearer* (Random House, 2024), Chapter 1.
460 Ben Turner, "Humanity faces a 'catastrophic' future if we don't regulate AI, 'Godfather of AI' Yoshua Bengio says," *Live Science*. October 1, 2024.
461 Joe Rogan Experience #2091 - Diana Walsh Pasulka. PowerfulJRE YouTube Channel. June 27, 2024.
462 Bill Gates on possibility, AI, and humanity. Reid Hoffman YouTube Channel. October 30, 2024.
463 https://openai.com/index/planning-for-agi-and-beyond/.
464 https://openai.com/about/.
465 https://www.reddit.com/r/ChatGPT/comments/1ggixzy/comment/luq8i5s/.
466 At UN, robot Sophia joins meeting on artificial intelligence and sustainable development. UN.org.
467 Yuval Harari. Read Yuval Harari's blistering warning to Davos in full. WEForum.org. January 24, 2020.
468 Yuval Noah Harari YouTube Channel. "Humanity is not that simple | Yuval Noah Harari & Pedro Pinto." June 2023.
469 Yuval Harari, *Nexus* (Random House, 2024), Prologue.
470 Ibid, Ch. 6.
471 https://x.com/lexfridman/status/1638908038928293892.
472 Sam Altman: OpenAI CEO on GPT-4, ChatGPT, and the Future of AI | Lex Fridman Podcast #367. Lex Fridman YouTube Channel. March 25, 2023.
473 https://twitter.com/lexfridman/status/1653051310034305025.

[474] Henry Kissinger, Craig Mundie, and Eric Schmidt, *Genesis* (Little, Brown and Company, 2024), Chapter 3.
[475] Ibid.
[476] Ibid.
[477] Ibid.
[478] Ibid, Conclusion.
[479] Ibid.
[480] Rod Dreher, *Living in Wonder: Finding Mystery and Meaning in a Secular Age* (Zondervan, 2024). Kindle Edition, page 118.
[481] D. W. Pasulka, *Encounters: Experiences with Nonhuman Intelligences* (St. Martin's Publishing Group, 2023), Kindle Edition. Pages 149-150.
[482] Rod Dreher, *Living in Wonder: Finding Mystery and Meaning in a Secular Age* (Zondervan, 2024). Kindle Edition, p.117.
[483] https://christogenesis.org/about/.
[484] https://christogenesis.org/.
[485] Rod Dreher, *Living in Wonder: Finding Mystery and Meaning in a Secular Age* (Zondervan, 2024). Kindle Edition, Chapter 10.
[486] Ibid.
[487] Ibid.
[488] Ibid, page 323 .
[489] Ilia Delio, *The Not-Yet God: Carl Jung, Teilhard de Chardin, and the Relational Whole* (Orbis Books, 2023). Kindle Edition, p. 318.
[490]AI Meets Religion: Dr Ilia Delio on AI and Human Evolution. Tech Innovations with Dr T YouTube Channel. July 7, 2024.
[491] Ilia Delio, *The Not-Yet God: Carl Jung, Teilhard de Chardin, and the Relational Whole* (Orbis Books, 2023). Kindle Edition, Chapter 12.
[492] Kevin Kelly, "The future will be shaped by optimists," TEDMonterey. August 2021. Ted.com.
[493] https://kk.org/interviews-matrix.
[494] https://kk.org/thetechnium/the-internet-is/.
[495] Kevin Kelly, *The Inevitable* (Viking, 2016), Chapter 2.
[496] Ibid, Chapter 12.
[497] Robert Geraci, *Apocalyptic AI*. (Oxford University Press, 2010), Appendix 1.
[498] Ibid, Chapter 4.
[499] Ibid.
[500] Ibid, Footnote 51.
[501] Anne Foerst, *God in the Machine* (Dutton Adult, 2004), Page 84.
[502] Ibid, Page 189.
[503] Ibid.
[504] Ibid.
[505] Edmund Furse, "The Theology of Robots." *New Blackfriars*, Vol. 67, No. 795 (September 1986), pp. 377-386.
[506] Ibid.
[507] Ibid.
[508] AI promoters themselves also decry the name. Yet whereas I am condemning calling such things "intelligence," they—disclosing the depths of their delusion—are condemning calling AI "artificial."
[509] Plato, *Phaedo*.
[510] Stanley Jaki, *Brain, Mind and Computers* (Herder and Herder, 1969), Epilogue.
[511] Matthew J. Gaudet, *Encountering Artificial Intelligence: Ethical and Anthropological Investigations* (Theological Investigations of Artificial Intelligence Book 1. Pickwick Publications, an imprint of Wipf and Stock Publishers). Kindle Edition, Chapter 5.
[512] Erik Larson, *The Myth of Artificial Intelligence* (Harvard University Press, 2021), Chapter 18.
[513] Quoted in Stanley Jaki, *Brain, Mind and Computers* (Herder and Herder, 1969), Ch 1. Footnote 35.
[514] By this claim, Hobbes won for himself much well-deserved ridicule from his peers; indeed, even something so simple as geometrical reasoning cannot be explained thus, much less any other human thought.

[515] Kester Brewin, *God-like: a 500-year history of Artificial Intelligence in myths, machines, monsters* (2024). Act 1. Pages 28-30.
[516] https://www.vox.com/climate/2024/3/28/24111721/climate-ai-tech-energy-demand-rising.
[517] Quoted in "The Radical Movement to Worship AI as a New God." *The Daily Beast*, Tony Ho Tran. February 25, 2023. TheDailyBeast.com.
[518] Arvind Narayanan & Sayash Kapoor, *AI Snake Oil* (Princeton University Press, 2024), Chapter 4.
[519] Imane El Atillah, "Man ends his life after an AI chatbot 'encouraged' him to sacrifice himself to stop climate change." Euro News. March 31, 2023.
[520] Ibid.
[521] Ibid.
[522] Kevin Roose, "Can A.I. Be Blamed for a Teen's Suicide?" *The New York Times*. October 23, 2034.
[523] Miles Kruppa, "Google Paid $2.7 Billion to Bring Back an AI Genius Who Quit in Frustration." *The Wall Street Journal*. September 25, 2024.
[524] Tom Singleton, "How a chatbot encouraged a man who wanted to kill the Queen," BBC News. October 6, 2023.
[525] Ibid.
[526] Arvind Narayanan & Sayash Kapoor, *AI Snake Oil* (Princeton University Press, 2024), Chapter 4.
[527] As the tests I provide in this book become outdated, I will provide new ones at www.DSDOConnor.com.
[528] Note that I actually wrote out the thousand 2s; so there was no question that ChatGPT simply misinterpreted the question due to my own phrasing of it. I entered in an entirely straightforward equation.
[529] Arvind Narayanan & Sayash Kapoor, *AI Snake Oil* (Princeton University Press, 2024), Chapter 4.
[530] Noreen Herzfeld, *The Artifice of Intelligence* (Fortress Press, 2023), Chapter 7.
[531] Jonathan Vanian, "Meta's AI chief doesn't think AI super intelligence is coming anytime soon, and is skeptical on quantum computing." CNBC News. December 3, 2023.
[532] https://x.com/ylecun/status/1727727093671145978?lang=en.
[533] I know that such software is not infallible; but when I combine my own gut sense with that software's output, and when I further prompt ChatGPT to generate content on the same topic, only to see it deliver remarkably similar results, I can combine these insights and others to conclude quite confidently that the content was AI-generated.
[534] The company says it "transforms the online presence" of whoever is willing to pay the price. They will generate articles, for a fee, that make some individual, or business, or what have you, appear more important than it is. The company boasts that, thanks to the torrent of articles generated by its employees, they accomplish this "transformation" through generating massive viewership.
[535] www.DSDOConnor.com.
[536] Aatish Bhatia, "When A.I.'s Output Is a Threat to A.I. Itself." *The New York Times*. August 25, 2024.
[537] Ibid.
[538] Wallace Witkowski, "'Moore's Law's dead,' Nvidia CEO Jensen Huang says in justifying gaming-card price hike." Market Watch. September 22, 2022.
[539] Rachel Meetz, Shirin Ghaffary, Dina Bass, and Julia Love, "OpenAI, Google and Anthropic Are Struggling to Build More Advanced AI." *Bloomberg News*. November 13, 2024.
[540] Spencer Van Dyk, "'I warned you guys in 1984,' 'Terminator' filmmaker James Cameron says of AI's risks to humanity." CTV News. July 2023.
[541] Elon Musk tells Tucker potential dangers of hyper-intelligent AI. Fox News YouTube Channel. April 17, 2023.
[542] Gary Smith, *The AI Delusion* (Oxford University Press, 2018), Chapter 12.
[543] J. W. Rinzler, *The Making of Star Wars: The Empire Strikes Back (Enhanced Edition)*, 2013. Page 25.
[544] Dean Kowalski, *Steven Spielberg and Philosophy: We're Gonna Need a Bigger Book (The Philosophy of Popular Culture*. University Press of Kentucky, 2011. Pages 89-90.
[545] "The 21st Century's 100 greatest films." BBC News. August 22, 2016.
[546] Armond White, "A.I. Is the Best Film of the 21st Century," *National Review*. June 30, 2021.
[547] Ibid.
[548] Reminding us that his film is indeed a conjuring of the Pagan demons, names the movie's female robot "Galatea." This name comes directly from Greek mythology. In those myths, Pygmalion was

a king who fashioned a statue of a woman. He falls in love with this creation of his, and in response to his prayers, the goddess Aphrodite (Venus) breathes life into the statue. The statue-person's name? Galatea.

[549] Quoted in Gary Bates, *Alien Intrusion* (Creation Book Publishers, 2010), Chapter 2.

[550] Isaac Asimov, *Asimov's Guide to the Bible* (Wings, Reprint ed. 1981), on Matthew.

[551] Ibid, on Luke.

[552] Ibid, on Revelation.

[553] Matthew J. Gaudet, *Encountering Artificial Intelligence: Ethical and Anthropological Investigations* (Theological Investigations of Artificial Intelligence Book 1) Pickwick Publications, an imprint of Wipf and Stock Publishers. Kindle Edition, p. 143.

[554] St. Augustine, *City of God.* Book 8, Ch. 26.

[555] Ibid.

[556] Ibid.

[557] Robert Geraci, *Apocalyptic AI* (Oxford University Press, 2010), Appendix 1.

[558] Ibid.

[559] Ibid.

[560] Ibid.

[561] Ibid.

[562] Anne Foerst, *God in the Machine* (Dutton Adult, 2004), Pages 38-40.

[563] Ursula Bielski, *The Devil in Dreamland: Catholic Faith, UFOs, the Occult and the End of the Supernatural* (Gadarene Press, 2023). Kindle Edition, pages 343-344.

[564] Carl E. Schmoger, C.SS.R., *The Life and Revelations of Anne Catherine Emmerich* (TAN, 2012), Volume II. Ch.3.

[565] C.S. Lewis, *That Hideous Strength.* Chapter 8.

[566] Antje Jackelen, "*The Image of God as Techno Sapiens.*" In *Zygon*, Volume 37, Issue 2. June 2002. Pages 289-302.

[567] Ibid.

[568] Jamey Keaten, "'AI Jesus' avatar tests man's faith in machines and the divine." AP News. November 28, 2024.

[569] Christopher Partridge, *The Occult World* (Routledge, 2016), Page 202.

[570] St. John Henry Newman, *Discussions and Arguments. II: The Patristical Idea of The Antichrist.* Lecture 4: "The Persecution of the Antichrist," §2 and §5.

[571] The World Government Summit Day 1 is Now Live. World Governments Summit YouTube Channel. February 12, 2023.

[572] Jeff Seldin, "Russia Sees Artificial Intelligence as Key to World Domination." VOA News. September 1, 2017.

[573] Joanna Ng, "How Artificial Super-Intelligence Is Today's Tower of Babel." Christianity Today. June 17, 2020.

[574] Thérèse of Lisieux, *Story of a Soul: The Autobiography of Saint Thérèse of Lisieux* (Washington, DC: ICS Publications, 1996), 102.

[575] Charles Arminjon, *The End of the Present World and the Mysteries of the Future Life.* Pages 56-58.

[576] Ibid.

[577] For the sake of preventing this discussion from becoming too lengthy, I am abridging this passage to focus on the "thousand years" itself, and not the mysterious "first resurrection" of Revelation 20:5. Suffice it to say that this verse certainly must not be taken as a reference to the actual General Resurrection itself, as that would be heretical. *The* Resurrection of the Dead can only occur at the end of time. Rather, this passage could refer to a number of martyrs who had been killed during the persecution of the Beast being miraculously raised from the dead by God. No Christian should scoff at this. Obviously, the dead have often been raised by God, both in Scripture and in Church History. It can happen again. At Christ's crucifixion, in fact, this happened in very large numbers: "*And Jesus cried again with a loud voice and yielded up his spirit. And behold, the curtain of the temple was torn in two, from top to bottom; and the earth shook, and the rocks were split; the tombs also were opened, and many bodies of the saints who had fallen asleep were raised, and coming out of the tombs after his resurrection they went into the holy city and appeared to many.*" (Matthew 27:50-53)

[578] St. Irenaeus, *Against Heresies.*

[579] Consider that St. Augustine did not have access to these prophecies, almost all of which came

after his own lifetime.

[580] Here, Dyakonov points out: "[Because] in the physical world, continuous quantities … can be neither measured nor manipulated exactly. That is, no continuously variable quantity can be made to have an exact value, including zero. To a mathematician, this might sound absurd, but this is the unquestionable reality of the world we live in, as any engineer knows."

[581] Mikhail Dyakonov, "The Case Against Quantum Computing." IEEE Spectrum. November 15, 2018.

[582] These include the Kyakonov-Perel spin relaxation mechanism and Dyakonov-Shur plasma wave instability.

[583] Jonathan Vanian, "Meta's AI chief doesn't think AI super intelligence is coming anytime soon, and is skeptical on quantum computing." CNBC News December 3, 2023.

[584] Edd Gent, "Quantum Computing's Hard, Cold Reality Check." IEEE Spectrum. December 22, 2023.

[585] Except, perhaps, "evil qua evil."

[586] Michael Crowe, *The Extraterrestrial Life Debate, 1750-1900* (Dover Publications, 2011), Part 3.

[587] Ibid.

[588] Ibid. Part 2.

[589] Ibid.

[590] Ibid. Part 1.

[591] Ibid. Fr. Burque, however, also noted that, hypothetically, if aliens were proved to exist, it would have no negative effect on Catholicism; and for this simple observation, some wrongly cite him as a pluralist.

[592] Ibid. Part 3.

[593] *Lettres à un matérialiste sur la pluralité des mondes habités et les questions qui s'y rattachent.*

[594] Ibid.

[595] Joel Achenbach, "The Outre Limits." *The Washington Post.* March 19, 1997.

[596] Ibid.

[597]https://en.thebongiovannifamily.it/index.php/biography/832-the-story-of-giorgio-bongiovanni-1d-part.

[598] Ibid.

[599] en.thebongiovannifamily.it.

[600] Ibid., "The New Theology." Part VII.

[601] 2011 Messages: Reincarnation.

[602] Although some philosophers of science after Popper offered legitimate critiques that the latter had focused too exclusively on falsifiability as *the* (sole) criterion of science, Popper's insights nevertheless remain invaluable. I also hasten to add that, unfortunately, Popper's own (non) religious ideology compelled him to succumb to the usual fallacies of the secularists in that arena. Nevertheless, his criteria for distinguishing pseudoscience from authentic science are often quite helpful.

[603] Jacques Vallée, *REVELATIONS: Alien Contact and Human Deception* (Anomalist Books, 2008), Conclusion.

[604] Ibid.

[605] Jacques Vallée, "*Messengers of Deception: UFO Contacts and Cults* (Daily Grail Publishing, 2008 edition), Foreword.

[606] David DeSteno, "A new religion has Americans looking to the stars." *The Boston Globe.* March 27, 2024.

[607] Ben Zeller, *Handbook of UFO Religions.* Brill Handbooks on Contemporary Religion, Volume: 20. (Brill, 2021), Introduction.

[608] Paul Thigpen, *Extraterrestrial Intelligence and the Catholic Faith* (TAN, 2022), What About UFOs?

About the Author

Daniel O'Connor is an adjunct professor of philosophy and religion at a State University of New York Community College, where he has taught for eight years. For over a decade serving as an educator, he is honored to have instructed thousands of students in courses on philosophy, existentialism, Catholic catechesis, comparative religion, mathematics, physics, and more. Originally a mechanical engineer, Daniel's first job after graduating RPI (Rensselaer) had him running high-tech experiments at the GE Global Research headquarters aimed at achieving breakthroughs in carbon-nanotube based quantum electron emissions. There, he ran a test resulting in a world record, but realized—after defending the right to life on a radio show debate later that same day—that he could do infinitely more good for the world by speaking up for the truth than by contributing to the world's technological advancement. In 2013, Daniel obtained a master's degree in Philosophical Theology from Holy Apostles College and Seminary in Connecticut, and has been teaching ever since. He has thus far completed several years of post-masters studies toward his PhD in Philosophy at the University at Albany. Daniel is the author of several books, and he writes for his personal website (www.DSDOConnor.com) where his writings, talks, videos, and other works may be found. Daniel lives in New York with his wife, Regina, and their five children, Joseph, David, Mary, Luisa, and John Paul.

***

Have you found this book helpful?

Kindly consider leaving a review for it on Amazon:
Amazon.com/author/DSDOConnor

Are you interested in discounted bulk orders or in other versions of this book? (Paperbacks, Hardcovers, Ebooks, and Audiobooks are available.)

Please see:
TheFirstAndLastDeception.com, or

DSDOConnor.com

Made in the USA
Las Vegas, NV
19 September 2026

c790b732-0de0-459d-a32f-475d15b49834R01